The Snow Geese of La Pérouse Bay

Oxford Ornithology Series

Edited by C. M. Perrins

1. *Bird Population Studies: Relevance to Conservation and Management* (1991) Edited by C. M. Perrins, J.-D. Lebreton, and G. J. M. Hirons
2. *Bird–Parasite Interactions: Ecology, Evolution, and Behaviour* (1991) Edited by J. E. Loye and M. Zuk
3. *Bird Migration: A General Survey* (1993) Peter Berthold
4. *The Snow Geese of La Pérouse Bay: Natural Selection in the Wild* (1995) Fred Cooke, Robert F. Rockwell, and David B. Lank

The Snow Geese of La Pérouse Bay

NATURAL SELECTION IN THE WILD

FRED COOKE
ROBERT F. ROCKWELL
and
DAVID B. LANK

Oxford New York Tokyo
OXFORD UNIVERSITY PRESS
1995

Oxford University Press, Walton Street, Oxford OX2 6DP
Oxford New York
Athens Auckland Bangkok Bombay
Calcutta Cape Town Dar es Salaam Delhi
Florence Hong Kong Istanbul Karachi
Kuala Lumpur Madras Madrid Melbourne
Mexico City Nairobi Paris Singapore
Taipei Tokyo Toronto
and associated companies in
Berlin Ibadan

Oxford is a trade mark of Oxford University Press

Published in the United States
by Oxford University Press Inc., New York

A catalogue record for this book is available from the British Library

Library of Congress Cataloging in Publication Data
Cooke, F. (Fred)
The snow geese of La Pérouse Bay: natural selection in the wild /
Fred Cooke, Robert F. Rockwell, and David B. Lank.
(Oxford ornithology series; 4)
Includes bibliographical references (p.) and indexes.
1. Snow goose—Manitoba—La Pérouse Bay. 2. Snow goose—Manitoba—La Pérouse Bay—Evolution. 3. Natural selection. I. Rockwell, Robert F. II. Lank, David B. III. Title. IV. Series.
QL696.A52C66 1995 598.4'1—dc20 94-5300

ISBN 0 19 854064 7

Typeset by Footnote Graphics, Warminster, Wiltshire
Printed in Great Britain on acid-free paper by
Biddles Ltd, Guildford & King's Lynn

To Graham Cooch and Hugh Boyd for inspiration and guidance
To Sylvia, Pat and Connie for understanding and patience

Preface

It is a daunting task to synthesize the results of 25 years of research into the confines of a single volume. Since the beginning of the study in 1968, the La Pérouse Bay Snow Goose project has become one of the largest field studies of a bird population in the world. Over 300 biologists and others have helped to gather the data and more than 100 scientific papers have been published about the work. When we decided to write this book, we had to decide how to package our material. We did not want simply to write a monograph of the snow goose, valuable as such a book would be. Our emphasis from the outset was to look at the Snow Geese we were studying through the eyes of both population geneticists and population ecologists.

Excellent studies of avian populations have been carried out with a strong ecological focus, but few of these also considered the genetic and the evolutionary framework of their study species. On the other hand, population geneticists often regard the ecological environmental variation as a handicap in their attempts to understand the underlying genetic processes driving the evolutionary processes they are investigating. Vertebrate population ecologists tend to think in terms of 'ecological' time, covering perhaps a few decades at most, whereas population geneticists think more in terms of 'evolutionary' time, typically spanning millennia. In our study, we have tried to show that evolutionary processes also occur within the framework of 'ecological' time. Our study is not the only one to take this approach. Peter Grant and his colleagues have carried out a remarkable series of studies of microevolutionary studies of the Darwin's Finches of the Galapagos, and John Endler and associates have taken a similar approach with wild populations of Guppies.

The environment acts directly on animals by influencing an individual's growth rate, survival probability, and reproductive rate. This 'natural selection' by the environment systematically affects a population's size and genetic structure if particular segments of the population differentially survive and/or transmit their genes into the next generation, producing microevolutionary change. In this book, we examine the complex interaction between individuals with particular attributes, the environment, and effects on the population of geese which we were studying. Hence the sub-title of our book '*Natural Selection in the Wild*'.

In writing this book, we are responding to our many colleagues who have asked us to summarize our studies in one readily accessible form. This has not always been an easy task, and our understanding has continued to evolve as we probe further into the data. To write a book while the study is still continuing produces conflicts between our desires to study the problems yet further and the need to summarize what we believe that our data already show. We hope we have found a suitable balance between the need to provide a clear picture for the general reader and the need to explain the complexity of some of our analyses, which are not always those in widespread use in our field.

It would have been impossible to have carried out this work and written this book without the help of a large number of people. Dr Graham Cooch of the Canadian Wildlife Service (CWS) has a very special place in this endeavour. He first took the decision to send a naïve young Englishman into the wilds of the Canadian Arctic and, we suspect, was surprised when he came back alive. He and Mr Hugh Boyd, also of the CWS, have kept up their association and enthusiasm for the project, have helped in its financing, and, we believe, have shown the value for wildlife management policies in having a strong link between curiosity-based research on evolutionary processes and the more mission-orientated research needs of wildlife management.

Graham Cooch's son Evan has also played an important role. Now an Assistant Professor of Wildlife Ecology at Simon Fraser University, he has been associated with Snow Geese for most of his life. He has assisted in the collection and analysis of much of the data, and most recently provided essential technical support in the production and presentation of this complex manuscript. His work on Chapter 13 deserves particular mention.

Alex Dzubin, now retired but formerly with the CWS also deserves special mention. He taught us the importance of and methods for measuring geese, and gave up many of his holidays to collect data and revive our sometimes flagging spirits during the annual banding of the geese at La Pérouse Bay.

Botanical research on the salt marshes of La Pérouse Bay has been principally carried out under the superb direction of Bob Jefferies, who, with his team of former and present graduate students, enlightened us about the intricacies of the interactions between Snow Geese and salt marshes they exploit for food.

So many students, post-graduate, graduate and undergraduate, have participated in the project and given several summers to collecting data at La Pérouse Bay that it is perhaps unwise to pick out a few names for special mention for fear of unwittingly appearing to undervalue the contribution of others. Some who deserve special mention for their leadership roles and/or contributions to the intellectual development of the study, in approximately chronological order, are: Ken Ross, Paul Mirsky, George Finney, Peter Boag, Laurene Ratcliffe, Ken Abraham, Pierre Mineau, Roger Healey, Scott Findlay, Marjorie Bousfield, Jean Hamann, Chris Davies, Cheri Gratto-Trevor, Kathy Martin, David Hik, Tom Quinn, Charles Francis, Tony Williams, Matthew Collins, and Barbara Ganter. Others contributing notably to the compilation and organization of the data at Queen's University and Simon Fraser University include Lauraine Newell, Anna Sadura, Liz Maskell, Karen Williams, and Connie Smith. The photographs were taken by Lauraine Newell, Pierre Mineau, Barbara Ganter, Jack Hughes, Mike Carter, Greg Robertson, Ian Craine, and Fred Cooke.

Many of our friends and colleagues have braved the mosquitoes and Polar Bears and have suffered the torments of arctic gales, Hudson Bay fogs, and even torrential rainstorms, to extract information about Snow Geese and their world. We are also extremely grateful to the people who live and/or work in Churchill, not only for their logistical help but also for their friendship. They have accepted us into their community and taught us the value of being part of a vibrant Northern settlement. Our particular thanks go to Bonnie and Al Chartier, Bill and Diane Erickson, Lindy and Andrew Lee, Clifford Paddock, Pat Worth, Bud Amireau, Ray Ramos, Steve and Janet Miller, Lauraine Brandson, Bob Taylor, and the late Omer Robidoux.

To our academic colleagues, we owe a major debt as well. Ian Newton and the late Alan Pakulak played major roles in defining our field approaches. We have also benefited in major ways from our discussions with Marvin Seiger, John Ryder, Dave Ankney, Charlie MacInnes, Jean Bédard, Dick Kerbes, Jim Sedinger, Gilles Gauthier, Don Rusch, Bruce Batt, Lew Oring, Peter Grant, Arie van Noordwijk, Kate Lessells, Raleigh Robertson, Patrick Colgan, Allen Keast, Dolf Harmsen, Bob Montgomerie, Brad White, Don Rainnie, Harry Lumsden, Chris Perrins, Evgeni Syroechkovsky, Kostya Litvin, Przemek Majewski, the late Klaus Immelmann, Jacques Blondel, Jean-Dominique Lebreton, Bruce Risska and Howard Levene.

Finally we must thank the various granting agencies who have all helped to make this research possible: the Natural Sciences and Engineering Research Council of Canada (NSERC), the Canadian Wildlife Service (CWS), the Department of Indian and Northern Affairs (DINA), the Manitoba Department of Natural Resources, Ducks Unlimited Inc. and Ducks Unlimited Canada, the Mississippi and Central Flyway Councils, and to our present or former universities, Queen's University at Kingston, City University of New York, and Simon Fraser University.

January 1994

Burnaby, BC	F.C.
New York	R.F.R.
Burnaby, BC	D.B.L.

Acknowledgements for tables and figures

The following are reproduced by permission of Allen Press and the editor of *The Auk*: Fig. 3.2 from Cooke, F., MacInnes, C. D. and Prevett, J. P. (1975). Gene flow between breeding populations of lesser snow geese. *Auk*, **93**, 493–510. Material in Fig. 3.3 from Cooke, F., Parkin, D. T., and Rockwell, R. F. (1988). Evidence of former allopatry of the two colour phases of lesser snow geese (*Chen caerulescens caerulescens*). *Auk*, **105**, 467–79. Material in Fig. 7.5 from Francis, C. M., Richards, M. H., Cooke, F., and Rockwell, R. F. (1992). Changes in survival rates of lesser snow geese with age and breeding status. *Auk*, **109**, 731–47.

The following are reproduced by permission of Blackwell Scientific Publications Limited: material in Fig. 2.11 from Williams, T. D., Cooch, E. G., Jefferies, R. L., and Cooke, F. (1993). Environmental degradation, food limitation and reproductive output: Juvenile survival in lesser snow geese. *Journal of Animal Ecology*, **62**, 766–77. Material in Figs 6.1, 6.4, and 6.5 from Cooch, E. G., Lank, D. B., Rockwell, R. F., and Cooke, F. (1989). Long-term decline in fecundity in a snow goose population: evidence for density dependence? *Journal of Animal Ecology*, **58**, 711–26. Figs 7.1, 7.2, and 7.3 from Rockwell, R. F., Cooch, E. G., Thompson, C. B. and Cooke, F. (1993). Age and reproductive success in female lesser snow geese: experience, senescence, and the cost of philopatry. *Journal of Animal Ecology*, **62**, 323–33. Material in Figs 13.2 and 13.3 from Cooch, E. G., Lank, D. B., Rockwell, R. F. and Cooke, F. (1991). Long-term decline in body size in a snow goose population: evidence of environmental degradation? *Journal of Animal Ecology*, **60**, 483–496.

The following are reproduced by permission of the Ecological Society of America: Figs 6.6 and 6.10 from Francis, C. M., Richards, M. H., Cooke, F., and Rockwell, R. F. (1992). Long term changes in survival rates of lesser snow geese. *Ecology*, **73**, 1346–62. © 1992 by the Ecological Society of America. Reprinted by permission. Fig. 13.1 from Cooch, E. G., Lank, D. B., Dzubin, A., Rockwell, R. F., and Cooke, F. (1991). Body size variation in lesser snow geese: environmental plasticity in gosling growth rates. *Ecology*, **72**, 503–12. © 1991 by the Ecological Society of America. Reprinted by permission.

The following are reproduced by permission of the University of Chicago Press: material in Fig. 7.7 from Barrowclough, G., and Rockwell, R. F. (1993). Variance in lifetime reproductive success: estimation based on demographic data. *American Naturalist*, **141,** 281–95. © 1993 by the University of Chicago. All rights reserved. Material in Figs 10.1 and 10.2 from Rockwell, R. F., Findlay, C. S., and Cooke, F. (1987). Is there an optimal clutch size in lesser snow geese? *American Naturalist*, **130,** 839–63. © 1993 by the University of Chicago. All rights reserved.

The following are reproduced by permission of Munksgaard International Publishers Ltd and the editor of *Oikos*: Fig. 12.1 and Tables 12.2, 12.3, and 12.4 from Williams, T. D., Lank, D. B. and Cooke, F. (1993). Is intraclutch egg-size variation adaptive in the lesser snow goose? *Oikos*, **67,** 250–6.

The following are reproduced by permission of Birkhauser Verlag AG: Fig. 6.7 from Francis, C. M. and Cooke, F. (1993). A comparison of survival rate estimates from live recaptures and dead recoveries of lesser snow geese. In: *Marked individuals in the study of bird populations*, (ed. J.-D. Lebreton and P. M. North), Birkhauser Verlag, Basel. Figs 6.8 and 6.9 from Cooke, F. and Francis, C.M. (1993). Challenges in the analysis of recruitment and spatial organization of populations. In: *Marked individuals in the study of bird populations*, (ed. J.-D. Lebreton and P. M. North). Birkhauser Verlag, Basel.

The following is reproduced by permission of the New Zealand Ornithological Congress Trust Board: material in Fig. 12.4 from Cooke, F., Rockwell, R.F. and Lank, D.B. (1991). Recruitment in long-lived birds: genetic considerations. *Proceedings of the XXth International Ornithological Congress*, 1666–77.

Contents

1 Introduction

Natural selection is one of the primary mechanisms moulding the phenotypic characters of a population. This central tenet of evolutionary biology is well established and it is widely accepted that past and present environments have interacted and continue to interact in complex ways with the genetic constitutions of organisms to produce the populations we see today. While not discounting the importance of stochastic events in the evolutionary processes, natural selection is seen by many evolutionary biologists as a key to understanding the form, function, and demography of populations. There are manifold ways in which complex environments can alter organisms, both directly in shaping the appearance of individuals and indirectly through selective changes that modify populations' gene pools. Through detailed investigations of populations in their natural environments, we have begun to understand some of the ways in which natural selection operates. Classic studies by Clutton-Brock and colleagues (1982) on Red Deer (*Cervus elephas*), by Newton (1986) on Sparrowhawks (*Accipiter nisus*) and by many workers on Great Tits (*Parus major*) (for example McCleery and Perrins 1989), have helped clarify how natural selection influences both demographic and other phenotypic attributes of these populations.

Our aim is to examine the process of natural selection in a wild population of birds. Our rationale is that only by detailed examination of a group of interbreeding organisms throughout its natural habitat can we hope to discover the range of selective pressures under which they live. This in turn may allow us to deduce or test general principles providing broader understanding of natural selection. Such a study requires the recognition that animals move during their lives and that the selective factors influencing them at one stage may be quite different from those at another stage. Not only do many animals have seasonal patterns of movement, but certain individuals also move from their place of birth to their place of breeding. Only by monitoring gene flow and patterns of philopatry can the contributions of different environments on the organisms be assessed.

1.1 *Studying natural selection in Lesser Snow Geese*

We studied a breeding colony of Lesser Snow Geese (*Anser caerulescens caerulescens*) a *c.* 2 kg, arctic nesting, migratory goose. Our study site was along the shores of La Pérouse Bay (58°44′N, 94°28′W), a remote part of the Hudson Bay coastline in northern Manitoba, Canada. We have followed this population for 25 years under a range of environmental conditions. The Lesser Snow Goose is a challenging species on which to study natural selection in field conditions. It is a large bird which is justifiably wary of humans, relatively difficult to capture at most times of the year, breeds in remote areas, migrates twice yearly between arctic and south temperate latitudes, and intermingles on the wintering ground with birds breeding in colonies thousands of kilometres distant.

The evaluation of fitness ideally requires that each stage of the life cycle be examined, and an assessment of the genetic consequences of fitness requires knowledge of patterns of gene flow. Contrast studying such questions in Snow Geese versus doing so in a bird with a lifestyle such as that of the Large Cactus Finch *Geospiza conirostris*, the subject of an 11-year microevolutionary study by Grant and Grant (1989). The population of *G. conirostris* studied lives on the *c.* 16 km^2 island of Genovesa in the Galapágos. We do not wish to contrast the logistical difficulties of conducting research in cactus-covered tropical islands with those of working in snow-covered tundra, but to consider instead the intrinsic demographic sampling problems for the two species. The Large Cactus Finches of Genovesa comprise a virtually closed genetic population, although rare immigrants or hybrids may have profound long term impacts on genetic structure. The spatial stability of individuals and small population size enabled researchers to document complete life histories of the survival and reproductive success of individual birds, and build their analyses of natural selection and genetic consequences of it directly around these data. This approach is impractical in a species with a Snow Goose's population size and annual migrations. Robust samples of complete individual life histories cannot be obtained. Instead, our analyses of natural selection in Snow Geese primarily rely on combining less complete information which was collected from larger numbers of different birds at different stages of the life cycle. The pitfalls to such an approach are well recognized by workers in this field of study. One must address the problem that biased samples may be used to estimate particular components. We have adapted and developed methods which permit us to analyse the workings of natural selection to a considerable degree despite lacking complete data on individuals.

The analysis has proved tractable in this species for the following reasons. Snow Geese breed in colonies. Nests are close together and easy to find, thus many can be sampled with relative ease. This allows collection of sufficient data to measure relatively small differences in fitness, distinguish selection from random environmental noise, and have some confidence that a failing to find statistical differences truly represents the population, rather than reflecting a lack of power to detect differences. At the start of our study, the La Pérouse Bay Snow Goose colony consisted of 2000 pairs, large enough to provide sufficient sample sizes, but small enough to be able to mark and monitor a significant fraction of the individuals and nests. We have now collected basic nest and egg survivorship data from over 50 000 nests since 1968, and individually marked over 112 000 hatching goslings since 1971.

The Lesser Snow Goose has a conspicuous discrete, genetically-based plumage dimorphism (Fig. 2.1; Cooke and Cooch 1968), which provided the initial impetus for the study. Data collection quickly expanded to include phenotypic variation for quantitative characters such as body size, egg size, timing of breeding, clutch size, and other demographic fitness components. As well as being of intrinsic evolutionary and population genetic interest, the plumage polymorphism has proved to be a useful genetic marker for studying gene flow and aspects of social behaviour. Rates of intra-specific nest parasitism and extra-pair fertilization can be estimated using the plumage colour gene (Lank *et al.* 1989*b*), thus we can evaluate the accuracy of our assignment of paternity and maternity without resorting to biochemical techniques.

The large size and conspicuousness of Snow Geese makes them easy to observe at most stages of their life cycle. Our field work involved studying the birds during the breeding season, from their arrival at the breeding colony through the fledging of their young. Most of the classic long term studies of birds have focused on species with altricial or semi-precocial young. Such species provide obvious advantages when studying the role of the parents in providing nutrients and care during the totally dependent stages for their offspring. As a precocial species, the Snow Goose provides an interesting contrast. While young geese feed themselves, parents must defend them from predators and shelter them from foul weather for several weeks. In contrast to many precocial species whose young are difficult to follow after leaving the nest, we can observe brood structure, movements, feeding, social behaviours, and predator activities, using towers built on salt marsh feeding flats.

About three weeks after eggs hatch, large numbers of geese can be captured, measured, and marked because adults are flightless during a post-breeding wing moult and the young are not yet fledged. We have marked over 100 000 adults and young with numbered metal wildlife service bands, most of which were also given engraved coloured plastic leg bands. The plastic bands allowed for individual visual identification at nests, while feeding, or even at remote migration or wintering areas. Recaptures of marked individuals provide information on gosling survival and recruitment into the breeding population, adult local survival, and data for estimates of population size (Seber 1982). The metal bands allowed us to obtain information about the movements and survivorship of our marked birds away from our field site and outside of our field season, from reports of band numbers of individuals shot by hunters.

The Lesser Snow Goose is a heavily hunted species. Hunters produce useful data because large numbers of wildlife service bands from shot birds are reported to the Canadian Wildlife Service and the US Fish and Wildlife Service, who subsequently notify us. We thereby know the place and time of the death of 12 per cent of the birds we have banded. This band recovery rate is substantially higher than rates for non-hunted species in North America, which are typically much less than 1 per cent for small passerine birds, and for most of the other species of birds studied. Accurate methods for estimating true survival rates may be used when there is a reasonably high probability that death will be detected (Brownie *et al.* 1985). This situation occurs in Snow Geese and a number of other waterfowl species in which large numbers of individuals have been banded, shot, and the bands reported to wildlife agencies by hunters.

Many survival estimates in migratory birds are based on the return of animals to a study area in future years, but this partially confounds survival with local philopatry. By resighting and recapturing birds annually, we can compare survival rates based on band recovery data with those based on the re-encounters at the breeding colony (Lebreton *et al.* 1992), and thus differentiate between survival *per se* and local philopatry.

If hunting is a major source of adult mortality, are we studying natural selection 'in the wild'? Although native hunters have probably killed Snow Geese since arriving on the American continent, as evidenced by stone corrals constructed by Inuit to capture geese, hunting by immigrants from Europe commenced only in the sixteenth century. The expansion of the immigrant human population had numerous effects on Snow Geese survivorship. The populations of competing natural predators were reduced. More importantly perhaps, farming, ranching, and forest clearing provided rich new feeding opportunities for geese along the migration routes and on the wintering grounds. Corn, pasture, winter wheat, and rice are major new food sources. Most estimates suggest that the Snow Geese of the Mississippi and Central Flyways are currently an expanding population (Francis *et al.* 1992*b*). For the past 60 years, one could argue that a continent-wide experimental manipulation of Snow Goose population size has been attempted as wildlife managers attempted to provision geese on migration and regulate annual hunter kill by setting bag limits for the species. We cannot assess how these countervailing factors balance out with respect to Snow Goose demographics, and their relative importance has undoubtedly changed over the past 200 years. Clearly, however, selection regimes on particular traits at certain parts of the life cycle may have changed substantially in the past 20 generations of geese.

The breeding grounds have been much less affected by human activity than the migration and wintering grounds have been. An advantage of our remote location is that the site was unlikely to be drastically altered, facilitating a long term study. The vegetation of the Hudson Bay lowlands comprises low arctic willow and salt marsh plants, little affected by humans since emerging from the ancient Tyrrell Sea in the late Pleistocene. Snow Geese are basically grazers, and it has been possible to study the autecology and community ecology of their principal food plants in some detail and to characterize the nutritional needs of the geese themselves (for example Jefferies *et al.* 1979; Bazely and Jefferies 1986; Jefferies 1988*a*; Hik *et al.* 1992; Frey *et al.* 1993). On the other hand, the remoteness of the location presents problems. To provision a field station for up to 20 biologists for 4 months each year has been a major logistic and financial challenge. In addition there are several natural hazards to ourselves, such as a large population of Polar Bears (*Ursus maritimus*) in the later part of our field season.

Snow Geese are socially monogamous and pair for life. The species is somewhat sexually dimorphic in size, and sex differences in juvenile growth rates provide useful tools for investigating selection on growth rates. Sex ratios seem to be similar at all life-cycle stages and for this reason most geese are probably successful in acquiring a mate. As such, sexual selection, which has been the focus of large numbers of studies of reproductive success, is less important in Snow Geese at least as far as mate acquisition itself is concerned. The type of mate acquired may be important, but this is effectively examined by looking for patterns of non-random mating.

The ability to observe and document so many stages of the life cycle represents one of the main advantages of working with a large, conspicuous vertebrate, but the long generation time means that many years must elapse before useful genetic data are obtained, a disadvantage compounded by difficulties in obtaining genetic data from controlled aviary crosses.

Because Lesser Snow Geese have commercial and recreational importance, research funds are available for addressing both theoretical and managerial questions. However, after all the rationalizations as to why this species is appropriate for our study, the pleasure of working with a beautiful bird in a remote and exciting part of the world is probably a more important reason for choosing to work with Snow Geese than we would like to admit.

1.2 *Research approach*

Our basic research philosophy stems from a belief that evolutionary processes can only be understood by a combination of genetic and ecological approaches. Too often, studies are limited by emphasizing only one of these approaches. Ecologists tend to stress the immediate importance of environmental variation, without considering the phenotypic and genotypic responses to this variation. Geneticists, on the other hand, concentrate on the phenotypic and genotypic response to environmental change without considering the more proximate effects of the environment. We will illustrate these contrasting approaches in several chapters.

We have been influenced by the work of Endler (1986). In *Natural Selection in the Wild* he stresses three requirements for an adequate examination of the pattern and process of natural selection. First, one must document phenotypic variability in the population and for this we chose both morphological and demographic characters. Of the former, we chose plumage variability, a character with a reasonably simple genetic basis, along with body and egg size, quantitative characters for which the relationship between genotype and phenotype is more complex. Of the latter, we chose the timing of reproduction and clutch size, although as we shall see later, these characters appear to be interrelated. Each of the characters illuminates features of natural selection that were not clear to us at the beginning of our work and provide new insights into the complex interactions between environment and the phenotypic expression of the organism.

Secondly, one must examine the genetic basis for variation in the characters. A genetic analysis requires pedigree studies and the need to follow the population through at least two generations. Parent–offspring comparisons require known individuals and for this reason it was essential to mark individually a large number of birds in the population and to monitor individual family histories. We also needed to examine phenomena such as extra-pair copulation and intra-specific nest and brood parasitism to ascertain whether the putative parents in a family situation were indeed the true parents. Genetic studies are much more difficult to carry out in the field than in the laboratory, and yet they are essential for a study of natural selection. After all, it is the total variation, expressed under the natural range of environments, that is subject to natural selection. Delineating the apparent relative contributions of genotypes and environments to phenotypic variation is essential to our work.

The third part of a study of natural selection is monitoring of possible changes in the distribution of the phenotypes with time. Changes could occur as a direct result of changes in the environment and we document some of these in this book. They could also occur as a result of changes in the composition of individuals comprising the population. For example, if there were a massive influx of birds with a different phenotypic composition to that of the resident birds, we would expect a change in the pattern of variability of the population. Gene flow is, therefore, one of the possible mechanisms resulting in changes in the phenotypic composition of the population and must be carefully assessed when studying natural selection. Finally, changes in the patterns of variability might indicate that natural selection is occurring. If phenotypic frequencies of the character distributions differ at different stages of the life history, or if the frequencies differ from one generation to the next, this may be evidence of some form of selection acting on the population. Detecting selection depends upon the careful measurement of differences in the relative fitness of different phenotypes in the population, and this is the major aim of the book.

Our analyses of selection are basically done on a univariate basis; that is, we consider one character at a time. Although we routinely use multivariate statistical models to isolate fitness consequences of particular traits while controlling for phenotypic values of other traits, this differs conceptually from analysing *selection* on particular traits while controlling for *selection* on correlated traits (Lande and Arnold 1983). The applicability of the genetic assumptions behind multivariate selection models for quantitative characters is a subject of lively debate (for example Mitchell-Olds and Shaw 1987; Grafen 1988; Schluter 1988), but our principal reason for adopting a univariate approach is the composite nature of our data. It is difficult to apply multivariate models because we have limited overlap in our data sets for different attributes and components of fitness among individuals, in contrast to analysing the attributes and complete life histories of single birds. In short, our data are not well suited to formal analyses using this approach. However, we can and do think of and discuss our findings in a multivariate, integrated context.

One would like to be able to go beyond an analysis of selection, and predict from the patterns of selection documented what the future evolutionary trajectory of a population might be. For the one trait where we have a clearly understood genetic mechanism, namely variation in plumage colour, we can and do make and test such predictions. For our other traits, however, doing so goes beyond the boundaries of this study. In the past 20 years, evolutionary biologists have become increasingly sophisticated in their conceptions of the relationships between phenotypic variation, genetic variation, natural selection, and what is expected as an evolutionary response from a population, that is changes in its genetic constitution. For traits with a strong relationship between genotype and phenotype, the expected response is clear. All else being equal, selection on phenotype should produce changes in genotypes. The classic studies of natural selection in changing environments illustrate this nicely (for example Kettlewell 1973). Even for traits for which the relationship between genotype and phenotype is robust with respect to environmental variation during development, however, all else may not be equal. Variation at one genetic locus may affect several traits simultaneously, a phenomenon known as pleiotropy. In this case, to predict changes in allele (gene) frequencies one must consider how selection is acting on all affected traits, and determine a net effect. An allele which stimulated the growth of longer legs during development, for example, might also result in the growth of longer arms. One would have to assess the fitness impact of phenotypic variation in both leg and arm length, and know that genes at one allele affected both characteristics, to determine the net fitness of particular genotypes.

Such considerations may be unimportant much of the time. As kangaroos demonstrate, a population can evolve developmental mechanisms which allow for the growth of long legs and short arms, thereby uncoupling the hypothetical pleiotropic effects imagined above. To do so, genetic variation at additional loci must affect the eventual phenotype produced. The joint action of several loci in affecting a single trait is called epistasis. As the number of loci affecting development increases, the relationship between genetic variation at particular loci and the resulting phenotypes becomes weaker. This is not simply a numerical dilution. Concomitant with increasing epistasis is an increasing opportunity for non-genetic environmental variation to affect development. Since the expression of any allele depends on input from its environment, including both the 'external' environment and input from other loci, epistatic systems provide more potential pathways for 'environmental' variation to affect gene expression and phenotypic development. Eventually, our only hope of analysing the genetics of traits is by using the techniques of quantitative genetics, in which we estimate the relative effects of the total 'additive genetic variation' on phenotypic characters by examining the covariance of the phenotypic values of parents and offspring or among other close relatives.

To predict how a population might evolve, that is to change its genetic makeup, due to selection on a quantitative genetic trait requires information not only about the selective regimes on correlated characters, but also on the genetic correlations involved, what is usually referred to as the genetic variance–covariance matrix (Lande and Arnold 1983; Arnold and Wade 1984*a*, *b*). Obtaining such information is not as difficult as it might seem. One obtains values by calculating covariances across traits and generations, for example correlating parent's arm length on offspring leg length and vice versa. While these statistical procedures are reasonably well established, reliable calculations require a database beyond that which we have generated. Generating sufficient data to investigate the general genetic architecture of Snow Geese, other than that of the plumage polymorphism, goes beyond the goals of this research project.

1.3 *Measurements of fitness*

Fitness differences among different segments of the population are the key to an investigation of natural selection, but fitness itself cannot be easily measured. All we can hope to do is partition fitness into its various components, examine as many of them as possible and infer the overall fitness differences among segments of the population from these estimates. To completely measure fitness for a segment of the population would require knowledge of gamete production, zygote production, survival of those zygotes through the various stages up to reproductive maturity, mating and mate acquisition ability, breeding propensity and survival. A complete investigation of all components is probably a practical impossibility since not all stages of the life cycle are available for measurement. Ideally, however, the organism chosen for study should be one in which as many as possible of these components of fitness can be reliably estimated. Our basic approach follows Prout (1969) and we describe how the fitness components are defined and measured in Chapter 4.

1.4 *Organization of the book*

The organization of the book is as follows. In Chapters 2 and 3 we describe the basic natural history of the species. First we describe the actual population with which we work and then show its relationship to the broader distribution of the species. This is necessary because gene exchange is important to an overall understanding of evolutionary processes. In Chapter 4 we present our model of fitness components. We document the various hazards the birds face at different stages of their lives and assess the relative importance of these to overall fitness variation. In Chapter 5 we consider the consequences of mating and reproductive decisions, both in terms of choice of mate and choice of reproductive tactic. In Chapters 6 and 7 we describe the variation in fitness components in relation to time and age. These differences mainly relate to non-genetic changes but are essential for understanding the ways in which the genotypic endowment of a bird can result in different phenotypic expressions depending on the age of the bird and on the particular environment in which it finds itself. In Chapter 8 we outline our approaches to analysing the genetic component of variation in our population. In particular, this chapter examines the uses and limitations of heritability measures for the quantitative characters we are studying.

In the remaining chapters, we concentrate on specific traits to understand the selection pressures that the population may be subject to. Through these analyses, we hope to understand why the population is the way it is. In Chapter 9 we consider plumage polymorphism. Such polymorphism is rare in birds, but provides a relatively simple case for examining possible fitness consequences of a conspicuous physical difference between segments of the population, a difference that has a simple genetic basis. The final chapters consider characters that vary continuously rather than discretely. In Chapter 10 we examine variation in clutch size, a trait that is unique among those we evaluated since it is itself a component of fitness. In Chapter 11 we consider variation in the timing of nesting. While it turns out that this trait is correlated with clutch size, it is not a fitness component *per se*. In Chapters 12 and 13 we examine egg and body size variation respectively. Variability in each of these traits has been the subject of many studies and our intensive examination has led to some rather unusual findings. The last chapter summarizes the specific findings in the book, but more importantly outlines what we have discovered about natural selection by using both genetic and ecological approaches.

2 *The Lesser Snow Geese of La Pérouse Bay*

We originally chose to work with Lesser Snow Geese, *Anser caerulescens caerulescens*, the smaller subspecies of the Snow Goose, because it has a conspicuous plumage dimorphism. The Latin nominate refers to the blue phase, formerly known as the Blue Goose, with grayish brown feathering on most of the body and a white head. The English nominate refers to the white phase which is all white, except for its black primary feathers (Fig. 2.1). We were initially interested primarily in how this polymorphism was maintained. We were also interested in the fact that the population showed positive assortative mating, which we hypothesized to be a result of birds choosing mates of the same plumage colour as their parents. To examine this hypothesis, we needed to find a small (by Snow Goose standards), self-contained, and yet relatively accessible colony where we could embark on a long

Fig. 2.1 Plumage dimorphism in the Lesser Snow Goose, represented by a mixed morph pair. Blue phase is on the left, white phase is on the right. The blue phase bird is a probable homozygote for the dominant blue allele, and would score as a '6' on our belly plumage scale (Chapter 9).

term study, using birds marked for individual recognition. The La Pérouse Bay colony suited our purposes ideally.

This chapter describes the La Pérouse Bay Snow Goose colony, the annual cycle of the birds which breed there, and our research activities at the colony. We review events at the breeding grounds: arrival, nest initiation, incubation, hatching, and fledging. We then examine the non-breeding season by following the geese through fall migration to their wintering grounds and back to the breeding grounds the following spring. This natural history overview provides essential background for understanding the manifold ways in which natural selection can influence Snow Geese. We discuss the birds breeding at La Pérouse Bay as if they were a closed biological population, but this is by no means true. The following chapter places the La Pérouse Bay Snow Goose into context within its breeding population and species.

2.1 *The La Pérouse Bay colony*

The Snow Goose colony at La Pérouse Bay, in northern Manitoba, Canada, is located 30 km east of the town of Churchill, and 15 km west southwest of Cape Churchill (Fig. 2.2).

2.1.1 History

The colony itself had been censused by aerial photography since 1963, and contained approximately 2000 breeding pairs when we began our study. The colony is well separated from other colonies. To the northwest, the next colony is more than 200 km away, where the Tha-Anne River meets the Hudson Bay coastline. To the southeast, the nearest colony is at Cape Henrietta Maria, more than 750 km away.

Nesting at La Pérouse Bay was documented first in 1953 by Wellein and Newcomb (1953) and 3 years later by Foster (1957), who reported finding 15 nests on a foggy day. Prior to that, H. Lumsden (personal communication) had walked along the coast in July 1943, but saw no evidence of nesting. Flights along the coast in the region of Cape Churchill and La Pérouse Bay in the years 1949–1952 (Hawkins 1949; Hawkins *et al.* 1951) had failed to detect nesting geese, although the observers had commented on the suitability of the areas as a Snow Goose nesting area. Inquiries among long-time residents of the Churchill area failed to uncover evidence of nesting

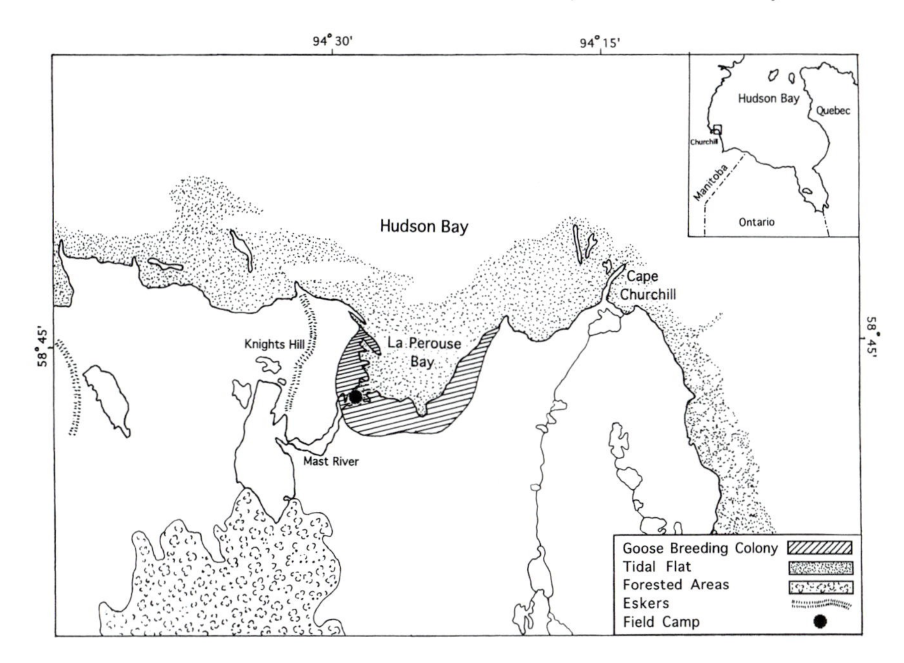

Fig. 2.2 La Pérouse Bay, and its location relative to Cape Churchill, on the shore of Hudson Bay in northern Manitoba, Canada (*inset*). Snow Goose nesting areas, tidal salt marsh flats, and forested areas are indicated.

earlier than the 1950s, and searches of the archives of the Hudson Bay Company in Winnipeg provide no evidence of nesting in previous centuries. Hanson *et al.* (1972) reported no nesting in 1962, despite careful searching, but at least 5623 birds in 1963. Successful nesting has been documented every year since then.

The colony thus originated in the early fifties, with nesting intermittent until 1963. It probably arose when birds were prevented from reaching their more northerly breeding colonies by adverse weather conditions. This type of 'short stopping' has been postulated to explain occasional large increases in the number of breeding birds (Geramita and Cooke 1982).

In 1968, the nearly 2000 pairs of geese nested primarily along the west coast of La Pérouse Bay. During the next 10–15 years, the population grew at about 8 per cent annually (Fig. 2.3) and the nesting area expanded north along the eastern coast of La Pérouse Bay and 2–3 km inland towards the southeast (Abraham 1980*a*). In those years, population sizes estimated by direct nest counts were reasonably concordant with estimates based on Jolly–Seber capture–recapture models of data collected in annual banding drives (for example Seber 1982). Since 1982, however, the colony has expanded

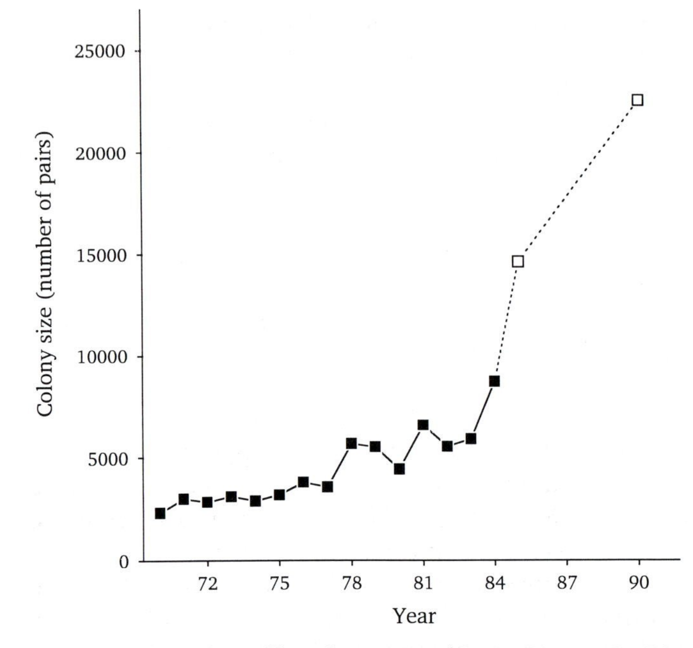

Fig. 2.3 Estimated number of breeding pairs of Snow Geese at La Pérouse Bay, 1968–1990. Data for 1968–1984 are population estimates based on Jolly–Seber analysis of data from banding drives. Points for 1985 and 1990 are extrapolated from analysis of aerial photographs of the colony (R. H. Kerbes, personal communication).

in number and geographic area to such an extent that neither of these methods are reliable. Large numbers of geese now nest along the smaller bay east of La Pérouse Bay and lower density nesting occurs as far as Cape Churchill (Fig. 2.2). High-density nesting is also found up to 5 km inland with sparser nesting extending up to 10 km. Aerial flights over large parts of the colony by the Canadian Wildlife Service resulted in estimates of 14 200 nesting pairs in 1985 and 22 500 pairs in 1990 (R. H. Kerbes, personal communication). The areas originally occupied in the central part of the colony are now sparsely used. The geographical pattern of colony growth is analogous to that of fairy ring fungi (for example *Marasmius oreades*), whose mycelia grow at the edges and decay in the centre.

2.1.2 Nesting and foraging habitat

The breeding habitat comprises several vegetational assemblages that change as one moves from coastline to the more elevated inland areas. Starting towards the shore, the coastal salt marshes on which geese feed, but do not nest, are comprised mainly of *Carex subspathacea*, a small tufted sedge, and goose grass, *Puccinellia phryganodes* (Fig. 2.4(a)). This zone is followed by a short grass area consisting mainly of *Calamagrostis deschampsioides*, *C. neglecta*, and *Festuca rubra*. Inland from this, the main vegetation type is dominated by the short shrubby willow *Salix brachycarpa* (Figs 2.4(b,c)) and the coarse sand dune lyme grass, *Elymus arenarius* (Fig. 2.4(d)). Most geese nest in this habitat. In the northwestern part of the colony there are some areas which are almost pure stands of lyme grass. Even further from the shoreline, other willowy shrubs such as *Betula glandulosa*, *Myrica gale*, *Salix planifolia*, *S. alexensis*, and *S. calcicola* appear. The vegetation here may be up to 2 m tall. Geese do not nest as densely in this habitat.

The coastal salt marshes have supported brood rearing by geese, and are undoubtedly responsible for the successful colonization of the site by Snow Geese. The growing, above-ground shoots of leaves of *P. phryganodes* and *C. subspathacea* provide the major forage. Light to moderate grazing by geese enhances the productivity of the salt marsh by 40–100 per cent above that of ungrazed control plots, as the faecal matter produced by geese increases the rate of nitrogen cycling (Cargill and Jefferies 1984; Bazely and Jefferies 1985; Jefferies 1988*a*,*b*). Both the quality and quantity of forage for geese improves because grazed plants respond by producing a higher proportion of young, protein-rich leaves. Since at least 1979, however, this positive feedback cycle has broken down because of over-exploitation by Snow Geese during pre-nesting grubbing and overgrazing during brood rearing periods (Fig. 2.4(e,f); Hik and Jefferies 1990; Hik *et al.* 1991). The area of intact salt marsh sward decreased by *c.* 12 per cent per year between 1985–1992, and the productivity of the remaining habitat has decreased by about half between 1979 and 1991 (R. L. Jefferies, personal communication; Williams *et al.* 1993*b*). The effect of geese on the vegetation is dramatically illustrated by constructing exclosure plots which prevent foraging over one or more seasons. The dramatic difference in vegetation within and outside of one such plot is shown in Fig. 2.4(g). In the summer of 1992, the fencing of six such plots was removed to allow foraging by geese. These became sites of intense social competition for a day or two, after which the vegetation had been grazed down to biomass levels similar to nearby control areas (Mulder *et al.* 1994). As traditional habitat has become degraded, the geese have made more use of inland meadows and upland areas. More recently, an increasing number of parents and their broods travel south and east to the portion of the Hudson Bay coastline that extends south from Cape

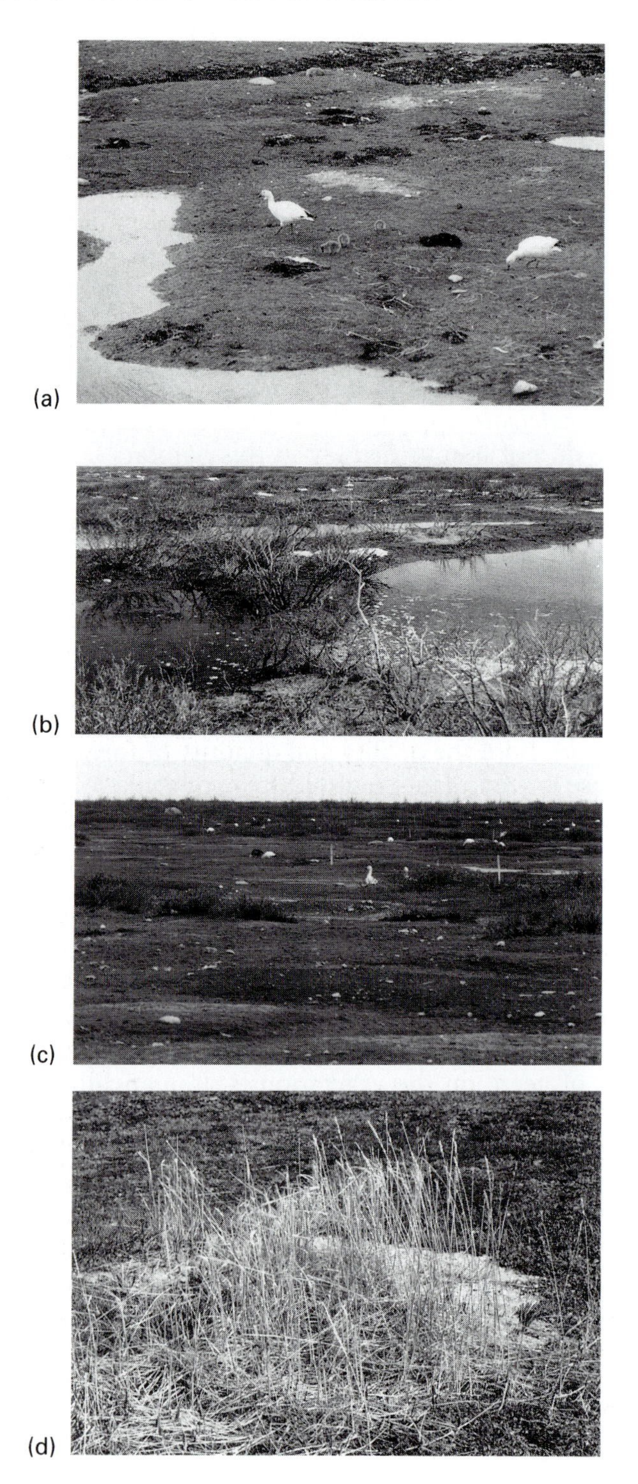
(a)
(b)
(c)
(d)

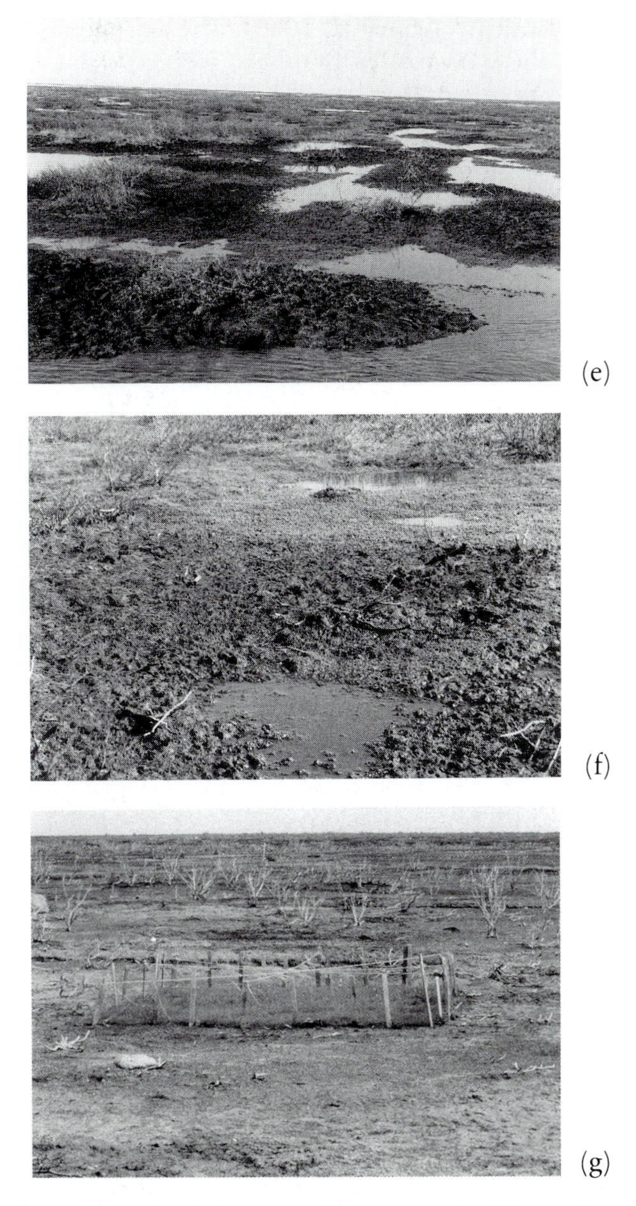

Fig. 2.4 Feeding and nesting habitats used by geese at La Pérouse Bay. (a) Salt marsh feeding flats. (b) Willow nesting habitat. (c) Willow nesting habitat. (d) Lyme grass (*Elymus arenarius*) nesting habitat. (e) Pond edge which has been 'grubbed' by geese in the spring, routing out rhizomes and stolons of monocotyledonous plants, such as *P. phryganodes*, *E. arenarius*, *Equisetum* spp., *Triglochin palustre* and *C. subspathacea*. (f) Close-up of 'grubbed' pond edge. (g) The effects of grazing by geese on salt marsh vegetation, as illustrated by vegetational differences within and outside a goose exclosure.

Churchill (Cooch *et al.* 1993). The salt marshes there are narrower than those originally found in La Pérouse Bay but extend at least 50 km south. Families nesting in La Pérouse Bay have been recaptured 5 weeks after hatch at least 30 km down that coastline.

2.2 *The breeding season*

2.2.1 Our field season at La Pérouse Bay

Each year since 1969, we have arrived in the Arctic and set up camp prior to the arrival of the geese. Our current base camp is a collection of small huts built on an island in the Mast River (Fig. 2.2). Before the geese lay, and usually while the area is still snow-covered, we divide *c.* 50 ha of nearby nesting areas into three to six study plots. During initiation and laying, each plot is visited daily by one researcher, who follows a standard procedure to search for and codify information about nests as they are established and progress. We try to follow between 100 and 500 nests from the laying of the first egg each year. An additional 1000–2500 nests containing two or more eggs are monitored each year during and after the laying period in expanded study plots. Information from these nests increases our sample sizes for estimating components of fitness from hatching onwards. This is essential because we are less likely to ascertain the fate of mobile geese than we are that of static eggs and nests. Nests are visited less frequently during the incubation period, and again daily during hatch, at which time hatching goslings are individually marked with monel numbered web tags. Birds are disturbed as little as possible during the 4–5 weeks brood rearing period. Before the geese fledge, while the parents are also flightless due to a complete primary molt, we round up families in mass banding drives.

2.2.2 Spring arrival

The first Snow Geese typically arrive with the first southerly winds in May to find the La Pérouse Bay area still covered with snow and ice. These early birds often temporarily disappear. As the season progresses, a succession of cold and warm periods occur, and increasing numbers of geese are seen with the southerly winds associated with the passages of these weather systems. The arrival peak (or peaks) is not closely correlated with the first arrival, but depends on the rate and pattern of subsequent warming and snow melt.

The arrival pattern at La Pérouse Bay is consistent with visual and radar observations of the influence of weather on the timing of migratory departures of Snow Geese from staging areas in southern Manitoba and North Dakota (Blokpoel 1974; Blokpoel and Gauthier 1975). Major flights occur when the passage of frontal systems brings southerly and southwesterly winds. Radar traces show geese travelling day and night heading towards the Hudson Bay or James Bay coastline.

Conditions on the spring staging areas in southern Manitoba and North Dakota influence the date the first geese depart to the north. Lacking specific data on the departure dates of Snow Geese from the staging areas, we correlated mean April temperatures in Winnipeg with the first arrival of Snow Geese at La Pérouse Bay (Fig. 2.5). Higher April temperatures are associated with earlier first arrivals at La Pérouse Bay. Higher temperatures presumably cause earlier snow disappearance and forage availability, allowing geese to acquire nutrients necessary for earlier departure to the north.

The temperature on the staging areas is not the only determinant of arrival on the breeding grounds. April temperature in Winnipeg is not correlated with peak arrival (as opposed to first arrival), which is much more dependent on weather conditions further north. The James and lower Hudson Bay coastlines provide important sources

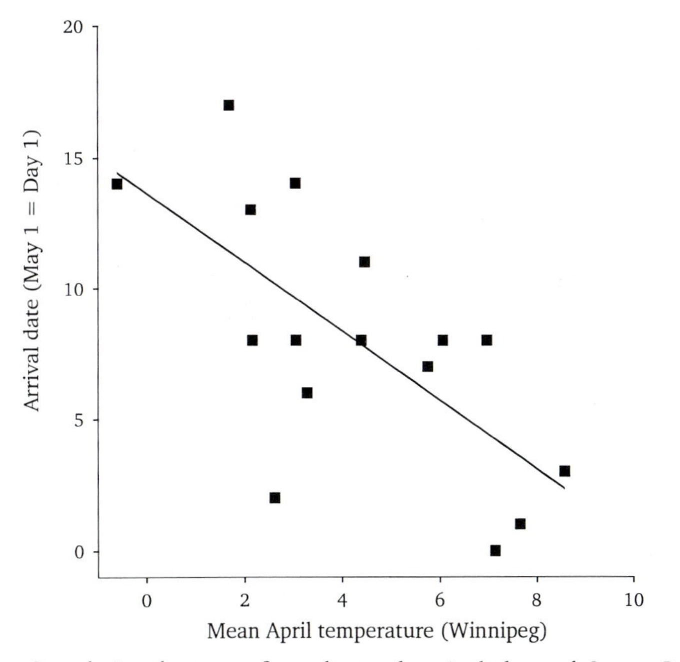

Fig. 2.5 Correlation between first observed arrival date of Snow Geese at La Pérouse Bay and mean April temperature (°C) in Winnipeg, 1972–1987 ($r = -0.68$, $n = 20$, $P = 0.004$).

of nutrients, particularly protein, for the geese en route to breeding colonies (Wypkema and Ankney 1979). Geese can feed only after snow and ice have disappeared at least sufficiently to expose some of the perennial vegetation, mainly rhizomes and stolons of monocotyledonous plants, such as *P. phryganodes*, *E. arenarius*, *Equisetum* spp., *Triglochin palustre* and *C. subspathacea*. Northerly progress up the coast is more directly influenced by snow disappearance than by temperature *per se*. Large numbers of geese are seen at the colony almost as soon as suitable vegetation emerges from beneath the snow.

Geese actively feed upon arrival. In years when the number of geese is large relative to the available vegetation, considerable damage can be done to the salt marsh feeding areas. Plants are uprooted, and areas where they grew are converted to anaerobic ponds which do not support the growth of higher plants. In places, huge areas of the salt marsh have been changed into plantless, foodless, areas, as has happened at other Snow Goose colonies (Kerbes *et al.* 1990). This loss of vegetation is detrimental to the geese in the long run, and restricts the use of such areas by goslings and adults in the critical post-hatch period (Jefferies 1988*a*; Cooch *et al.* 1993).

2.2.3 Nest initiation and egg laying

Field work at initiation involves slogging through partially flooded icy tundra, searching likely spots where females may be building nests and laying eggs. Unlike many birds, geese construct nests during the laying period, with nest size increasing as eggs are added. This both saves time and prevents the waste of down feathers if a nest is abandoned, as many are. Figure 2.6 shows the development of a nest and its clutch. Nest sites are marked with a wooden stake placed approximately 1 m from the nest.

Nests found at the one-egg stage are referred to as 'intensive' nests. These provide us with a reasonably complete picture of the initial components of reproductive success needed for this study (Chapter 4), including: the date of nest initiation; the sequence, size, and total number of eggs laid; and the frequency of nest failure prior to the start of incubation.

What determines when nest initiation begins? Successful nest initiation requires bare ground which is not subject to flooding, and is thus also regulated by local weather conditions. First and mean initiation dates (the dates of the first egg of the first nest and the average of nests sampled) vary widely from season to season (Table 2.1). One can predict mean initiation date well from two pieces of

Fig. 2.6 The development of a Lesser Snow Goose nest and clutch. (a) A typical one egg nest. (b) Two eggs with some down. (c) A typical nest site, with a four-egg nest. (d) A five-egg clutch showing numbered eggs. (e) A nest covered with down, as a goose would normally leave it unless suddenly disturbed. (f) A nest following a snow storm, showing the imprint of the incubating female.

Table 2.1 Annual variation in nest initiation and hatching dates of Lesser Snow Geese nesting at La Pérouse Bay, 1973–1992

	Initiation				Hatching		
Year	First	Mean	SD	*n*	Mean	SD	*n*
1973	24 May	28.5 May	2.0	273	24.3 June	1.9	1179
1974	27 May	31.0 May	2.5	512	26.5 June	1.9	1123
1975	19 May	25.7 May	2.4	438	22.2 June	2.3	972
1976	18 May	21.9 May	2.1	325	19.2 June	2.5	1286
1977	14 May	19.3 May	2.5	190	15.0 June	2.3	1509
1978	30 May	4.0 June	2.7	274	1.3 July	3.0	1088
1979	29 May	31.7 May	1.6	71	26.8 June	1.7	981
1980	12 May	18.7 May	2.8	405	14.2 June	2.9	1613
1981	23 May	28.2 May	2.0	453	24.1 June	2.1	1811
1982	13 May	21.7 May	2.5	473	18.0 June	2.4	1802
1983	8 June	11.2 June	1.6	316	6.8 July	1.6	2246
1984	21 May	30.3 May	3.1	324	25.4 June	2.5	2316
1985	17 May	23.9 May	2.8	520	19.8 June	2.9	2482
1986	23 May	26.1 May	1.6	260	20.3 June	1.7	2011
1987	26 May	31.0 May	1.6	165	26.0 June	1.7	1273
1988	26 May	30.0 May	1.9	307	25.7 June	1.9	1804
1989	23 May	2.3 June	3.5	223	27.8 June	3.6	1355
1990	24 May	29.1 May	1.6	315	23.6 June	1.5	1610
1991	19 May	24.0 May	1.7	80	19.5 June	1.7	515
1992	3 June	4.4 June	1.4	57	29.1 June	2.0	230

information: the mean April temperature in Winnipeg (T_{AW}), Manitoba, and the amount of snow on the ground as of the first of May at Churchill (S_{MC}) using the equation:

$$\text{mean initiation date} = 26.2 - 0.80 \times T_{AW} + 0.24 \times S_{MC}$$

counting days from May 1 ($r^2 = 0.84$, $n = 16$, $P < 0.0001$; E. G. Cooch, unpublished analysis). Snow Geese thus breed later in years when it is warm on the staging grounds and the snow is deep on the nesting grounds. We have enough faith in this equation to use it as the basis for making airline reservations to bring up a crew for our banding operation which will be held 3 months later.

It would be valuable to know the time interval between an individual bird's arrival and first laying. We lack individual arrival information because extensive flooding limits our mobility at this time and because local breeders are not readily found among the much larger numbers of birds using the area as a stopover on migration to colonies further north. However, in years with clear peaks of first arrival ($n = 10$), the interval between arrival and the peak of nest initiation is 11.0 ± 1.7 days, ranging from 7 to 13 days. This interval coincides with the time necessary for a Snow Goose to develop fully a ripening egg follicle into an egg. The number of yolk rings can be counted using a staining technique developed by Grau (1976). Snow goose eggs have an average of 12 rings (F. Cooke, unpublished data). Assuming deposition of one ring per day, in most years females begin rapid yolk formation around the time of arrival at the breeding colony.

The initiation and laying period at the colony is synchronized, with the last geese beginning to lay their eggs 11.9 ± 3.1 days, ranging from 7 to 18, after the first geese begin. The initiation date distribution varies annually (Fig. 2.7; Findlay and Cooke 1982*a*,*b*), although, in general, it exhibits the slight positive skewness characteristic of most nesting curves (Gochfeld 1980). Annual differences in the pattern of snow and melt water disappearance account for much of the annual variation in the shape of these curves.

Nest sites are chosen primarily by the females, who are strongly philopatric. First-time breeding females nearly always return to their natal colony. Small groups of first time breeders often clump

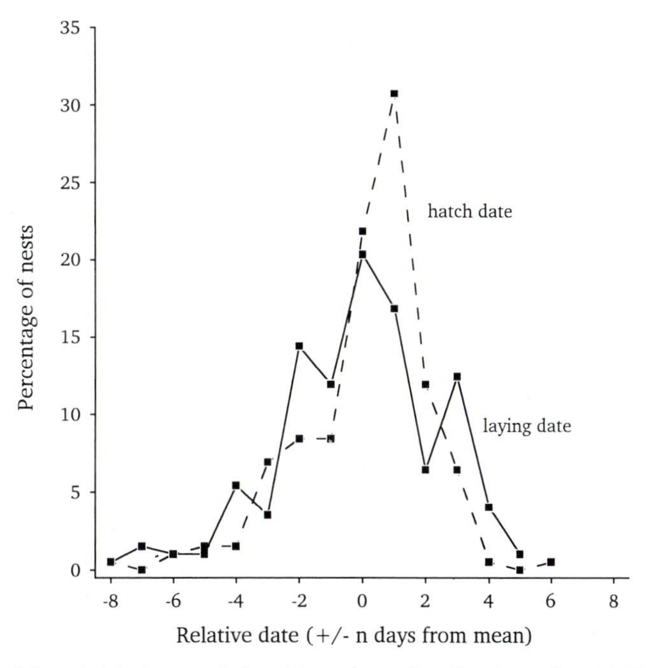

Fig. 2.7 Nest initiation and hatching date distributions for 1978, plotted relative to mean laying and hatch dates. The initiation date distribution is wider than that at hatch because geese laying smaller clutches initiate later than those laying larger clutches, resulting in a more synchronized hatch than laying period.

together, either in areas peripheral to the existing colony, in previously unused areas within the colony or, to a lesser extent, in recently vacated areas within the colony (Cooke *et al.* 1983). They often select a nest site in that part of the colony closest to the area they used for feeding when they were goslings. Once a female nests in an area, she tends to reuse it in successive seasons. Most re-encountered females nest within 50 m of their previous nest site (Cooke and Abraham 1980). Such local fidelity occurs not only among older birds, but also in 3-year-old birds that have bred only once before in that location. There is no evidence that successful young birds move into 'better' nest locations as they grow older (Cooke *et al.* 1983). These facts suggest that birds are just as tenacious to a location when they are second time breeders as when they are older breeders, and points to the fact that previous occupancy of an area is a major predictor of future nest location. Reproductive performance also plays a role in site fidelity, in that less successful females are more likely to move to a different part of the colony (Hik 1986).

Precise nest site location is influenced by the patterns of snow cover, melt water, and vegetation, as well as the presence of old nest cups. In years when early nest sites are scarce due to snow or water cover, there may be considerable conflict between pairs over the site. Even so, a pair may be absent from a site for much of the time during the early stages of egg laying, and two (or more) females may choose nest sites within 1 m of each other, or even lay at the same site. Nests in such close proximity probably occur because females lay at different times of day and may be unaware of each other until the site is occupied more continuously. Pairs do not incubate nests closer than 1.5 m from one another, and so one of such nests is usually abandoned, often at the one- or two-egg stage.

Eggs are laid at intervals of approximately 33 h (Schubert and Cooke 1993), although there may be variation in this pattern. Regular incubation commences at the laying of the penultimate or of the last egg and occupancy of the nest by the female is more or less continuous from then until hatch. Clutch size is thought to be determined through allocation of a female's limited nutrient reserves, most of which she transported to the colony on migration, between egg production and an incubation reserve (Ankney and MacInnes 1978, Chapter 10).

We classify causes of nest failures during the laying period as abandoned, depredated, or either. Cold eggs remaining in a nest where no new eggs are laid for two or more consecutive days are considered abandoned. Predation is assumed if there is any direct evidence of severe egg damage, missing eggs, or predator presence at nests known to be occupied the previous day. When no eggs remain in an abandoned nest, it is often difficult to tell whether the eggs were depredated before or after the nest was abandoned, and such cases are coded as 'predated or abandoned'. Eighty-eight per cent of 560 nests abandoned from 1975 to 1982 were abandoned at the one- or two-egg stages (Hamann 1983), and most of these occurred during the first few days of egg laying in the colony. Birds which lose their first one or two eggs to predators nearly always abandon a site (Collins 1993). When predation is suspected, there is a 16.5 per cent probability of nest loss in the pre-peak period, 5.2 per cent during the peak period, and 1.5 per cent in the post-peak period. Predators therefore are more likely to cause pre-incubation failure in early-nesting birds. The primary predators during this time period are Herring Gulls (*Larus argentatus*), Arctic Foxes (*Alopex lagopus*) and Caribou (*Rangifer tarandus*). Arctic wolves (*Canis lupus*) and Ravens (*Corvus corax*) may also depredate some nests during this period.

We have little direct information on the subsequent behaviour of birds which lose eggs before completing their clutch. Since the birds are wary of people, it is difficult to read the rings that would individually identify them. The principle options are to continue laying elsewhere, to lay parasitically in the nests of others, or to resorb remaining follicles and opt out of laying for the season. At least two out of three marked females whose nests were experimentally destroyed at the one egg stage successfully established a new nest in which they laid additional eggs (Ganter and Cooke 1993). An indirect analysis compared annual rates of pre-incubation failure with mean annual clutch size. Increased clutch sizes would be expected if all birds became parasites, while decreased clutch sizes would be expected if all birds laid continuation clutches. No relationship was found. Individuals may pursue either strategy, or resorb follicles, perhaps depending on the number of eggs remaining to lay at the time of failure and/or their body condition.

2.2.4 Incubation

The incubation period commences with the laying of the last or penultimate egg and ends with the hatching of the eggs. The female only incubates for an average of 23.6 days, counted from the date eggs are consistently warm when the nest is visited at the end of laying through and including the date of the first pipped egg. Larger clutches have slightly longer periods than smaller ones (23.3 for two-egg clutches versus 23.6 for six-egg clutches). Controlling for clutch size, incubation periods are 0.1 day longer for nests initiated at the end of the laying period versus those at the beginning.

During the incubation period, females have been thought to remain on the nest almost continuously. However, B. Ganter (personal communication) observed that some females may leave their nest for up to an hour at a time and up to 3 h per day.

Nests may be abandoned by nesting geese if weather conditions are particularly harsh. Following a period of inclement weather towards the end of incubation, emaciated females may be found close to their abandoned nests or even found dead on or near the nests. Dead females are likely to be lighter than average (Ankney and MacInnes 1978). Such events are rare at La Pérouse Bay, however, and most females successfully hatch their eggs.

The behaviour of the male during incubation was studied in 1976 and 1977 by Mineau (1978). Approximately 20-30 per cent of the male's time is spent feeding at the beginning of incubation, increasing to more than 30 per cent by the end. As feeding increases, the time spent resting or alert decreases. Most of the time the male is in close proximity to the nest and Mineau estimated that it is absent from the nesting area less than 10 percent of the time. Feeding generally occurs within the area of the nest. Males are *c.* 19 per cent lighter at the end of incubation than at the start, a loss about half that of nesting females (Ankney 1977).

During incubation, the birds may defend their nest and clutch against a variety of predators. Total nest loss can result from groups of Herring Gulls, Caribou, Arctic and Red (*Vulpes fulva*) Foxes, Arctic Wolves, and Polar Bears. The latter two, and Bald and Golden eagles (*Haliaeetus leucocephalus* and *Aquila chysaetos*) occasionally depredate the incubating female as well as the clutch of eggs but such events are rare. Partial clutch loss is usually the result of predation by lone Herring Gulls, Parasitic Jaegers (*Stercorarius parasiticus*), Ravens, Sandhill Cranes (*Grus canadensis*) and both Arctic and Red Foxes. The extent of loss may depend more on the predator(s) than on the behaviour of the pair. Approximately 13.4 per cent of nests which hatch have some level of partial clutch loss (M. Collins, personal communication).

2.2.5 Hatching

Hatching begins when the first signs of a star-shaped crack appear on the egg. The gosling gradually creates a small hole which is then extended around the widest part of the egg until the entire blunt end

Fig. 2.8 A mixed colour phase brood of 'fluffy' goslings. Blue phase on the left, white phase on the right.

of the egg is pushed off. The gosling kicks itself out of the shell completely and within a few hours has completely reabsorbed the yolk sac and the down feathers have dried (Fig. 2.8). The downy goslings leave the nest usually within 24 h of hatching and are taken by their parents to the brood rearing areas.

The hatching period is one of intense activity for us as well as the geese, and extra researchers are brought into camp. Each is assigned one of six to eight plots containing approximately 300 non-intensive nests. These, plus the four to six intensive plots are visited daily from 2 days before the first nest on the colony has hatched until all the nests are hatched or unsuccessful. The status of each nest is recorded, goslings present are weighed and classified as blue or white by down and foot colour, and sometimes sexed by eversion of the cloaca. A small individually numbered tag is placed on the web of one foot for identification if the bird is re-encountered (Fig. 2.9). When possible, we record which gosling came from which egg, to relate back to laying sequence in the intensive nests. If not noted on previous nest visits, we record parental colour phases, and examine legs for the presence, colour, location, and alphanumeric codes of any bands on their legs, to determine parental age or individual identity. Hatch is the easiest time to gather these data because parents are most attentive to the nest at this stage.

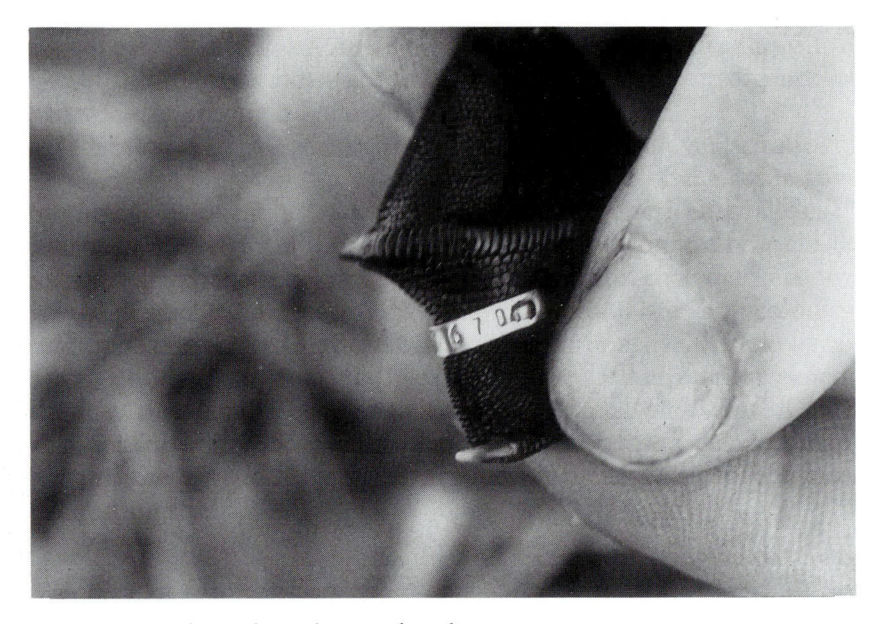

Fig. 2.9 The foot of a web tagged gosling.

Pairs vary in their nest tenacity and defensive behaviour towards us. Some parents stay extremely close to the nest and may even attack the researcher (Fig. 2.10). Others others fly more than 100 m from the nest when it is visited. We quantify parental 'stickiness' as closest distance to the nest by males and females when web tagging goslings. First time breeders are less tenacious to the nest, but become more so as they age (Cooke *et al.* 1981).

Eggs within a clutch generally hatch synchronously, through a mechanism involving an interaction between parents and eggs (Davies and Cooke 1983*b*). Unincubated eggs added to a clutch 1–3 days after the onset of incubation are able to hatch together with existing eggs. Despite this synchrony, Syroechkovsky (1975), on Wrangel Island, and Cargill and Cooke (1981), at La Pérouse Bay, found that hatching sequence correlated with laying sequence. This implies that some development of embryos occurred throughout the laying period.

There is synchrony of hatch at the colony level as well. The whole process, from hatching of the first nest to hatching of the last one, usually occurs within a 2 week period. Factors influencing this synchrony have been examined by Findlay and Cooke (1982*a*,*b*). The timing of hatch obviously follows from the timing of laying and incubation onset. Since birds laying larger clutches generally initiate earlier in the season, but have longer laying periods than those with

Fig. 2.10 Lesser Snow Geese threatening a field worker at a nest.

smaller clutches, the hatching is more tightly synchronized than laying. A typical pattern of laying and hatch from the same sample of nests for a season at La Pérouse Bay is shown in Fig. 2.7. Table 2.1 shows the annual mean and variation in hatching dates for our whole colony sample from 1973 to 1992. Fitness consequences associated with timing are considered in Chapter 11.

Losses during the hatching period are of eggs or newly hatched goslings. Eggs may fail to hatch because they are infertile, because the embryo died or because the embryo's development is retarded relative to the others in the nest. A small number of the embryos are severely deformed. We record the types of hatching failure where possible. An egg becomes a gosling, for our purposes, when it reaches the star pipped stage described above. This definition is precise and measurable, if arbitrary. Losses at the gosling stage occur if the gosling is unable to extricate itself from the egg, predated, trampled by the female or abandoned by parents and more developed siblings. Some abandoned goslings have morphological abnormalities and, as a group, have a higher level of developmental instability (Rockwell *et al.* in preparation).

While gosling loss is not extensive during this period (less than 5 per cent), this is a time when our activities may measurably affect hatching success and subsequent gosling survival. Our nest visits to weigh and web tag goslings disturbs the colony and distracts adults from their parental duties. Avian predators such as Herring Gulls and Parasitic Jaeger are better able to remove eggs or newly-hatched goslings from adjacent nests where we are not present to deter predators. We may also cause the parents to take their goslings away from the nest early, leading to an excess of goslings abandoned at the nest. Disturbance may occasionally cause goslings that are already mobile to leave their nest and attach themselves to another family. We probably also reduce the frequency of goslings that are unable to extricate themselves from the egg, because our web tagging procedure, which may be done at the pip stage, may indirectly assist goslings having difficulty. This is especially true for rare cases where embryos are reversed within the egg and pipping begins at the narrow end. Without our intervention, such goslings would die.

2.2.6 Brood rearing

Brood rearing begins when the goslings leave the nest with their parents, and ends when they acquire their primaries, learn to fly and begin the migration south. We collect data in two ways during this period. Throughout the period, we observe family groups from towers situated on the coastal marshes of La Pérouse Bay. Near the end of the period, we capture large numbers of birds in our annual banding operation.

Families leave the nesting areas soon after the goslings are dry, and move onto the coastal marshes. Families remain together as a 'unit' and join loose assemblages of 20–100 families. During the 4–6 week brood rearing period, the birds spend most of their time grazing on sedges and grasses. The goslings must grow from their average hatching weight of 85 g to a fledging weight of near 1600 g for males and 1500 g for females. They replace their downy feathers with juvenile plumage and grow their flight feathers. Adults provide parental care during this period by brooding the goslings during rain and cold weather and by warning and defending against predators, but they too must feed extensively. Both sexes undergo a complete remigial moult and the female regains some of the weight lost during incubation. Particularly early in the period, the male is the primary vigilant parent.

Gosling losses during the brood rearing period are of two types, with different consequences for parents. Parents may lose their entire brood, an event most likely to occur shortly after the family leaves the nest. As with total nest failure, most of this loss results from the action of groups of herring gulls. Parents that fail totally at this stage may join those that failed completely during egg laying or incubation

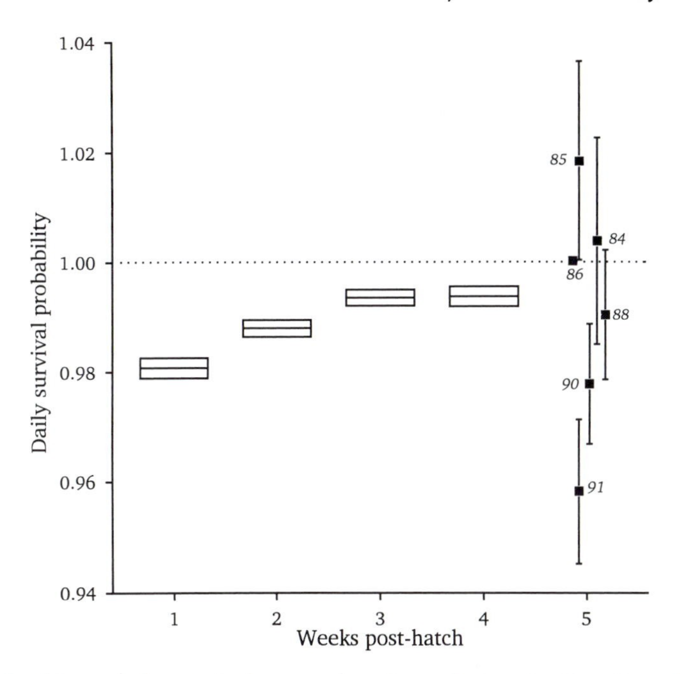

Fig. 2.11 Mean daily survival rates of goslings during the first 5 weeks after hatching, based on repeated observations of individually identified broods. Total brood loss is not included. (Adapted with permission from Williams *et al.* 1993*b*).

and leave the La Pérouse Bay area (see below). Most parents lose only a part of their brood before the goslings fledge. As with total brood loss, most of this occurs shortly after leaving the nest.

We examined the temporal changes in gosling survival probability based on changes in the brood sizes of individual families seen from observation towers (Fig. 2.11; Williams *et al.* 1993*b*). Survival rates were lower at the start of the period, when goslings are small and quite dependent on parental care, and increased throughout the following 3 weeks. In recent years, daily survival rates have declined at the end of this period, suggesting stress on the goslings as foraging habitat quality has declined. The main predators causing partial loss are Herring Gulls, Parasitic Jaegers, Ravens, Arctic and Red Foxes, Arctic Wolves, and Polar Bears.

Just before adults regain their ability to fly, a team of 12–20 persons and a helicopter round up goose families in our annual banding operation. Breeding adults with their surviving offspring are in loose assemblages on the coastal marshes. A few non-breeding yearlings are caught each year and, very rarely, larger groups of non-breeding birds are caught with the moulting flocks. A 'catch crew' in the helicopter locates adjacent flocks that total *c.* 1000-1200 individuals. Members of the crew are put on the ground and herd the flocks together. Once the round up begins, the helicopter ferries a 'net crew' to a nearby dry site where they erect a keyhole-shaped net enclosure plus two post-processing holding pens with water and forage for the birds. The combined flock is slowly driven in through the keyhole entrance, the opening is closed, and the flock is processed for the remainder of the day. The annual banding operation usually lasts 5–6 days.

During processing, 50–100 birds are separated from the flock in the main net and driven into a small catch pen. We catch and band the goslings first, to minimize their time in the main net and maximize their time in the pen with food and water. Male and female goslings are banded with numbered, wildlife service bands and females are also marked with year specific alphanumeric bands. Sex, colour, and web tag number (if present), are recorded for each bird. Web tagged goslings are also weighed and measured. Unbanded, unknown-aged adults of both sexes are banded with both wildlife service and year-specific alphanumeric bands, the location of which distinguishes them from the known-aged goslings. The bands of recaptured adults are recorded and as many as possible are weighed and measured. When the entire flock has been processed, the birds are released *en masse*.

2.2.7 Non-breeding birds

The salt marshes at the periphery of the nesting colony, as well as areas within the colony, are used extensively throughout the laying and incubation period by non-breeding birds. They associate in small groups of 20–50 birds. Many of these birds are yearlings that have a plumage visibly distinct from the adults. Yearlings return to the colony with their parents but when the parents begin nesting, they generally leave the nesting area and associate with other non-breeding birds. However, yearlings occasionally visit the nesting area and smaller numbers are seen near their parents. Some even accompany the new family to the feeding flats.

Not all non-breeding birds are yearlings. Birds of age 2 years and older are not distinguishable from one another on the basis of plumage, but colour bands on the legs of some of the birds allow us to determine their age. While we do not have extensive data on non-breeders, some generalizations are possible. Non-breeding birds on the colony can be fit into three categories: (1) pre-breeders, those not yet old enough to breed; (2) failed breeders who have started a breeding attempt but failed to complete the incubation; and (3) non-breeders, birds which have bred in a previous year but 'opted-out' of nesting in the year of observation. It is unfortunately easier to define these categories than to assign adult-plumaged birds to them. In 1977, Mineau (1978) observed non-breeding birds prior to hatch from an observation tower. Fifty-three per cent of 113 adult-plumaged birds carrying colour bands were 2-year-olds, far higher than their frequency among the banded breeding population. Thus non-breeding is more frequent among younger birds (Chapter 7).

Yearlings are unpaired in the non-breeding flocks, in contrast to most adults. Colour-banded females return to their natal colony far more frequently than males. Sulzbach (1975) sighted 15 per cent of 185 yearling banded females compared to only 8 per cent of 225 yearling banded males in 1974. This ratio may differ from the overall sex ratio of male and female yearlings, although survival analyses do suggest that the mortality of males in their first year is higher than that of females (Francis *et al.* 1992*a*). Higher female than male natal philopatry is characteristic of Snow Geese at other stages of their life. But why should this be manifest in the first year of life, when yearlings generally are thought to return with their parents? Young birds may become separated from their families and return north on the spring migration without them. In such cases, females might home to their natal areas despite the lack of parental guidance, while males might more readily accompany the other geese to different Snow Goose colonies.

The asymmetry in philopatry of the sexes is even stronger in the 2-year-olds. Many 2-year-old females are seen among the non-breeding flocks but no 2-year-old males have been observed. The paired status of 2-year-olds shows that pair formation first occurs sometime between the departure of the geese as yearlings and their arrival on the breeding grounds the following season.

Most non-breeding birds do not remain within the geographic bounds of La Pérouse Bay for the entire breeding season. Once hatching begins for the nesting birds, there is a mass exodus of failed and non-breeders. The departure usually takes place when winds are from the south and often coincides with a moult migration of Canada Geese (*Branta canadensis*) that come from southern Manitoba and are on their way north along the Hudson Bay coast. A moult migration of Snow Geese was documented for La Pérouse Bay in 1977 by Abraham (1980*b*). Non-breeding yearlings and older birds left La Pérouse Bay and were seen flying north. Some of those birds were later captured in association with non-breeders from the much larger Snow Goose colonies in the McConnell, Tha-Anne, and Maguse

River area during a massive banding effort by the Canadian Wildlife Service. One female, regularly observed as a 2-year-old non-breeder at La Pérouse Bay from 14 May to 4 June 1977, was captured in the CWS banding operation on 23 July 1977. The following year she bred at La Pérouse Bay.

2.2.8 Effects of our activities

Our research activity undoubtedly affects the things we measure. Our presence has not prevented the colony expanding from 2000 pairs to more than 22 500 pairs, but would the expansion have been even greater if we had not disturbed the birds?

We have no way of assessing the impact our presence has on the birds. Early nesting attempts are often abandoned, but it is not clear whether the rate is higher because of our work. Losses during incubation are relatively low (Chapter 4), thus we cannot be substantially influencing fitness measures at this stage. Our greatest impact may well be at hatch, as discussed above. Observers in the towers have reported an increased number of goose families with day-old goslings leaving the colony immediately before the daily arrival of researchers. Our activities may both increase the rate of abandonment of late-hatching goslings and increase the mixing of goslings among families (Chapter 5). Finally, our banding drives are stressful, especially for goslings (Williams *et al.*, 1993*d*).

Our general presence in the area may deter certain predators, such as wolves, Caribou and foxes. We can state with certainty that the researchers themselves are at times deterred from the colony by the presence of other predators, such as Polar Bears and Black Bears (*Ursus americanus*)!

Our measures of reproductive failure probably differ from those in an undisturbed colony, but these differences should not greatly influence within-population analyses in which we contrast fitness parameters for different segments of our population (for example blue versus white morphs). Observer effects will be a problem in this context only if they are biased with respect to the categories of birds being compared.

2.3 *The non-breeding season*

2.3.1 Fall migration

Once goslings fledge and their parents have regained the ability to fly, families make frequent sorties from the colony area and may be seen as far away as Churchill (50 km). While we do not have precise data on the timing of departure from La Pérouse Bay, substantial numbers of the geese have left by mid to late August. Once the birds leave La Pérouse Bay, we can only trace them indirectly. Hundreds of thousands of Snow Geese concentrate at suitable locations along the Central and Mississippi Flyways. The arrivals and departures of these flocks are monitored by federal, provincial and state agencies whose reports give us a general picture of the migration patterns on an annual basis. For more precise analyses, we use band recovery data that provide information on location, date and cause of death. Figure 2.12 shows the spatial and temporal distribution of band recoveries of La Pérouse Bay Snow Geese from August 1977 through February 1978, a typical year.

In September, most birds are still to be found along the coast of Hudson Bay and James Bay, though some have already moved to southern Manitoba and North Dakota. Most birds from La Pérouse Bay migrate directly from Hudson Bay rather than via James Bay (Francis and Cooke 1992*a*). Figure 2.12 probably underestimates the number of geese still in the north, because hunters are infrequent along the Arctic coast and many of these are Cree who have less opportunity to report the bands. By October most birds have left the north and are concentrated in southern Manitoba and North and South Dakota, and a few have moved down the Missouri River valley in Iowa, Nebraska, Kansas, and Missouri. A southward movement continues into November with fewer birds in the northern states and provinces, and more in the central states. Some birds have already moved to the wintering areas of Texas and Louisiana. There is major direct flight from the central states to the Gulf coast, a distance of some 800 k. Most geese overfly the states of Oklahoma and Arkansas, though recently more have been staying there. Most birds are on the traditional wintering grounds along the Louisiana and Texas coasts by December.

The geese feed on a variety of marsh grasses and sedges as they move down the Hudson and James Bay Bay coasts (Alisauskas 1988). Once they reach the northern and central states, corn (*Zea mays*) left in the fields after harvest forms a major portion of the diet. Those birds that move further south for the winter, feed in rice growing areas and in coastal salt marshes. Currently, there is every indication of food abundance as a result of human agricultural practice, although projected declines in agriculture and continued urbanization could have future negative effects on the goose population.

Outbreaks of diseases are occasionally reported among migrants.

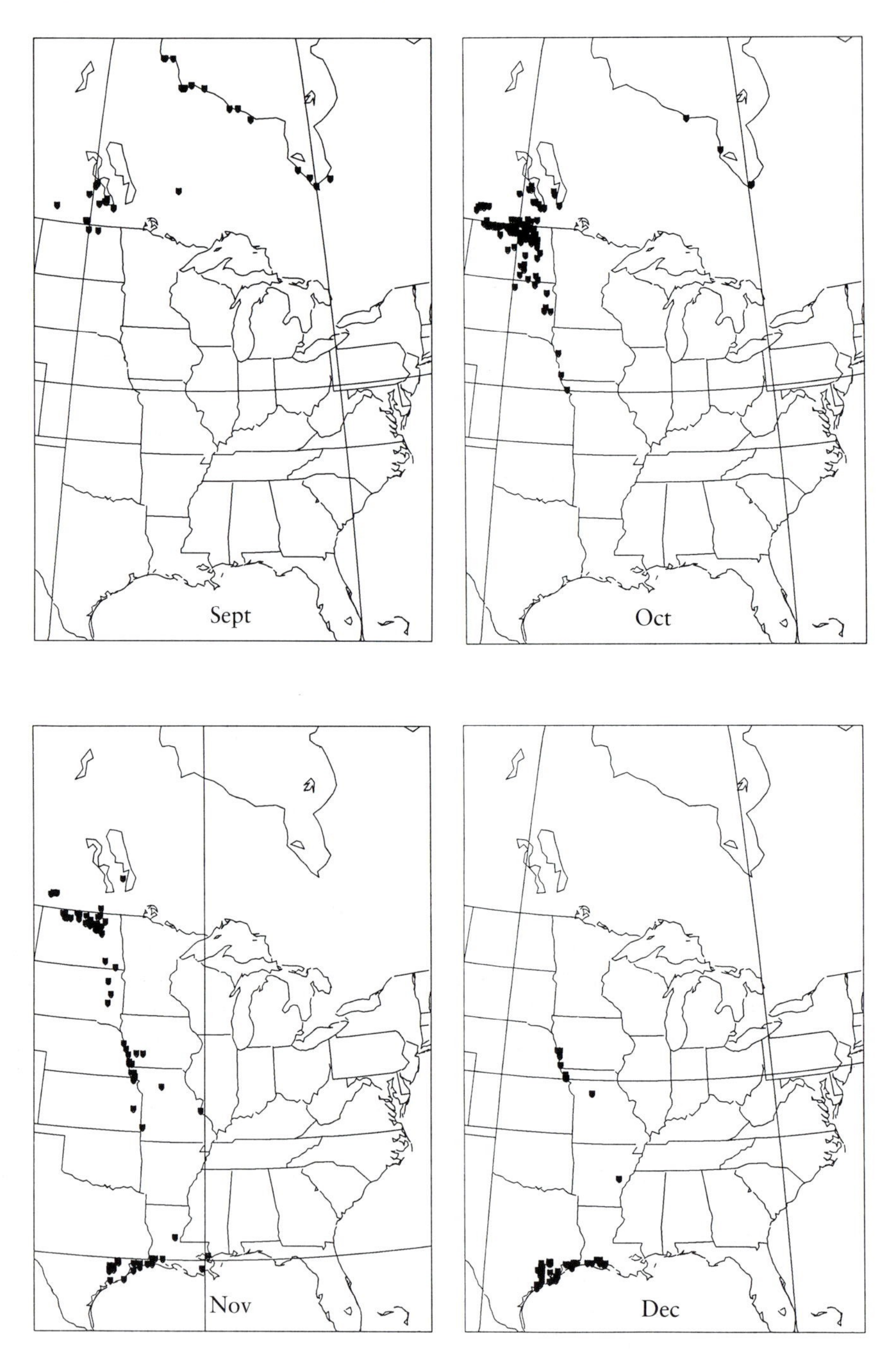

Fig. 2.12 Fall and winter band recovery locations of Lesser Snow Geese marked at La Pérouse Bay. Recoveries were reported between September 1977 and January 1978; the February plot pools data from 1970–1983.

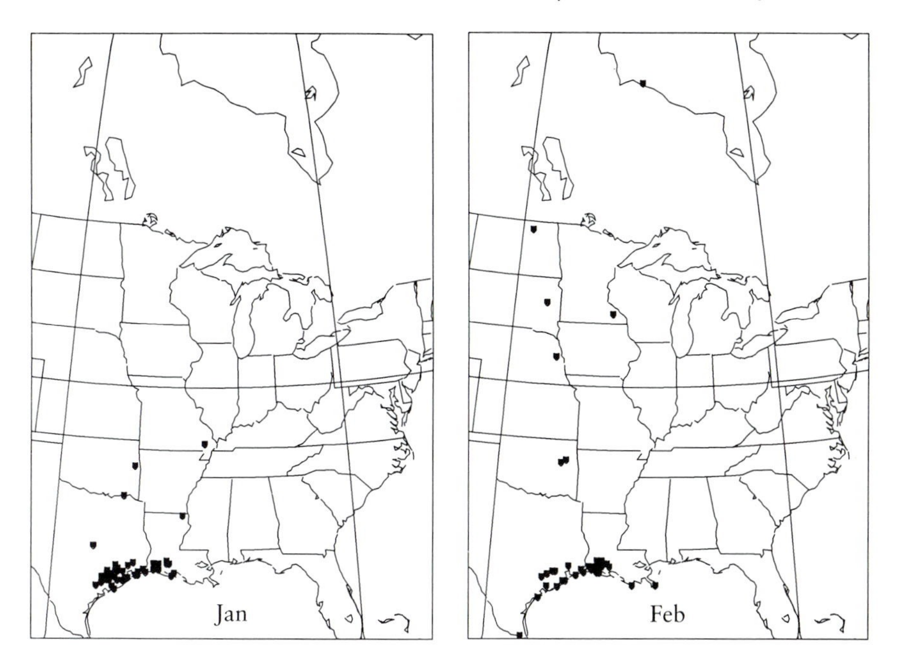

These have resulted in the deaths of hundreds of individuals in areas of major Snow Goose concentration. Avian cholera, new duck disease and duck viral enteritis are all diseases which have been reported (Wobeser 1981). However, since hundreds of thousands of geese may be present at the locations where outbreaks are observed, disease is not often a major cause of mortality during migration.

By far the major reported cause of mortality is shooting. Man is undoubtedly the main predator and 97 per cent of our band recoveries are of hunted birds. Using band recovery data obviously overrepresents the hunting as a cause of mortality, since birds shot by hunters are far more likely to be reported to the banding office than birds dying from other causes. Most birds shot are retrieved and can be reported, whereas birds dying from other causes may never be found. Hunters have been the main cause of mortality among adult geese, but not necessarily among immature birds (Francis *et al.* 1992*a*). While immatures are more vulnerable to hunters than adults (Boyd *et al.* 1982), many appear to die before ever reaching areas where hunters are frequent. This has been particularly true in recent years for the La Pérouse Bay population and in years when overall immature mortality is high. Years with higher immature mortality actually have a lower proportion of immatures among the geese reported shot by hunters, the opposite of what is predicted if hunters

Table 2.2 Causes of mortality of reported with band returns from Lesser Snow Geese banded at La Pérouse Bay, 1969–1992

Cause of death	Frequency
Shot	10 302
Found dead	281
Disease-related	39
Caught by miscellaneous animals	14
Found dead: skeleton or bones only	12
Collected as scientific specimen	12
Illegally taken	7
Caught due to injury	6
Weather-related mortality	6
Found dead on highway	4
'Obtained' no information	4
Caught by dog	4
Hit power lines or towers	3
Caught by hand	3
Caught due to lead poisoning	2
Miscellaneous recovery	2
Caught by raptor	1
Found in a building	1
Mortality when banded	1
Oiled feathers	1
Poisoned	1
Joined captive or domestic flock	1

were responsible for the high immature mortality (Chapter 7; Cooke and Francis 1993). It seems more reasonable to assume that in years when immature mortality is high, many birds die of other causes (starvation, disease or predation) prior to reaching the more southerly locations. Many of these deaths may occur along the Hudson Bay coast, or immature birds may fail to survive the migration across the boreal forest.

Other types of death reported from the band recovery data are listed in Table 2.2. Often the causes of death are not clear, particularly when the birds are found dead. They may have been injured by a hunter and subsequently died. Predators are not identified in the above list, but there are records of coyotes, foxes, and eagles killing geese during fall migration. There is one interesting death worth noting. A flock of geese was seen flying in southern Texas during a violent thunderstorm. Several were struck by lightning and some of the birds fell to the ground, dead. One of these birds was carrying a band which had been applied at La Pérouse Bay.

2.3.2 Winter

Most of the La Pérouse Bay geese winter in the Mississippi or Central Flyway (Fig. 2.12). Historically, most of the birds wintered primarily in Louisiana, Texas and the adjacent part of Mexico. Since the early to mid 1980s, increasing numbers have been remaining in Iowa, Nebraska, Missouri, and Arkansas. A handful of birds have been re-encountered away from this area. Westward, single birds were recovered in California and New Mexico and three in the western part of Mexico. Eastward, two birds were recovered in Quebec, single birds in New Jersey and New York, two from Maryland, and one each from North Carolina and Georgia. The most far-flung re-encounter was a live bird identified in the Netherlands! A 4 year old male was part of a flock of 18 Snow Geese sighted in April 1980 (Blankert 1980) which had presumably wintered in Europe. This bird provided the first direct evidence that wild Snow Geese from North America occasionally appear in Europe.

Wintering Snow Geese utilize two major habitats in the Gulf coast region. Coastal salt marsh is the traditional habitat. McIlhenny (1932) stated that in Louisiana Snow Geese were never found feeding more than 8 miles from the salt beaches. He recorded them as grubbing rather than grazing, feeding primarily on the rhizomes of *Scirpus robustus*, *S. americanus*, *Spartina patens*, and the stolons of *Sagittaria platyphylla*. They also feed on the seeds of the grasses *Leptochloa fascicularis* and *Echinochloa crus-galli*. Sections of the marsh that have been freshly burned over are favoured when available. In Texas they feed in salt marshes and wet prairies close to the sea, feeding on *Spartina patens* and *Scirpus oneyi* (Stutzenbaker and Buller 1974). They also feed in the adjacent wet prairies dominated by *Andropogon* spp. grasses.

Within the past 50 years, the rice growing prairies have become the second major gulf coast wintering habitat of Snow Geese. In these regions, rice is grown on a rotation system, with rye grass (*Lolium* sp.) planted for two or more years after the rice crop. This is for cattle grazing, but it also provides a valuable food supply for grazing Snow Geese. These areas, mainly west of Houston, are now a major wintering areas for Snow Geese.

Birds from La Pérouse Bay are found throughout suitable areas of Louisiana and Texas (Fig. 2.12). A few are found as far east as the Delta National Wildlife Refuge at the Mississippi delta. More commonly they occur in the Rockefeller and Sabine Refuges of western Louisiana. In Texas, large numbers are found at the Anahuac National Wildlife Refuge and along the coastal marshes west of Galveston into eastern Mexico. Large numbers share Arkansas National Wildlife Refuge with wintering endangered Whooping Cranes (*Grus americana*) and occur in the Corpus Christi area. Inland, the prairies of Katy and Lissie in the heart of the rice growing regions are particularly favoured by Snow Geese.

Although there are some small scale movements within the wintering grounds, the birds show considerable winter philopatry (Smithey 1973). Prevett (1972) reported seeing neck collared birds in identical fields in successive years. Some birds from La Pérouse Bay have been caught and released in successive years during the banding operations carried out by the US Fish and Wildlife Service at Sabine Refuge in Louisiana. Further studies of local movements need to be conducted during winter, but it seems likely that juveniles arriving with their parents learn the particular wintering grounds from them and return to those areas in the following years after they have become independent.

The recent northern extension of the wintering range has resulted from both the construction of wildlife management areas, such as the Squaw Creek National Wildlife Refuge, Missouri, and changes in agricultural practices that have made these areas more attractive to the geese.

Established pairs of Snow Geese remain together on southward migration and throughout the winter. Unpaired geese form new pair bonds on the wintering grounds and/or on northward migration (Cooke *et al.* 1975). Since members of new pairs may well have hatched at different colonies, establishing a pair bond away from the breeding grounds leads to potential conflict between pair members as to where to return and breed in the spring. As in most other species of North American waterfowl (Anderson *et al.* 1992), this conflict has been settled by males following the female back to her natal colony, producing a strong pattern of natal female philopatry and male dispersal (Table 2.3). This process has genetic consequences which will be examined in Chapter 3.

2.3.3 Spring migration

Snow Geese leave the wintering grounds in early to mid-March, and move north as melting snow and ice make food and roosting areas available. Since no legal hunting occurs in the spring, our band recovery records are too limited to document the northward movement with much precision. However, by pooling data from several years, some indication of timing can be obtained (Fig. 2.13). As during the fall migration, La Pérouse Bay birds travel with the birds from other western Hudson Bay colonies. Most birds are in Iowa and Missouri

Table 2.3 Re-encounter frequencies as 2-year-olds or older of female and male fledglings banded at La Pérouse Bay, 1969–1992

	Females		Males	
Cohort	Re-encountered	Not re-encountered	Re-encountered	Not re-encountered
1969	12	243	0	292
1970	98	1218	3	1352
1971	45	364	1	417
1972	85	1017	0	1301
1973	135	1020	2	1409
1974	61	700	5	788
1975	109	1012	6	1415
1976	147	1127	4	1334
1977	140	1074	6	1346
1978	82	1002	4	1029
1979	217	1894	13	2291
1980	230	1656	10	2122
1981	125	1762	4	1954
1982	160	1544	10	1630
1983	66	1276	5	1285
1984	140	1383	3	1491
1985	105	1834	3	1687
1986	37	1180	0	1238
1987	58	1729	0	1840
1988	26	1363	0	1301
1989	3	438	0	412
1990	1	423	0	435
1969–90	2082	25 259	79	28 369

in late March and early April, moving to the Dakotas and southern Canada, their final staging area, in early April.

The birds feed all along their northern migration, making use of many of the same sources used in the fall as well as newly emergent winter wheat. They gain weight as nutrient reserves are stored for the breeding season, and there is a recrudescence of the gonads. Departure from the prairie regions occurs when meteorological conditions for northward migration are favourable (Blokpoel and Gauthier 1975) and sufficient reserves for migration and breeding have been accumulated (Alisauskas 1988).

2.4 *Summary*

1. The Lesser Snow Goose is an excellent species for studying natural selection in the wild because it is relatively easy to

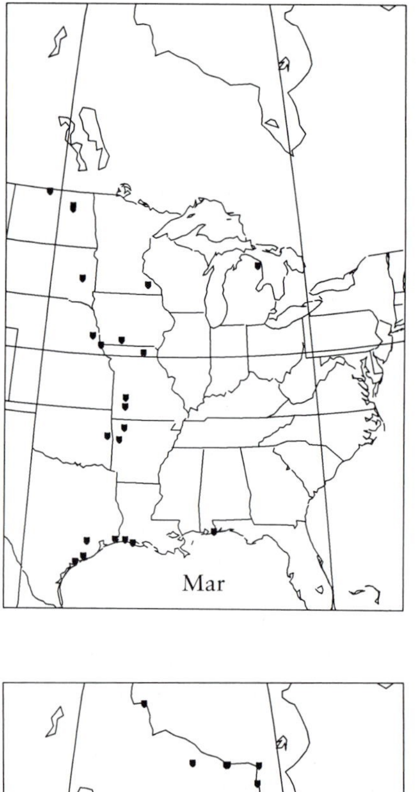

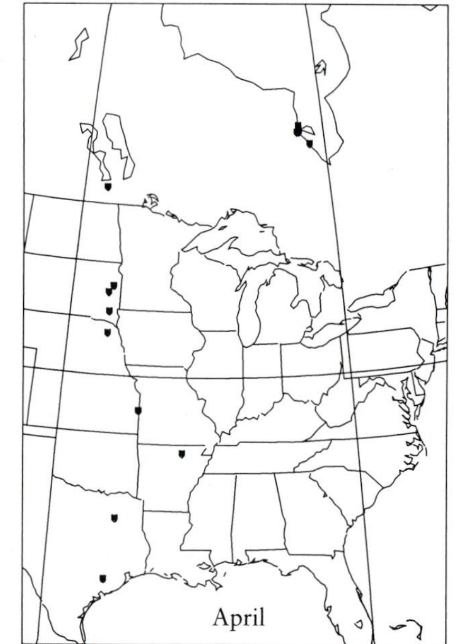

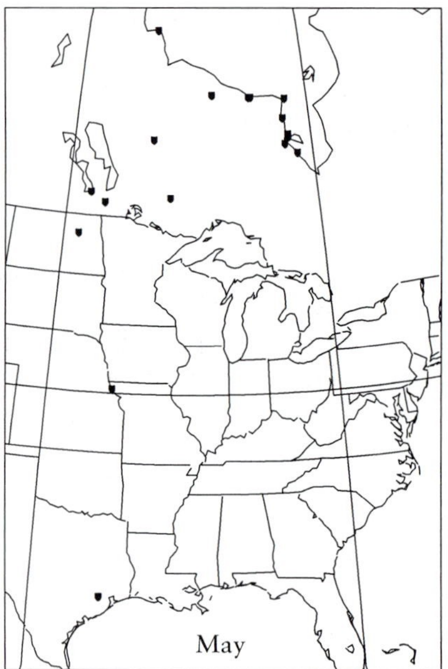

Fig. 2.13 Spring band recovery locations of Lesser Snow Geese marked at La Pérouse Bay. Recoveries were reported between March and May, 1970–1983.

study at all stages of its life cycle. Large samples allow us to document the patterns of selection which operate at the different stages.

2. The breeding colony at La Pérouse Bay, Manitoba, Canada, a remote region of the Hudson Bay coastline, has been studied since 1968. The colony was established in the 1950s and since the study began has grown from 2000 to more than 20 000 breeding pairs.
3. The nesting area is located in shrubby willow habitat adjacent to salt marshes which provide nutrients for geese and goslings during the breeding season.
4. The arrival of geese at the breeding colony generally coincides with the first disappearance of snow cover, but date and pattern vary among seasons. Geese grub heavily on salt marsh vegetation prior to nesting, which has led to a deterioration in the availability of nutrients during the study.
5. Nest initiation is highly synchronized and occurs soon after arrival, as suitable nest sites become available. Eggs are laid at approximately 33 h intervals. Large numbers of nests are abandoned or predated during the egg laying period. We gather information on several hundred nests during laying.
6. Incubation begins after the laying of the last or penultimate egg. Only females incubate, for an average period of 23.6 days. Nest and egg predators include gulls, jaegers, foxes, wolves, caribou and ravens.
7. Hatching is highly synchronized both within and among nests. Most eggs hatch. Families leave the nest within a day of hatching and move to salt marshes to feed. We gather information on several thousand nests at hatch.
8. During the 4–6 week brood-rearing period, goslings grow rapidly and parents become flightless during their annual wing and tail moult. Towards the end of this period, we capture 4000–8000 geese each year and mark them for individual identification. Marking provides useful information on migration routes, survival rates, and colony size, and identification if the birds return again in subsequent years.
9. Non-breeding birds, mainly younger birds, frequent the salt marshes adjacent to the nesting area for the early part of the nesting period but leave on a predominantly northward moult migration during the hatch period.
10. Fall migration begins in late August with birds moving south to other parts of Hudson or James Bay, or moving directly to southern Manitoba and the Dakotas. Later as winter con-

ditions set in they move further south to Missouri and Iowa and many finally winter in the southern parts of Texas and Louisiana. Hunting is the major cause of mortality among adult geese, but at least in some years large numbers of immatures also die soon after fledging before they reach areas frequented by hunters.

11. The geese winter in salt marshes along the Gulf coast and in inland areas to the north, where cultivated crops are available. There is evidence for increased use of inland areas in the past 40 years.
12. Pair formation occurs on the wintering grounds or on spring migration, resulting in a strong pattern of natal female philopatry and male dispersal.
13. Spring migration occurs from mid-March to mid-May along a similar route to that taken in fall. Much of the nutrients required for nesting are acquired in the Dakotas and Southern Manitoba prior to their final major flight to the Hudson or James Bay coast.

3 *Population structure and gene flow*

The Snow Geese nesting at La Pérouse Bay are not an isolated biological population. Pair formation occurs away from the breeding grounds, where La Pérouse Bay geese mix with those breeding at other nesting colonies. The level of gene exchange among colonies needs to be assessed if we are to infer potential local evolutionary consequences of the selective regime operating among the birds nesting at La Pérouse Bay. We aim to assess how both selection and gene flow can act to modify the genetic structure of our population. This chapter puts our colony into perspective by examining the population structure of Snow Geese and their close relatives.

3.1 *Taxonomy*

Geese belong to the family Anatidae in the order Anseriformes. Most waterfowl biologists concur with the classification of Delacour and Mayr (1945) that true geese can be subdivided into two genera *Anser* and *Branta*, separated mainly by bill morphology and colour of soft parts and plumage. Shields and Wilson (1987), using divergence of the mitochondrial genome calibrated with fossil evidence, calculated that the genera split 4–5 million years ago. Some authors (American Ornithologists' Union (AOU) 1983) have separated those *Anser* species with considerable amounts of white in their plumage, including Snow Geese and Ross' Geese (*Anser rossii*), into the genus *Chen*, and Oberholser (1919) went as far as to put Ross' Geese into the genus *Exanthemops*, calling it an 'excellent genus'. However, there is frequent hybridization among Anser geese and hybrids are usually fertile. Even intergeneric hybrids occur, and the genetic similarity among geese species is high. In our opinion Snow and Ross' geese should be referred to as *Anser*.

At La Pérouse Bay, a few Ross' Geese nest among the Snow Geese, and mixed pairs and intermediates occur. Marked offspring from these mixed pairs have themselves produced offspring, so these taxa are extremely close genetically, as confirmed in two recent studies examining the DNA of nearctic white geese (Avise *et al.* 1992; Quinn 1992). Snow Geese have traditionally been classified into two subspecies: the colour monomorphic Greater Snow Goose (*Anser caerulescens atlantica*) and the plumage dimorphic Lesser Snow Goose (*Anser caerulescens caerulescens*). The subspecies are mainly differentiated by size. The Greater Snow Goose weighs from 2700 to 3500 g with a bill length of 60–70 mm, whereas the Lesser weighs from 2300–3100 g and has a bill of 52–60 mm. There is considerable overlap in size and the subspecific division has little utility. Originally, the two colour phases of the Lesser Snow Goose were regarded as distinct species (*Anser hyperborea* and *A. caerulescens*), but work by Cooch (1961) and Cooke and Cooch (1968) showed them to be a single interbreeding dimorphic species.

3.2 *Population structure within Snow Geese*

Snow geese nest in colonies from Greenland in the east, across Arctic and sub-Arctic North America, to Wrangel Island in northeastern Russia to the west. It is natural to think of these widely scattered nesting colonies as appropriate units of study, as we do in much of this book. Since Snow Geese pair on the wintering grounds or on migration (Cooke *et al.* 1975), genetic populations are most usefully viewed from this perspective, as this is where gene flow occurs. Breeding colonies are ecological entities, some of which include members of quite separate subpopulations. Banding studies have identified three distinct Snow Goose wintering areas: the Atlantic coast, the central and southern United States, and the Pacific and southwestern states (Fig. 3.1). Birds, and thus genes, mix extensively among colonies wintering within each area, but there is virtually no exchange among these three biological populations (Dzubin 1974).

Birds wintering in the Atlantic region nest primarily in the high Arctic islands of Bylot, Ellesmere, northern Baffin and Greenland. These birds are primarily in the size range of 'Greater Snow Geese', but some birds from the Hudson Bay–Foxe Basin nesting areas, including some blue phase birds, also winter on the Atlantic coast. Thus gene exchange may occur between Lesser and Greater Snow Geese.

The central and southern region is concentrated along the Gulf of Mexico coast of Louisiana, Texas, and Mexico, with birds also wintering inland in Arkansas, Missouri, Nebraska, and Iowa. This region is used by birds from all the colonies of the Hudson Bay and Foxe Basin region, and is the primary wintering grounds of the La Pérouse Bay birds (Francis and Cooke 1992*a*). Some birds breeding in the Queen Maud Gulf and Central Arctic region area of northern Canada also winter in the central region.

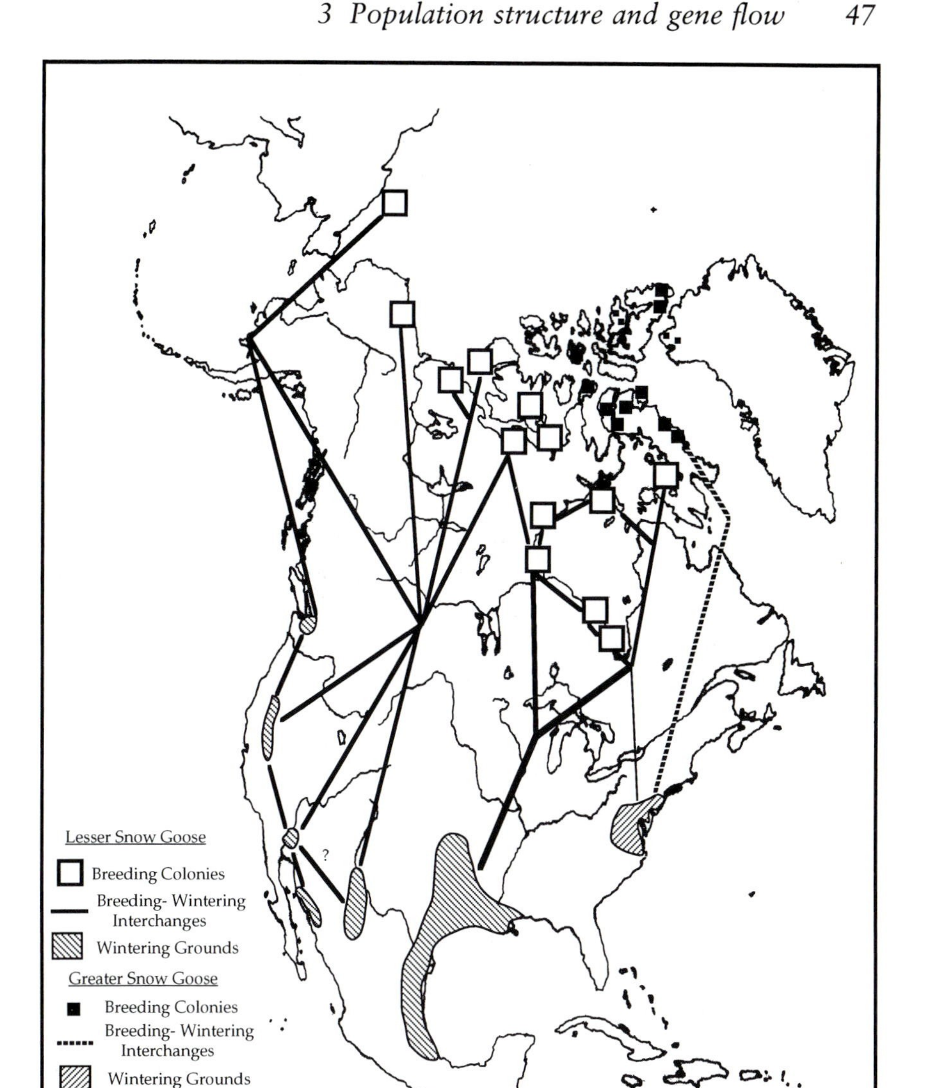

Fig. 3.1 The locations of breeding colonies and wintering areas of Lesser and Greater Snow Geese. The lines link breeding, migration, and wintering sites of populations.

The Pacific and southwestern region extends south from British Columbia to northern Mexico and inland to New Mexico. The main area of concentration is the Central Valley of California. Although wintering locations are geographically discrete, probably reflecting the distribution of suitable habitat in the region, there is some exchange of birds between these locations (Dzubin 1974). We refer to these areas as the western wintering area and ignore possible population substructuring among these wintering grounds.

Birds nesting on Wrangel Island, northern Alaska, the Yukon, and the western Northwest Territories use the western wintering grounds. Some birds from the Queen Maud Gulf and Central Arctic area also winter here, thus the birds from those breeding colonies winter in at least two distinct areas. However, since pair formation does not occur in the summer, this coincidence of breeding areas will not result in extensive gene exchange between the Pacific and Gulf Coast wintering populations.

Band recovery data indicate little exchange of birds among the three major wintering populations (Dzubin 1974). The lack of movement suggests that the three wintering populations may be genetically differentiated from each other, but Avise *et al.* (1992), using mitochondrial DNA (mtDNA), found no evidence of differences. However, Quinn (1992), using only a rapidly evolving region of the mtDNA, did find significant differences between the western and central populations. He did not test for differences between these and the Atlantic wintering population.

3.3 *Mid-continental Snow Geese*

Since La Pérouse Bay geese pair and exchange genes with geese from other nesting colonies which use the same wintering grounds, we will focus our attention on the central region. The breeding colonies whose birds use the central and southern wintering region, and their approximate sizes, are given in Fig. 3.2 (1–13 on map). Most of the colonies are close to the sea in areas of recent isostatic glacial rebound. Europeans had encountered Snow Goose broods on Southampton Island by 1904, and Inuit middens dated at 2000 BP contain Snow Goose bones (F. G. Cooch, personal communication). However, the nesting colonies were not located and described by scientists until Soper (1930) published reports on those of Baffin Island and Sutton (1931) described those on Southampton Island. More southerly colonies have been established more recently, and the expansion of geese from the central wintering area into the Queen Maude Gulf and the central arctic mainly has occurred since the start of our study.

3.3.1 Plumage dimorphism

The Snow Geese of the central region are dimorphic in plumage, but the morphs are not distributed evenly among colonies. The evolu-

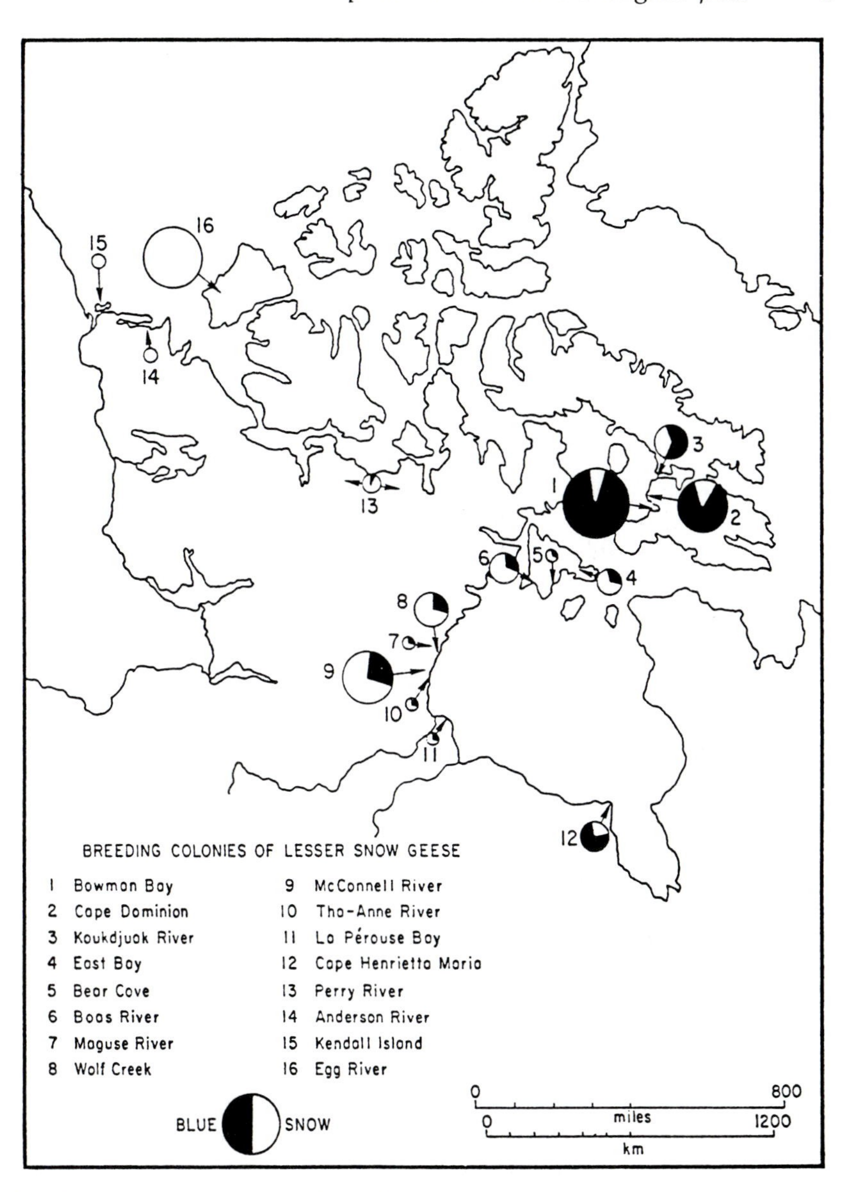

Fig. 3.2 Colony sizes and colour-phase ratios of Lesser Snow Goose colonies in the Canadian arctic. (Reproduced with permission from Cooke *et al.* 1975).

tionary origin and current status of the morph distribution is one reason why Lesser Snow Geese intrigued evolutionary biologists. Is the pattern stable or transient? If stable, what mechanism maintains it? If transient, what were the populations' previous histories and what is their expected evolutionary trajectory?

Estimates of the ratios of the two colour phases at breeding colonies are depicted in Fig. 3.2. The proportion of blue phase generally increases from west to east, with a maximum at the colony located along the coast line of the Great Plain of the Koukdjuak. North of this colony, the proportion of blue phase birds declines, as it does to the south and to the west. The blue phase ratio for the Queen Maud Gulf and Central Arctic region also decreases towards the west. The phase ratios in these colonies may not reflect their contribution to central wintering areas, since some of the overwhelmingly white phase Pacific-wintering geese also nest there.

The distribution of the colour phases is also non-random across wintering areas. Along the Gulf coast, the blue phase predominates in the east and the white phase in the west (Fig. 3.3*a*), roughly paralleling the longitudinal pattern at breeding colonies. There is a steep change in phase ratio close to the Louisiana and Texas border. In Texas, blue phase geese occur at higher frequencies in the inland regions than in coastal areas of the same longitude (Cooke *et al.* 1988).

The wintering ground phase ratio distribution is basically a result of parallel movements south from breeding colonies in the north, but a more active process of differential migration by colour phase is also involved (Cooke *et al.* 1975; Francis and Cooke 1992*a*). Among birds within breeding colonies, the two colour phases use different wintering areas, on average, with blue-phase individuals more likely to winter in the eastern part of the range and white-phase birds in the west (Table 3.1, Fig. 3.3*b*). The colour phase ratios in the band recovery data from birds marked at La Pérouse Bay are shown in Fig. 3.3*b*. To the east, these geese are whiter than the mean of the overwintering populations, perhaps reflecting the preponderance of white birds at La Pérouse Bay. Nonetheless, blue-phase recoveries from the colony are strongly biased towards the east and white phase recoveries towards the west.

3.3.2 Female breeding philopatry and male immigration

Lesser Snow Geese from many breeding colonies overlap in the wintering grounds. Geese from one colony thus encounter geese from other colonies, albeit with a minor colony-specific spatial bias (Francis and Cooke 1992*a*). Since pair formation is likely finalized during the winter and spring, this mixing results in the exchange of genetic material between breeding colonies, assuming that there is not strong genuine colony specific assortative mating. Evidence of large-scale genetic exchange was first provided by Cooke *et al.* (1975), using

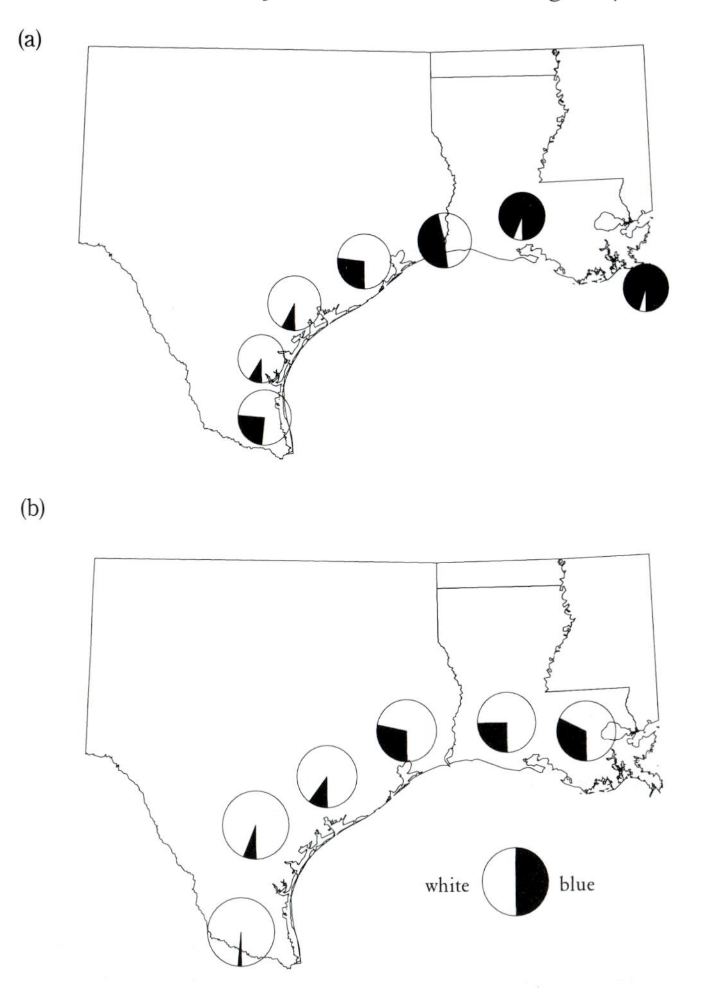

Fig. 3.3 (a) Colour phase ratios of Lesser Snow Geese wintering along the Gulf of Mexico coast, based on Audubon Christmas bird count data, 1979–1984. (Adapted with permission from Cooke *et al.* 1988). (b) Colour phase ratios of La Pérouse Bay Lesser Snow Geese wintering along the Gulf of Mexico coast, based on band recoveries.

data from both La Pérouse Bay and McConnell River, NWT. At La Pérouse Bay, we banded 27 341 female and 28 448 male fledglings between 1969 and 1990, of which 2082 females (7.6 per cent) but only 79 males (0.3 per cent) were re-encountered at the colony at ages 2 years or older, clearly confirming overwhelming female-biased philopatry (Table 2.3). Since male and female survival rates are approximately the same, apart perhaps for the first year of life (Francis *et al.* 1992*a*), and since the sex ratio of adults is approxi-

Table 3.1 Percentage of blue-phase Lesser Snow Geese among band recoveries in the Gulf Coast region (south of latitude 31°) banded at northern breeding colonies, longitudes 87–92° (data provided by F. G. Cooch, R. H. Kerbes, C. D. MacInnes and H. G. Lumsden). From Cooke *et al.* (1988)

	Recovery longitude						
Location	97	96	95	94	93	92	<92
Baffin Island	—	33	62	50	92	94	100
Cape Henrietta Maria	0	50	43	71	78	90	86
East Bay, Southampton Island	25	19	32	38	48	80	100
Boas River, Southampton Island	16	20	18	28	42	85	52
McConnell River	24	18	20	31	37	60	76

mately 1 : 1 at the time of banding, Cooke *et al.* (1975) concluded that nearly all of the adult male breeding geese at La Pérouse Bay must have hatched elsewhere, and that nearly all breeding male geese hatched at La Pérouse Bay emigrated and nested in other colonies. Although 0.3 per cent philopatric males is a tiny proportion, it is about eight times higher than we would expect if La Pérouse Bay females mated completely at random with males in the central wintering area, assuming that La Pérouse Bay is about 0.5 per cent of the total population and taking into consideration the rate at which we detect returning breeders. Spatial bias on the wintering grounds may account for the over-representation, but some of these 79 putative males may be incorrectly sexed females, errors which have come back to haunt us.

We have limited direct documentation on colony movements to and from La Pérouse Bay. The species has been banded intermittently at most of the other major colonies in the Canadian and Russian Arctic. Figure 3.4 plots the recapture locations of 10 males and three females banded at La Pérouse Bay that were encountered elsewhere during the breeding season. One male was found in the western Canadian Arctic, near Amundsen Gulf, but the rest fall within the Hudson Bay drainage. Figure 3.4 also plots the original banding locations of 15 foreign males, five foreign females, and three birds of unknown sex that were recaptured at La Pérouse Bay, all of which are from the Hudson Bay drainage. The sex ratios in these exchanges show that while natal philopatry is overwhelmingly female biased, females are not absolutely tied to their natal colonies (Geramita and Cooke 1982). As expected from the pattern of migration and mating described above, the primarily exchanges of breeding birds between La Pérouse Bay are with those of other colonies in the Hudson Bay

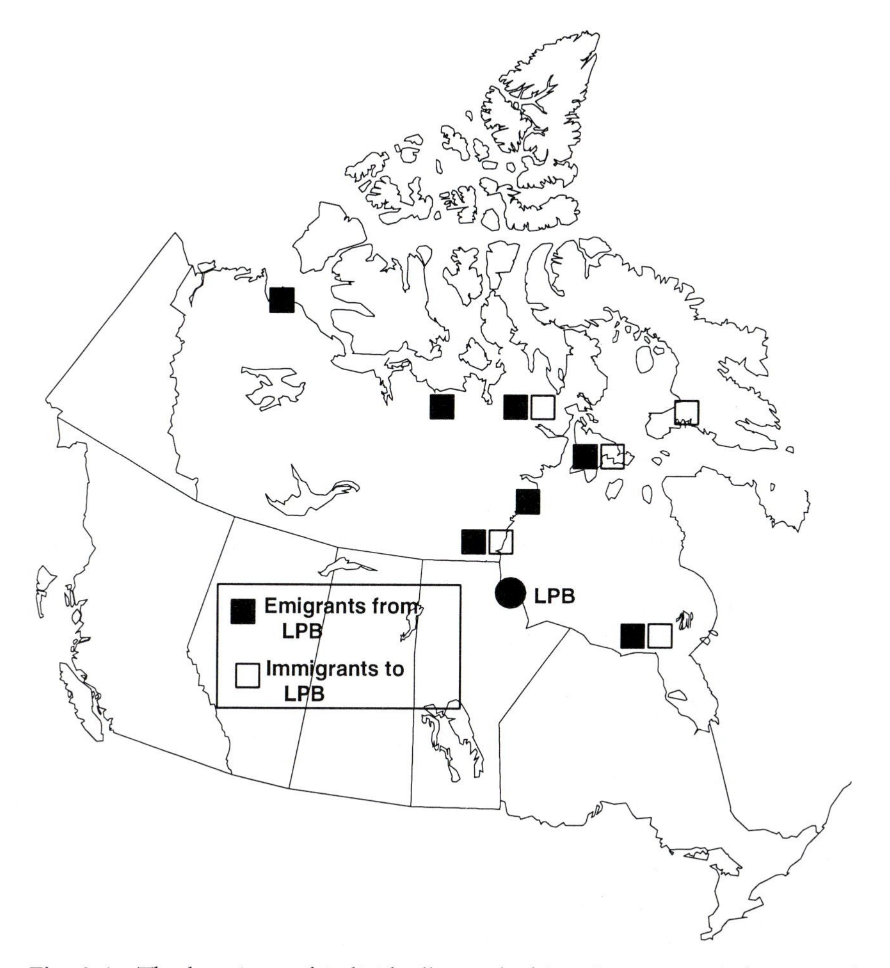

Fig. 3.4 The locations of individually marked immigrants (open boxes) and emigrants (closed boxes) documented moving between La Pérouse Bay (LPB) and other Lesser Snow Goose breeding colonies.

and Foxe Basin region. No exchanges yet have been detected between La Pérouse Bay geese and populations breeding in northern Russia.

The gene flow into La Pérouse Bay arising from the immigration of males from other breeding colonies is on the order of 50 per cent per generation (Rockwell and Cooke 1977). This estimate applies to all genes residing on any but the female W chromosome and those associated with mitochondrial and other extranuclear bodies. The amount of gene flow into a colony is inversely proportional to the probability that a female will chose a male hatched from her own breeding colony. This depends in part on the size of the breeding colony relative to the wintering population. Birds from a small breeding colony are much less likely to find a mate from their own breeding colony than those from a large colony. Assuming total overlap of birds from the various breeding colonies on the wintering grounds, Rockwell and Barrowclough (1987) developed a general model to calculate the amount of gene flow into each breeding colony. Under the conditions of their model, gene flow into La Pérouse Bay is 49 per cent, while gene flow into the large Baffin Island colony is less than 30 per cent. In reality the amount of gene flow due to male immigration is likely to be somewhat lower than predicted, since the wintering distributions of the birds from the various Hudson Bay breeding colonies, although broadly overlapping, are not identical. For example, geese from Cape Henrietta Maria, 1000 km to the east, had a more easterly wintering distribution than those from La Pérouse Bay (Francis and Cooke 1992*a*). Finally, even a low level of exchanges of females among colonies could be sufficient to minimize genetic differentiation among the three major wintering areas documented in the mtDNA studies (Avise *et al.* 1992; Quinn 1992).

3.3.3 Genetic colony structure within the central population

The gene flow between breeding colonies has several theoretical and practical implications which we must consider when interpreting the data gathered at La Pérouse Bay. Snow goose colonies differ greatly in mean laying dates and clutch sizes, with earlier laying and larger clutches in the more southerly colonies (Davies and Cooke 1983*a*). Suppose that the timing of laying had a high heritability, meaning that that variation in laying date predominantly reflected differences in genes. Were this the case, selection against late laying birds at Southampton Island would affect the gene pool at La Pérouse Bay colony, even if this selective regime did not occur locally. Males carrying alleles for late nesting would be less available to La Pérouse Bay females than if there were no such directional selection elsewhere. Similarly, we must consider whether the selective regime detected at La Pérouse Bay will apply to other colonies.

The genetic structure of populations is formally measured by F_{ST}, the among-population component of genetic variance. Species with little gene flow among strongly differentiated subpopulations have F_{ST} values which approach 1.0, while genetically homogeneous species with large amounts of gene flow have values near 0.0. Rockwell and Barrowclough (1987) used an indirect demographic method to calculate the expected F_{ST} values of the Hudson Bay Snow Goose breeding colonies. They applied the island model of gene flow (Wright

1943; Nei *et al.* 1977) to the central Hudson Bay Snow Geese, and estimated F_{ST} for the population at 0.0000142, an extremely low value. At equilibrium, allelic frequencies will be virtually identical among colonies. The rate of approach to equilibrium will be tempered by colony sizes, the initial state of the population, local selection regimes, and non-random mating. Since the last three variables will differ among loci, the rates of equilibration will also vary among loci. The colour locus itself is an example of such interplay. The previous history and positive assortative mating mean that it will take considerable time for the phase ratios in the Hudson Bay colonies to equilibrate.

3.3.4 Population consequences of assortative mating

The high rate of gene flow makes it difficult to envision adaptation to local conditions at La Pérouse Bay, except perhaps through genes located on the female W chromosome, or outside the nucleus (Rockwell and Cooke 1977). Given its small size, La Pérouse Bay colony gene frequencies should rapidly converge towards the mean of the central region birds. Even the larger colonies should converge genetically, albeit at a slower rate. Why then have colonies remained distinct with regards to colour phase, a conspicuous and autosomally linked character? The blue phase ratio of females captured in banding drives at La Pérouse Bay has increased slightly, but systematically, during our study (Fig. 3.5). Thus blue alleles are increasing in the philopatric core population of the colony. In contrast, the phase ratio of males, immigrants to the colony, has remained constant. Obviously, gene frequencies at the colour phase locus are not rapidly converging with that of the population as a whole. The trend in females is in the expected direction, but at a far slower rate than expected, and no trend at all is seen in the males.

This paradox is resolved because the species has a strong, but not perfect, pattern of positive assortative mating with respect plumage colour, due to imprinting on family colour when a gosling (Cooke *et al.* 1976). The behavioural mechanism and evolutionary history of these preferences are considered in Chapter 5. Here we consider the current consequences for population structure and gene flow of non-random mating with respect to colour phase.

Among females reared at La Pérouse Bay, 90.4 per cent with two white parents paired and returned with white mates, and 78.2 per cent of females with two blue parents returned with blue mates. Females from mixed colour-phased families obtained 62.5 per cent white and 37.5 per cent blue mates (Fig. 3.6; Cooke 1988). This

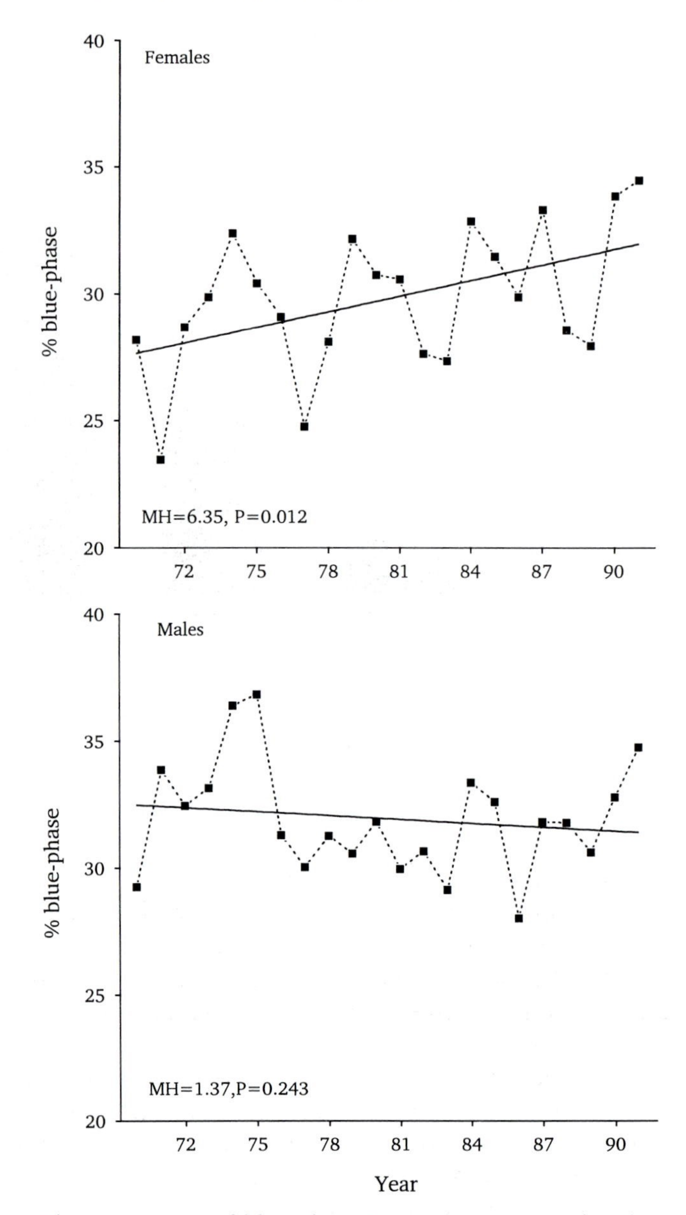

Fig. 3.5 The proportion of blue phase Lesser Snow Geese females (*top*) and males (*bottom*) encountered in banding drives at La Pérouse Bay (1969–1992). There is a significant increase in the frequency of blue females, but no change in that of males (a linear regression line is drawn, MH = Mantel–Haensel χ^2 test for linear trend in frequency data, d.f. = 1).

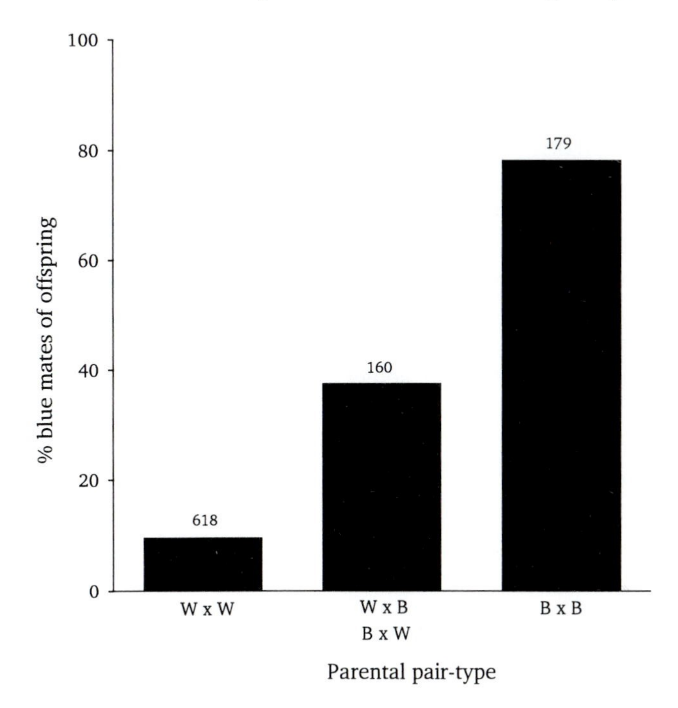

Fig. 3.6 Percentage of blue phase mates obtained by offspring from white × white (W × W), mixed (W × B, B × W), and blue × blue parental pairtypes, 1970–1984 (Cooke 1988).

analysis categorizes birds according to the colour of their parents only, rather than their whole family, which is not always possible, and sib colour has a minor effect on mate choice. Assortative mating impedes the movement of genes between colour phases of the population, and thus between subsets of breeding colonies wintering in the central region.

We explored the population consequences of different rules for assortative mating through detailed mathematical modeling (Geramita *et al.* 1982). Here we give a simplified version of the assumptions, techniques, results, and conclusions. Our aim was to explore the consequences of non-random mating and gene flow on the colour phase ratio at the La Pérouse Bay colony, and predict changes for other colonies. Starting with the phase ratio and pair-bond frequencies of 1969, we modeled changes expected over 4 non-overlapping generations under a variety of 'mating rules'. The model tabulated four generations of changes in three parameters: the simulated breeding colony's colour phase ratio, the proportion of mixed pairs, and the relative frequency of mixed pairs in which the male was blue. The last parameter is important since immigration to the colony is strongly sex biased.

Our model corresponded as closely as possible to the real world, apart from the mating rules which we were testing. We assumed that each breeding colony in the arctic started with a distinct phase ratio, as occurs in the Hudson Bay–Foxe Basin population. Females from these colonies migrated to the wintering grounds, selected a mate according to some mating rule, and returned with the mate to her natal breeding colony. The phase ratio in the wintering grounds was set to approximate the observed east–west cline (Fig. 3.3(a)). We assumed no overall selective differences between morphs (Chapter 9). The rules modelled were: (1) random mating with respect to colour; (2) mate choice under the assumption that birds always choose a mate of the same colour as their parents if from colour-matched pairs, or choose randomly if they are the offspring of a mixed pair; (3) as above, but allowing sibling colour also to influence choice; and (4) as above, but allowing for a range of levels of mate choice due to other criteria ('mistake making'). We evaluated which of these rules most closely predicted the changes which actually occurred at our colony.

At La Pérouse Bay, the frequency of blue females has increased slightly (Fig. 3.5), but that of males has not. The proportion of mixed pairs has remained relatively constant around 16 per cent, and within this group, the proportion of mixed pairs with blue males decreased from 1969 to 1980 but has since increased somewhat (Table 3.2). The first three mating rules failed to predict these outcomes. Random mating predicted a rapid rise in the frequency of mixed pairs which did not occur in the real population. Complete assortative mating based on either parental colour or parental plus sibling colour predicted a sharp decline in the frequency of mixed pairs, again in contrast with our data. Introducing a level of *c.* 15 per cent non-colour-based mating generated a pattern close to the population changes seen. This corresponds well with the average non-matching mating rate observed among returning females (Fig. 3.6).

The general prediction of our model is a slow equilibration of phase ratios and pair types at the different colonies within the Hudson Bay and Foxe Basin breeding area. This appears to be happening among females, but not among the immigrant males. While the local frequency of blue males should also slowly increase, it is a more complicated process. Females from blue families are less likely than those from white birds to pick matched colour mates, and birds from mixed families acquire mates biased towards white (Fig. 3.6). The colour morph frequency of males may in part reflect a bias in the

Table 3.2 The percentage of pairings with respect to colour-phase and sex of Snow Geese nesting at La Pérouse Bay, 1968–1991

Year	Pair type				*n* pairs
	blue female × blue male	blue female × white male	white female × blue male	white female × white male	
1968	18.2	5.2	13.8	62.9	385
1969	15.6	5.7	8.8	69.9	628
1970	17.5	5.9	10.0	66.7	928
1971	16.9	4.6	10.4	68.1	604
1972	18.0	6.2	9.7	66.1	1286
1973	19.1	6.2	10.0	64.7	1411
1974	19.8	6.5	9.8	63.8	1407
1975	19.6	6.9	9.5	64.0	1606
1976	19.0	8.2	9.4	63.4	1488
1977	19.8	8.0	10.3	61.9	1630
1978	18.3	8.2	9.3	64.2	1619
1979	18.2	7.6	9.9	64.3	1479
1980	17.6	8.3	9.0	65.1	1618
1981	18.7	7.8	8.4	65.2	1774
1982	19.7	8.0	8.8	63.5	1846
1983	20.2	8.0	8.7	63.2	1989
1984	20.2	7.6	9.3	62.9	2308
1985	18.2	7.5	10.0	64.3	2426
1986	19.6	7.0	9.0	64.4	1979
1987	20.1	7.6	9.4	62.9	1244
1988	21.2	7.1	9.4	62.3	1664
1989	20.2	6.6	10.2	62.9	1403
1990	22.5	7.0	9.3	61.2	1439
1991	20.5	6.0	8.4	65.2	885

prevalence of potential mates on the wintering grounds and/or effects of sibs, which also affect mate choice (Cooke *et al.* 1976). Sibs are more likely to not match parental colour in blue than in white families, both because two heterozygous blue parents can produce white offspring, and because white females are more frequent at La Pérouse Bay, intra-specific nest parasitism produces predominantly white goslings (Chapter 5).

Applying the model to a breeding colony where most birds were blue phase predicted a drop in the frequency of blue phase birds, which are in the minority within the wintering population, and a maintenance in the frequency of mixed pairs. Among mixed pairs, the colony would have had an initial excess of pairs with a white male and blue females which should decrease with time. A paucity of mixed pairs should remain so long as the pattern of mate choice remains the same. The positive assortative mating practiced by Snow Geese retards, but does not completely stop, the convergence of phase ratios among breeding colonies.

Gene flow would lead to a rapid equilibration of the phase ratios at all colonies if it were not for the retarding effect of the assortative mating. How does this restriction affect other genes? Only alleles in gametic phase disequilibrium with the colour locus itself would be affected by the mating system. For most autosomal markers, we would predict little genetic differentiation among breeding colonies, and this is confirmed by studies using both allozymes (Cooke *et al.* 1988) and nuclear markers (Quinn 1988). As other nuclear genetic markers become available in Snow Geese (Quinn and White 1987), perhaps including some linked to the colour locus, a detailed mapping of the genetic relationships among colonies will prove a most interesting study.

3.3.5 Evidence of former allopatry

If the breeding colonies of Lesser Snow Geese wintering in the central region are slowly becoming more similar in phase ratio, one can infer that they were formerly more distinct. Perhaps in the recent past, the two colour phases were completely allopatric. Extrapolation from the model of Geramita *et al.* (1982) suggested that the two phases may have been allopatric as few as 10 generations ago. If so, rather than the two colour phases of Lesser Snow Goose reflecting a stable balanced polymorphism, the evolutionary history may be that of a transient polymorphism in two formerly allopatric taxa which have recently merged to become sympatric. The present distribution of the phases should, under this explanation, be thought of more in terms of historical accident rather than in terms of differential selection pressures, although this of course begs the question of why a novel morph, presumably the blue phase, initially spread within the subpopulation in which it arose.

If the morphs were geographically separated as little as 10 non-overlapping goose generations ago (40 years ago), there should be historical evidence of this and of subsequent changes in distribution. Cooke *et al.* (1988) examined this question in detail and produced convincing evidence that there have been major changes in the distribution of the two colour phases in the past 300 years. We document some highlights of this evidence below.

Evidence of distributional changes on the breeding grounds is unavailable for previous centuries, since nesting colonies in North America were not discovered by scientists until this century. Bent

(1925) reports Lesser Snow Geese nesting from northern Alaska to Southampton Island, but the breeding grounds of blue phase birds was one of the mysteries of North American ornithology until Soper (1930) discovered nesting colonies on the west coast of Baffin Island, south of the Koukdjuak River. Although a few breeding white-phased geese were found as well, there was not a single mixed pair and all broods contained matched young. The following year, however, Sutton (1931) reported 2 per cent mixed pairs at Southampton and intermixing of nests of the two phases within colonies. By 1940, more mixing of the colour phases was occurring. Both Bray (1943) and Manning (1942) recorded both colour phases and mixed pairs at several northern colonies.

The frequency of mixed colonies increased with time, and within colonies, colour morph frequencies are converging. Cooch (1961) noted a gradual increase in the frequency of blue phase birds at all colonies between 1940 and 1959. We feel that Cooch's figures on phase ratios at the predominantly blue phase colonies on Baffin Island may have been collected in less than ideal conditions, since subsequent visits by Kerbes (1969), R. K. Ross (unpublished data), and L. Dupuis in 1979 (unpublished data) found a decline in the relative frequency of blue phase birds. In any event, the presence of both colour phases in the Hudson Bay breeding colonies is now universal (Fig 3.2), the frequency of whites having increased in the predominantly blue colonies and the frequency of blues having increased in the predominantly white colonies since they were first discovered or censused (Cooke *et al.* 1988).

Documentation of phase ratios on migration routes goes back much further in time. Cooke *et al.* (1988) infer that seventeenth century missionaries and Huron Indians encountered blue phase birds in Midland, Ontario. Midland is along a direct line if Blue Geese migrated from Baffin Island to the Mississippi Delta, but far from the route of the Greater Snow Goose or of white phase Lesser Snow Geese migrating from Hudson Bay to Texas. At the present time, no Snow Geese are common in central Ontario.

In the eighteenth century, the distributions of the two colour phases were better defined in the north. Andrew Graham spent many years of his life at Hudson Bay Company forts along the Hudson and James Bay coastlines. He reports an overwhelming predominance of migratory white phase birds at the westerly settlements and blues at the easterly settlements (Graham 1769). In the nineteenth century, the two phases were still geographically distinct in the Southern James Bay region, according to Barnston (1860), who wrote that in the fall 'Snow Geese' migrated from the north-west while 'Blue

Geese' came from the north-east. This distribution on the Hudson and James Bays persisted for at least the next 60 years (Saunders 1917), but changes were evident by 1941. Lewis and Peters (1941) reported almost as many blue as white phase geese in the fall on the west coast of James Bay, but state that local residents informed them that 20 years previously blue geese were rare there and had gradually increased in frequency. On the east side of James Bay, blue phase geese greatly outnumbered whites. By the 1970s the situation had changed even further, with 40–50 per cent blue phase birds at least as far west as the Manitoba–Ontario border (H. Lumsden and J. P. Prevett, unpublished data). These descriptions suggest slow but steadily increasing intermixing of the two phases over the earlier period and a more major change occurring some time after 1917.

Wintering records of Snow Geese show a similar separation in the early records and more recent merging of the colour phases (Cooke *et al.* 1988). K. C. Oberholser's careful records, examined on microfilm in the library of Texas A & M University, show that blue phase geese were extremely rare in Texas in the nineteenth century, while white birds were abundant. Early records of geese from Louisiana (McAtee 1910, 1911; McIlhenny 1932), by contrast, show that blues were overwhelmingly more frequent than whites.

Morph separation persisted for the early years of the twentieth century, but Audubon Christmas bird counts show that geographical overlap of the phases increased in the early 1930s. Stutzenbaker and Buller (1974) suggested that changes in agricultural practices may have caused this shift. From the late 1800s through the 1930s, prairie areas some 100–200 km inland from the traditional salt marsh wintering areas of the geese were gradually turned over to rice cultivation. After the rice crop was harvested, rye grass was seeded for two or more years for use as cattle pasture, which now attracts large numbers of wintering geese. The opening up of these novel feeding areas appears to have promoted mixing of the colour phases in winter, thereby increasing the opportunities for the formation of mixed pairs.

There have been no dramatic changes in phase ratios at central wintering locations where Audubon Christmas bird counts have been made regularly since the early 1950s. US Fish and Wildlife Service inventories of winter waterfowl in Texas and Louisiana also show no change over this period.

The wintering ground distribution is the critical one from the point of view of gene flow. Changes in wintering distributions of the colour phases in the twentieth century initiated the coincident changes in the distributions of the colour phases on the breeding grounds and on migration routes by changing the colour distribution of potential mates. As the white birds and the blue birds merged early in the twentieth century, there would be corresponding changes in the phase ratios in the breeding colonies. Historically, birds from white phase colonies wintered in Texas and birds from blue phase colonies wintered in Louisiana. As birds from the different wintering areas mixed due to the habitat changes, mates of the opposite plumage colour became available. This would allow some blue geese to enter the white colonies and some whites to enter the blue colonies. Because of female philopatry, the first immigrants of the opposite colour would be males.

In summary, all historical evidence, from the breeding grounds, migration routes, and the wintering grounds suggest that very little intermingling of the colour phases occurred prior to 1920. A major change occurred subsequently, probably triggered by a changes in food utilization on the wintering grounds. Prior to 1920, the two morphs were almost allopatric in both breeding and wintering areas. Historical evidence suggests the two morphs were essentially isolated from one another for at least 300 years prior to that, and we might speculate that the taxa were allopatric at least since the late Pleistocene. Ploeger (1968) reached a similar conclusion, but appeared to be unaware of the importance of wintering ground allopatry.

If the blue and white phases were isolated as far back as the Wisconsin glaciation, genetic differences between colour phases would have occurred at loci other than the colour locus. Although sympatry was restored within this century, we might still expect some residual genetic differences to be present. There is indeed some evidence for genetic differences in allozymes (Cooke *et al.* 1988), but not in analyses of mitochondrial DNA (Avise *et al.* 1992; Quinn 1992). More detailed genetic analysis using molecular markers with different rates of change is needed.

The historical and genetic evidence leads us to a revision of the taxonomy. Although it is correct to think of the Snow Goose as divided into three population units (Atlantic, Gulf Coast and Pacific) at the present time, it is clear that the Gulf Coast taxon represents a recent merging of two taxa, one white plumaged and the other blue plumaged. Thus, while current taxonomists sensibly define the blue and white plumages as a within population dimorphism, earlier taxonomists (prior to 1930) were equally correct to consider blues and snows as separate taxa, since they looked quite different and were allopatric. This is one case where the change in taxonomic status reflects a genuine biological change in the species itself.

3.4 *Summary*

1. The Snow Goose and closely related Ross Goose should be considered members of the genus *Anser*.
2. Lesser Snow Geese comprise three distinct populations from a genetic viewpoint. These are defined in terms of broad wintering areas—the western, the central and the Atlantic—between which there is little or no gene exchange at the present time.
3. Birds from the La Pérouse Bay colony winter in the central area, which comprises the Gulf Coast regions of Texas and Louisiana and states along the Mississippi valley to the north.
4. The La Pérouse Bay colony cannot be considered in isolation from the other colonies in the Hudson Bay, Foxe Basin, and Queen Maud Gulf areas of the Arctic. This is because birds from all the colonies from this region winter mainly in the Gulf Coast states of Louisiana and Texas. During this period pair formation occurs and birds originating from different breeding colonies frequently pair with one another.
5. Pairing on the wintering grounds results in large-scale gene exchange among breeding colonies, since the pair returns to the natal colony of the female. As a consequence, most males breeding at La Pérouse Bay and at many other breeding colonies have hatched elsewhere. Thus if we are to examine the effects of natural selection on the birds at La Pérouse Bay, we must take into account possible selection acting on males which hatch at other breeding colonies.
6. Although most gene exchange among breeding colonies occurs through the male, some movement of females among colonies has been documented.
7. The pattern of gene flow documented leads to the prediction that there should be little genetic differentiation among colonies within the Hudson Bay region.
8. Despite large-scale gene flow, breeding colonies differ considerably in the relative frequency of the two colour phases. The present-day distribution can be explained by the historical segregation of the phases and present-day strong assortative mating with respect to colour in the species.
9. There is evidence of a recent (early twentieth century) merging of two formerly allopatric populations, one blue and one white, which can account for the clinal distribution of colour phases both in the breeding and the wintering range. The breeding colonies are not yet in genetic equilibrium with respect to genes at the colour locus.

4 *Fitness components model of Snow Goose life cycle*

Closing his chapter 'Struggle for Existence', Charles Darwin (1859) gave his view of fitness by stating '. . . that the vigorous, the healthy, and the happy survive and multiply'. An individual's relative fitness is measured by its ability to pass copies of its genes to the next and future generations compared to other members of its population. While the concept is straightforward, many would agree with Lewontin's (1974) statement that 'Although there is no difficulty in theory estimating fitnesses, in practice the difficulties are virtually insuperable'. A major advance towards measuring fitnesses under real world conditions was made by Prout (1969, 1971) who partitioned fitness into its two basic components, namely, viability and fecundity, and evaluated each separately, as well as the interactions between the two. This approach brought analytical and empirical research on fitness back to Darwin's original duality of 'survive and multiply'. Bungaard and Christiansen (1972) and Christiansen and Frydenberg (1973) furthered this approach, adding components for sexual and gametic fitness and stressing the need to evaluate fitness with components that correspond to stages of the life cycle of the species being studied. Arnold and Wade (1984*a*,*b*) refined and extended the approach to polygenic phenotypes and stressed Prout's (1965) point that the components and stages must be assessed within a single generation. Hedrick and Murray (1983) provide a particularly lucid overview of both the basis and application of the components of fitness approach.

The fecundity or reproductive success of a member of one generation is its contribution to the next generation. This includes the production of offspring, their survival and, ultimately, their recruitment into the breeding population. The viability of that same individual is its probability of survival from the present to some time in the future, most usefully reckoned as the next opportunity for reproduction. In longer lived species, viability must be modulated by the probability that the age of first breeding may vary and that some reproductively mature individuals do not breed in some years. Fecundity, viability, and breeding propensity must be considered in the final evaluation of fitness, particularly since trade-offs among them may be crucial in the evolution of life history strategies (for example Williams 1966; Stearns 1992).

We have adopted a components approach towards measuring fitness in Lesser Snow Geese. We divide fitness broadly into fecundity, viability, and breeding propensity. Within these broader headings, we have established components that correspond to distinct stages of the life cycle of this species. The fecundity components are involved in the production of generation $n+1$ independent offspring from generation n adults. The viability components determine the survival of those offspring to reproductively mature adults and the continued survival of those breeding individuals. The breeding propensity components address both the age of first breeding and subsequent 'opting out' in particular years.

The boundary between fecundity and viability is not clearly delineated in any species. Ideally, it should mark a point where offspring reach an age of demographic independence, that is, an age where the offspring's continued survival depends exclusively on its own age and phenotype. In Snow Geese, goslings from one year often remain with their parents throughout their first winter and spring migration back to the breeding colony. Demographic independence is likely attained over this longer period with dependency on parents declining over at least 2 years. We have chosen fledging as a convenient boundary between fecundity and viability since pre-fledging survival is heavily dependent on parental phenotypes. Once the goslings can fly, they are much less susceptible to predators and require substantially less parental care.

Examination of this set of components provides a temporal picture of the constraints affecting an individual's fitness from its birth to its death. Much of the work we present will focus on potential fitness differences between phenotypic segments of the population. By using components of fitness rather than comparing total fitness, we increase the likelihood of detecting differences between the segments, particularly since there may be compensation among the components. For example, one colour morph might lay four eggs and hatch half of them, while the other morph might lay six eggs and hatch only a third of them. While both morphs hatch two eggs, they would attain this fecundity by different routes, following different life-history strategies, which might be compared with aspects of their evolutionary histories and ecological situations.

The Lesser Snow Goose life cycle can be viewed as a components model (Fig. 4.1) in which various stages of the life cycle are represented by state variables connected with transition probabilities. The reproductive output of a single nest flows through the system. This

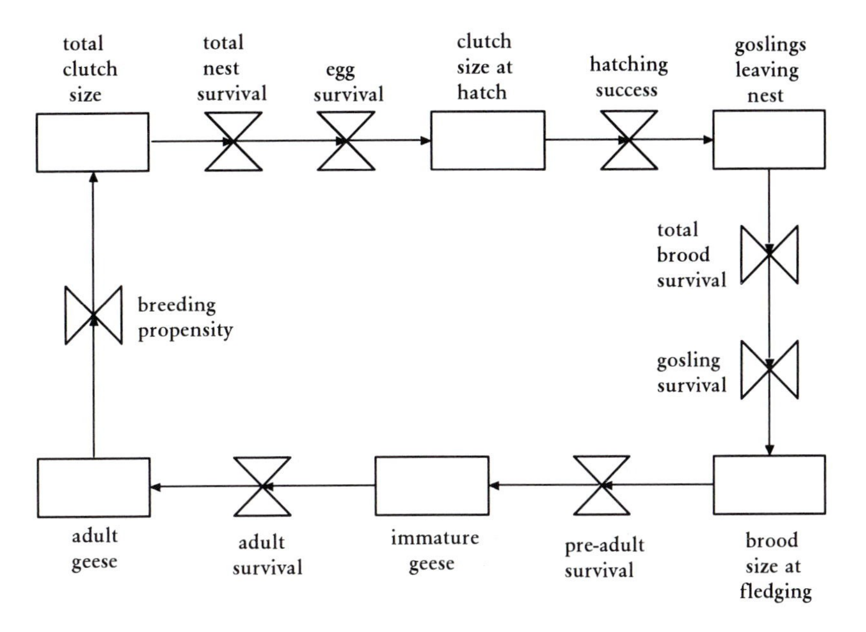

Fig. 4.1 Life cycle model of Snow Goose fitness components, showing the reproductive output from a single nest. State variables are indicated as boxes, transition variables by bow ties.

output is followed from the initial number of eggs laid to the recruitment of female offspring from the nest into the breeding population and ultimately to the death of those individuals. The nest can be associated with any phenotypic attributes of the attendant pair one wishes to examine. Table 4.1 presents global estimates of fitness component values for the La Pérouse Bay population. The values are the expectations for the average nest, means taken over all ages and years, without regard to any particular phenotype. Our fundamental approach is to compare the relative fitnesses of segments of the population with respect to phenotypic variables against these average expectations.

The remainder of this chapter provides a detailed description of our model of fitness components in Snow Geese. We relate the model and the estimates of state variables and transition probabilities to the species' natural history and to the manner in which data were collected. The presentation will allow the reader to assess precisely how the components of fitness were measured and provides an example of how such a study can be done. If the reader is examining an organism with the intent of assessing the role of natural selection in molding the organism, we recommend a detailed breakdown of the life cycle as we have done for the Snow Goose.

Table 4.1 Global expectations of the state variables and transition probabilities for the life cycle model of the Lesser Snow Geese at La Pérouse Bay

(a) Fecundity components. Standard errors are the root of the weighted variances among years for all but total nesting, hatching and brooding failure. Those are based on the standard binomial variance equation

State variable or transition probability	Mean	Standard error	Sample size
Total clutch size (*TCL*)	4.189	0.316	3559
Total nesting failure (*TNF*)	0.081	0.057	4839
Egg survival (*P*1)	0.949	0.014	3318
Clutch size at hatch (*CSH*)	3.921	0.265	27 780
Total hatching failure (*THF*)	0.010	0.008	32 146
Hatchability (*P*2)	0.928	0.020	22 074
Goslings leaving nest (*GLN*)	3.594	0.311	25 383
Total brooding failure (*TBF*)	0.085	0.068	1409
Gosling survival (*P*3)	0.755	0.057	2194
Brood size at fledging (*BSF*)	2.798	0.325	2194
(b) Viability components			
Annual first year survival from fledging	0.424	0.019	19 yr
Annual adult survival	0.816	0.016	19 yr

4.1 *Fecundity components*

Fecundity components start with the initial state variable clutch size and measure the survivorship of these eggs through fledging of the young.

4.1.1 Clutch size

The initial state variable in our model is the total clutch size. This is the number of eggs laid in a nest from initiation through the onset of incubation and can be viewed as a female's capital investment in reproduction for a given breeding season. The overall frequency distribution of this investment is summarized in Fig. 4.2.

Data are used only from nests in intensive study areas that were found at the one egg stage and for which incubation began. Eggs found outside of nests, or those thought to be laid by parasitic females, are included in total clutch size, since these are generally not clearly distinguishable from those laid by the nesting females. Nests containing more than seven eggs were excluded from all fecundity analyses for several reasons. First, such nests are rare, there having been only 45 of them in the 5624 intensively studied nests which

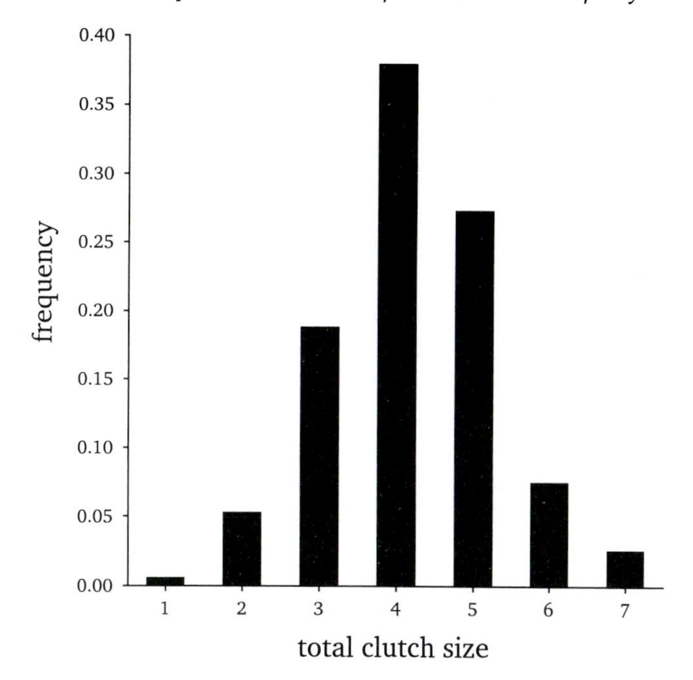

Fig. 4.2 Frequency distribution of total clutch sizes laid by Snow Geese at La Pérouse Bay. Only intensively studied nests found at the one egg stage are included. Data are pooled over 1973–1990, n = 3559.

could be used to estimate total clutch size. Second, such nests are all (or virtually all) the result of more than one female laying in the same nest. Evidence for this comes from behavioural observations (Mineau 1978), from the fact that dissected females seldom contain more than six post-ovulatory follicles (Ankney 1974; Hamann 1983; R. F. Rockwell unpublished data) and, most importantly, the observation that the frequency of goslings whose colour is genetically inappropriate for their parents, is higher in larger clutches (Chapter 5; Lank *et al.* 1989*b*).

4.1.2 Initiation and incubation components

Pre-incubation failure is the probability that a nest fails totally before the onset of incubation. It is the least understood fitness component for Lesser Snow Geese and the most difficult to contend with analytically, and not included in our general model shown in Fig. 4.1. As explained in Chapter 2, we do not know what proportion of females at such nests initiate and incubate a second reduced clutch, deposit their remaining eggs parasitically in other females' nests, or resorb remaining eggs and opt out of breeding for the season. We have direct evidence of renesting, but little opportunity for observing the other two options in the field. Since these possibilities have different effects on overall reproductive dynamics, we can not include pre-incubation failure directly in our composite estimates of reproductive success (see below). We can compare relative levels of pre-incubation failure among segments of the population for some phenotypic variables (for example white versus blue females), although not for others. We can not evaluate the relation of pre-incubation failure to clutch size, for example, with anything short of extensive clutch manipulation and capture and examination of failed females. Even for phenotypic variables where direct evaluation is possible (for example white versus blue females), our sample sizes are limited because pre-incubation failure often occurs so early during initiation that parents have not yet been identified.

Clutch size at hatch is the next state variable following total clutch size, and is defined as the number of eggs in the nest when the first egg begins hatching. It represents the return on a female's investment over the initiation and incubation period. As described in Chapter 2, both abandonment and predation can depreciate that investment. We evaluate such losses with separate probabilities for total and partial nest success, and separate transition probabilities which cover partial loss of eggs and complete hatching failure.

Total nest failure is the probability that a nest is completely predated or abandoned between the day incubation began and the time any hatching occurred. As explained in Chapter 2, total nest failure generally involves predator action (or abandonment) affecting all the eggs in the nest. Thus, the sampling variance of total nest failure is treated on a 'per nest' basis.

Our first transition probability is defined for those nests that do not fail totally. Egg survival is estimated for them as clutch size at hatch divided by total clutch size. Egg losses from 'successful' nests are likely related to single or, perhaps double, predation events or accidents, and thus sampling variance accrues primarily on a 'per egg' basis. One statistical reason for not treating total and partial losses with a single variable is that their sampling variances have different bases and their joint distribution is complex (Barrowclough and Rockwell 1993). A second is that the overall distribution of loss during this interval is bimodal (see below), and forming a single mean over such a distribution is questionable at best.

Different predators and predation strategies are primarily responsible for partial versus total loss, especially during the incubation period (Chapter 2). The impact of these differences on reproductive dynamics is seen in Fig. 4.3, where the frequencies of loss for several

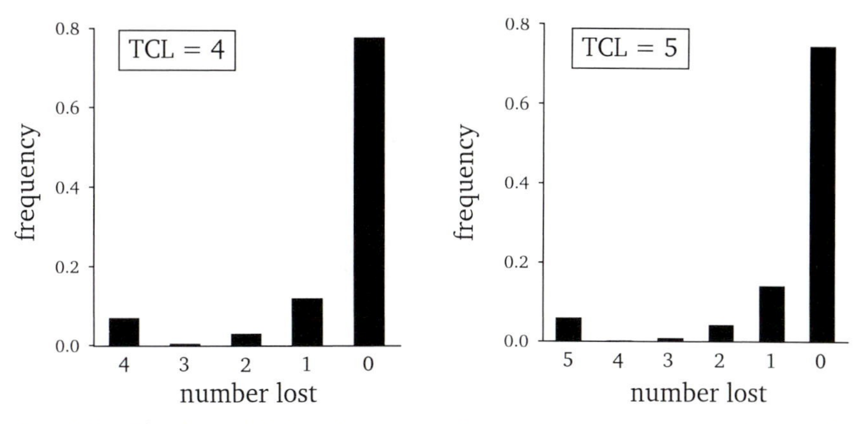

Fig. 4.3 The bimodal distribution of the number of eggs lost during the initiation and incubation periods for initial clutches of four (n = 1298) and five eggs (n = 857). Data are pooled over 1973–1990.

clutch sizes are presented. The asymmetrical bimodality is actually more extreme than depicted, since total clutch loss by definition occurs only during the incubation period, while partial clutch loss occurs over the longer period from initiation through incubation.

4.1.3 Hatching components

The next state variable in our model is the number of goslings leaving the nest. Losses during the hatching period can again be either total or partial. In contrast to incubation, the distribution of loss at this stage does not appear to be bimodal. Total hatching failure is evidenced when no goslings leave a nest for which at least one egg reached the pipping stage. Only 0.4 per cent of the 35 468 nests surviving through incubation between 1973 and 1992 failed completely, thus we generally have ignored this component. For the remaining 99.6 per cent of the sample, we define hatchability as the number of goslings leaving the nest divided by clutch size at hatch. As explained in Chapter 2, an egg 'becomes' a gosling when there is pipping, evidenced by the presence of a star-shaped crack in the shell. Losses during the hatching period can be related to problems with both eggs and goslings. The relative frequencies of such problems are summarized for several years in Table 4.2.

4.1.4 Brood rearing components

The last state variable in the fecundity portion of our model is the brood size at fledging. This is the number of goslings present with their parents during our annual banding operation. Of necessity, the

Table 4.2 Sources of loss in the transition from clutch size at hatch to goslings leaving nest for Lesser Snow Geese at La Pérouse Bay. The data are pooled over the years 1980–1984. The data are summarized both on a per nest basis and on a per egg basis. In both cases only nests for which clutch size at hatch >0 were considered

(a) Egg summary

Status of eggs in nest	Frequency	Percentage total	Percentage subtotal
Left as a gosling	6011	92.9	
Remained in nest	462	7.1	
	6473	100.0	
Remained as gosling	92	19.9	
Remained as egg	370	80.1	
	462	100.0	
Partially developed embryo	97	21.0	26.2
Rotten egg	132	28.6	35.7
Infertile or unknown	141	30.5	38.1
	370	80.1	100.0

Subtotal = 370 abandoned eggs.

(b) Nest summary

Contents of nest when family left	Frequency	Percentage total	Percentage subtotal 1	Percentage subtotal 2
No remaining eggs or goslings	1316	77.9		
One or more eggs or goslings	373	22.1		
	1689	100.0		
One or more abandoned goslings	84	5.0	22.5	
One or more abandoned eggs	296	17.5	79.4	
	380	22.5	100.0	
One or more partially developed embryos	85	5.0	22.8	28.7
One or more rotten eggs	117	6.9	31.4	39.5
One or more infertile or unknown eggs	113	6.7	30.3	38.2
	315	18.6	84.5	10.6

Subtotal 1 = 373 nests which had one or more remaining eggs and/or goslings.
Subtotal 2 = 296 nests which had one or more remaining eggs.
Note that percentages of subtotals reflect the fact that seven nests had both abandoned goslings and eggs and 19 nests had more than one type of abandoned egg.

banding operation occurs before the goslings can actually fly, so our estimate is taken slightly pre-fledging (Chapter 2). Because we are primarily interested in monitoring depreciation in the reproductive capital since the last state variable, we restrict our sample to those captured families where one or both parents was individually identified at the nest and for which all goslings that left the nest were web tagged. Since our nesting study plots include only a portion of the breeding colony and since the birds move considerably during brood rearing, the sample size of families meeting the criteria needed to estimate brood size at fledging is substantially smaller than for the nest-related components, averaging 249 broods per year (range 157–410) between 1973 and 1985, but fewer than 100 per year in the years since, reflecting dilution of our banding efforts as the colony greatly increased in numbers (Fig. 2.3).

We again separate total and partial losses primarily due to the different bases associated with their sampling variances and the fact that estimates of total loss are minima. The probability of total brood failure is estimated as the proportion of those nests where one or both parents were identified and whose goslings were all web tagged, for which the banded parents were captured at banding without any of their web tagged goslings. As explained in Chapter 2, if parents lose their entire brood before they have moulted their primary flight feathers, they may leave the La Pérouse Bay area. The likelihood of capturing parents that suffered total brood loss is thus lower than that for parents that did not fail completely during brood rearing.

The minimum nature of the estimate of total brood failure does not systematically hamper our ability to make relative comparisons among phenotypic segments of the population. Those comparisons are valid as long as it can reasonably be assumed that the conditional probability of a pair leaving the La Pérouse Bay area given that they failed totally is independent of the phenotypes being examined.

For those families that do not fail totally, we estimate gosling survival as the brood size at fledging divided by the number of goslings leaving the nest. The sample is again restricted as above. Loss of one or more web tagged goslings from a family may be due to one of three causes. First, the web tag may have been lost. We can measure the frequency of this by the presence of a distinctive small hole in the web, which occurs in about 4 per cent of encountered originally web tagged goslings (0.038 ± 0.007/year). Second, a gosling may have become detached from its own family and attached to another family which was not caught during the banding operation. The frequency of such 'fostering' may be as high as one in 20 (Chapter 5). Third, the missing gosling(s) may be dead. Predation is likely the major source of such death but both malnutrition and disease contribute directly and indirectly to gosling mortality, most of which occurs early in the brood rearing period (Fig. 2.11; Williams *et al.* 1993*b*).

4.1.5 Comparing transition probabilities

We have a methodological problem when comparing transition probabilities among segments of the population (Rockwell *et al.* 1987). This is readily seen when thinking about comparing the values of egg survival across clutch sizes. Since total nest loss is dealt with separately, egg survival values are constrained by clutch size. For clutch size 2, egg survival can only be 1.0 or 0.5, whereas for clutch size 3, it may be 1.0, 0.67, or 0.33, biasing this variable towards lower values. To adjust for this effect, transition probabilities are transformed to deviation scales that reflect departures from the value expected if there were no effects of the variable being analyzed and/or covariates, such as clutch size, year, or age, for example (Rockwell *et al.* 1987, 1993). When used with covariates, the approach assumes that any change in a transition probability is independent of the variable. Any departure from this should result in a significant interaction between covariates and variables in a factorial analysis of deviation scores. If no interaction terms are significant, we conclude that this assumption of the method has not been violated, and the interaction terms are eliminated from further analyses.

4.1.6 The composite estimate of fledging success

The total clutch laid is a capital investment in reproduction for a female. Her return on that investment, in terms of reproductive success, is the number of fledglings remaining after depreciation during initiation, incubation, hatching and brood rearing. For the 'average female', the expected value of that return can be calculated as a composite over the fitness components. We estimate this by multiplying total clutch size by the series of transition probabilities accounting for both total and partial success and thus make use of the maximum available sample size for each component (Rockwell *et al.* 1987).

Welsh *et al.* (1988) noted that the product of means is an unbiased estimator of the mean of products only if the components are not correlated. Our fitness components are not based on a single sample of individuals and, thus, covariances among the components can not be directly estimated. However, Rockwell *et al.* (1987) found that these fitness components are independent in Lesser Snow Geese. In accord with Welsh *et al.* then, our model is linear and we regard the composite to be an unbiased estimator of reproductive success in this species.

Formally, the expected brood size at fledging (E(BSF)) is estimated as:

$$E(BSF) = TCL \times (1\text{-}TNF) \times P1 \times P2 \times (1\text{-}TBF) \times P3$$

where TCL = total clutch size, TNF = total nest failure, $P1$ = egg survival, $P2$ = hatchability, TBF = total brood failure, and $P3$ = fledging success. These estimates are population values rather than representing the averages of values of a number of known individuals, and as such variances are derived not empirically but by Monte Carlo simulation which generates 95 per cent confidence limits for this estimate (Rockwell *et al.* 1993).

It should be noted that pre-incubation failure is not used in forming the composite. As discussed above, this stems from our lack of information as to what females suffering such failure actually do. If they resorb their remaining eggs and opt out of breeding for the season, then our composite is an accurate estimate for those females that begin incubating a clutch of at least one egg. It would overestimate the expected brood size at fledging for all females that initiated. On the other hand, if females either complete their clutch parasitically or initiate and incubate a reduced clutch, the composite is an unbiased estimate of the expected brood size at fledging for all females that initiated. Using the global estimates in Table 4.1, our general estimate for composite reproductive success is 2.34 (2.29–2.40) where the parenthetical values are the lower and upper 95 per cent confidence limits based on a harmonic mean sample size of 3011 (Rockwell *et al.* 1993).

4.2 *Viability components*

We estimate values of two viability components in our model: annual survival and breeding propensity, taking age into consideration.

4.2.1 Annual survival

Annual survival estimates are notoriously difficult to calculate in wild populations, even when sophisticated statistical techniques are available. We have used three measures of annual survival from fledging onwards. These are: simple re-encounter rates of individuals at the colony, formal models of annual local survivorship of a group based on re-encounters at the colony, and models of annual true survivorship based on reports of band recoveries from hunted birds. The latter are used in conjunction with the fecundity components defined above to produce accurate estimates of life history values. However, the first two measures may be used as relative indices for comparing segments of the population when true survivorship cannot be accurately calculated.

The first method is to simply tabulate 'return rates', defined as whether or not an individual is re-encountered at the colony in a subsequent year or window of years. We calculate return rates from resightings at nests or during the brood-rearing period, and/or recaptures during banding drives. This direct approach gives a survivorship value for particular individuals, and for some purposes provides a sufficient index of the relative survivorship rates of different phenotypic segments of the population.

Simple return rates suffer from two major problems. First, since birds may return but not be re-encountered by researchers, return rates are only a relative index of local survivorship and return, rather than accurate estimates of survivorship probabilities. The indices cannot be used in quantitative comparisons of the importance of survivorship differences relative to differences in fecundity, for example.

We can estimate true local survival rates for segments of the population with our second approach, namely using formal models of annual survivorship rates from capture–mark–recapture (CMR) data, as detailed below. The accuracy of these models depends on the degree to which we meet certain assumptions about our sampling regimes and the birds' philopatry patterns. But more fundamentally, all analyses of local re-encounters measure only 'local survival', defined as being alive in the area in which the study is carried out. As such, mortality is confounded with permanent emigration, defined as permanently leaving the population under study and not to be confused with annual spring and fall migrations. Survival calculated from recapture data thus underestimates true survival, unless there is no permanent emigration from the population under study, and such movements do occur (Chapter 3). Using re-encounter data to generate survivorship measures of male and female goslings, for example, would lead one to believe that few young male Snow Geese survive to breed at all! Where major differences in philopatry are not involved, however, re-encounter data are often our most powerful source of information on factors affecting variation in survival rates.

Our third method for measuring viability is using formal models of annual survivorship rates based on recoveries of bands from hunted geese, as detailed below. Recovery analyses rely on a completely separate data set from local re-encounters, involve a different set of assumptions, and provide our most accurate measures of true survivorship for large samples. However, the large variances in the parameter estimates produced by recovery analyses preclude their use for comparisons of survivorship rates among subsets of the population. We have used a mixture of techniques for different purposes. In theory, useful information can be derived by comparing values produced by recapture and recovery methods. Differences between recovery and recapture estimates, for example, provide information on emigration rates (Francis and Cooke 1993).

Recapture models of local survivorship

Local survival rates are estimated using procedures extended from Jolly–Seber CMR analyses (Seber 1982), with the capture and recapture data generated from banding drives. Recapture data allow the calculation of annual survival rates by comparing the number of marked birds from a particular cohort alive in year (x) with the number alive in year ($x+1$). The SURGE models developed by Lebreton and Clobert (1986) allow investigation of variation in local survival rates due to age, year, cohort or banding history, plus tests for differences between phenotypic segments of the population. A general account of these methods can be found in Lebreton *et al.* (1992). Cooke and Francis (1993) provide details on the application of these methods to Snow Geese.

Local survival estimates based on recapture data are subject to many biases, but we generally have larger samples than for recovery analyses, which means that sources of variation may be examined more effectively. One particular assumption which is violated in our data is the assumption of equal probability of recapture. We have found that *c.* 12 per cent of the adult geese banded at La Pérouse Bay were temporary immigrants, nesting for a single season and never returning, which therefore had a different probability of recapture from permanent residents. Removal of these temporary residents produces estimates of local survival with significantly better goodness-of-fit to the data.

Recovery models of 'true' survivorship

Annual survival rates based on recovery data estimate 'true survival', rather than 'local survival' in the area of study. Recovery analyses are based on band recoveries, which are a sample of the birds which were alive at the beginning of the hunting season, shot, and reported. These data are analyzed with models based on the methods of Brownie *et al.* (1985). In essence, the method compares the recovery rates of birds from cohorts x and $x+1$ over the time period between the banding of the second cohort and the end of the study. The ratio of the two recovery rates gives an estimate of the survival rate of the cohort x in the year between the banding of the two cohorts. If the birds in cohort x are goslings when banded, the survival rate calculated is the first-year, or immature, survival rate. If the birds are adults, then the year-specific annual adult survival rate is obtained. More complex models are available if yearlings are also banded in considerable numbers. Several models may be tested to investigate whether allowing annual survival rates to vary with year or cohort provides a better goodness-of-fit with the data. Recovery models provide perhaps the most reliable tool available for investigating variation in survival rates currently available to vertebrate evolutionary biologists. A good introductory account can be found in Nichols (1992) and a thorough presentation is found in Brownie *et al.* (1985). Details on the application of these methods to Snow Geese are presented in Francis *et al.* (1992*b*).

The recovery data analyses rely on a number of assumptions, summarized by Brownie *et al.* (1985), most of which appear to be met in the present study. Even when sample sizes are large, as in the present study, sampling variances are high due to the fact that ratios of estimates are being calculated. Fortunately, survival estimators are substantially more robust to the partial failure of assumptions than are the estimators of population size which also are produced by these methods (Lebreton *et al.* 1992). However, the wide variances mean that it is difficult to identify causes of variability in survival estimates. Annual, age-specific and cohort-specific variations in survival are difficult to test, for example, thus we use return rates or local survival estimates for these purposes. Finally, while far less sensitive than are recapture analyses, survival rates based on recovery data may also be influenced by patterns of dispersal under certain conditions (Francis and Cooke 1992*b*).

Age-specific survival estimates are particularly important for our calculations of overall fitness and are discussed in Chapter 7. Recapture data are not very reliable for age-specific survival estimates because (1) they confound mortality with permanent emigration and (2) the variable age of first time breeding weakens the reliability of the age-specific estimates. Models based on recovery data do not allow a detailed separation of age-specific survival either, but at least enable one to test the goodness-of-fit of one-age, two-age, and three-age survivorship models (Brownie *et al.* 1985). The two-age model, which assumed different survival rates for first year (that is from fledging to 1 year later) and older than first year birds, provided the best fit to the data, and those values are given in Table 4.1 as our best overall summary of survival rates of La Pérouse Bay Snow Geese. Using data from the much larger colony at McConnell River, Francis *et al.* (1992*a*) showed that 2-year-old birds (that is birds which are yearlings at the time of banding) had a survival rate of 91 per cent of that of older birds. Although we showed a significantly lower survival rate for yearlings than adults for the La Pérouse Bay birds also, the estimates of the differences are not very reliable. The existing models do not specifically test for senescence of survival rate, since all birds older than one are assumed to have the same survival probability. However, Francis *et al.* (1992*a*) developed a method combining the data from both recoveries and recaptures and showed that there were no detectable differences in annual survival from the age of 2 until at least 16 years.

4.2.2 Breeding propensity

The probability that a bird breeds is considered at two points in our model. The age of first breeding vector (b_x) gives the probabilities that an individual breeds for the first time at ages 1, 2 or 3 years. As near as we can determine, all individuals still living have attempted to breed by age 4 years, so the $b_x = 1$ for all birds 4 years old and older (but see below). The analytical details and general considerations of this set of probabilities are presented in Chapter 7.

Breeding propensity also enters the model for individuals that have bred at least once. Adult breeding propensity (b_a) is 1 minus the annual probability that an adult opts out of breeding for a year (Chapter 2). Unfortunately, we can not estimate this probability absolutely. The absence of a bird from our nesting sample in a given year may mean it has opted out, lost a clutch and not re-nested, or simply nested outside of our study plots. Since our nesting plots generally cover the same geographic portion of the entire colony each year the latter would likely reflect a change in nest site. Fortunately, we can make relative comparisons of breeding propensity among phenotypic segments of the population under the assumption that the probability of breeding and not being sampled is independent of the phenotype being examined. By examining all birds seen in years t and $t + 2$, we can divide our sample into two components; those seen also in year $t + 1$ and those not seen in the intervening year. If these proportions of those differ between the phenotypic segments phases, we might conclude that breeding propensity differs (Rockwell *et al.* 1985*b*).

4.3 *Summary*

1. We document the rationale for constructing a fitness components model.
2. Fecundity components are those involved in assessing the number of independent offspring produced by members of the parental generation.
3. Viability components are concerned with the survival of those offspring to become reproductively mature adults and the continued survival of those breeding adults.
4. The boundary between fecundity and viability components of fitness is not clearly defined theoretically, but may be done so operationally.
5. Global estimates for the components of fitness for the La Pérouse Bay population of Snow Geese are given.
6. Detailed descriptions of the fecundity and viability components are provided.
7. We compare the relative merits and applicability of three measures of survivorship: simple return rates, formal estimates of local survival, and estimates of true annual survival.

5 *Pairing, mating, and parental care*

The complexities of reproductive strategies in long term socially monogamous birds such as geese, in contrast to the more obvious ones in polygamous species, have only recently been appreciated by ornithologists. Young Snow Geese find mates during their second winter or spring, and establish pair bonds which nearly always are maintained until the death of a partner. The pair typically migrate to breeding grounds, breed, and migrate back south together for years. This chapter focuses on variation in five aspects of reproductive behaviour which themselves affect fitness, and which also potentially affect interpretations of the reproductive components of fitness defined in the previous chapter and used in subsequent analyses. We discuss mate choice, pair-bond duration, mate fidelity, intra-specific nest parasitism, and aspects of parental care.

5.1 *Mate choice*

Mate choice in biparental species should involve careful assessment by both sexes (O'Donald 1980; Kirkpatrick *et al.* 1990; Jones and Hunter 1993). Since the sex ratio of potential mates for Snow Geese forming pair bonds on the wintering grounds or on migration is approximately 1 : 1, based on approximately equal hatching sex ratios and sex-specific juvenile and first year survival rates (Cooch *et al.* submitted; Francis and Cooke 1992b), we expect no strong bias in intra-sexual competition by one sex or another (Emlen and Oring 1977). Pair bond formation has not been studied directly in wild Snow Geese. Instead, inferences about mate choice have been made by looking for assortative mating, defined as non-random pair formation with respect to phenotype, in wild birds, and by experiments with captive geese which included phenotypic manipulation of potential mate choice traits. The two phenotypic variables which we consider in Snow Geese are plumage colour and body size, both characters which could be detected and used by birds choosing their mates. Mate choice will not necessarily lead to assortative mating, nor does assortative mating in the wild necessarily demonstrate active mate choice (Crespi 1989). Cooke and Davies (1983) proposed that a rigorous demonstration of adaptive mate choice in pair-bonded species involved answering five cascading questions:

1. Is there non-random pairing with respect to some phenotypic character?
2. Does the non-random pairing demonstrate mate preference, or is it simply due to non-random prevalences of phenotypes at the time and place of mate choice?
3. Is mate choice based on the character itself or on correlated characters?
4. Is there genetic variability in the population for the character chosen?
5. Are the fitness consequences of choice consistent with the patterns of choice?

Table 5.1 summarizes the answers to these questions with respect to mate colour and body size, the data for which are reviewed below.

Table 5.1 Evidence for mate choice in Lesser Snow Geese with respect to colour and body size

Question	Colour	Body size
Assortative pairing?	strong	moderately
Active mate choice?	strong	none or weakly
Choice based on character?	yes	unknown
Significant genetic variability?	yes	yes
Fitness consequences	none or negative	none

Sources: Cooch and Beardmore (1959), Cooke and McNally (1975), Cooke *et al.* (1976), Ankney (1977), Cooke and Davies (1983), Davies *et al.* (1988), Cooch *et al.* (1992), E. G. Cooch (personal communication).

5.1.1 Mate choice with respect to plumage colour

Cooch and Beardmore (1959) found strong positive assortment with respect to plumage type within a mixed colony of Snow Geese nesting at Boas River, NWT. Approximately half as many mixed-colour pairs occurred as would be expected if the birds were pairing randomly with respect to colour, and mixed pairs consisting of blue-phased males and white-phased females considerably out-numbered the opposite mixed pair type. As we have seen in Chapter 3, this asymmetry helps explain the present day distribution of the colour phases. A similar pattern of non-random pairing occurs at the La Pérouse Bay colony (Table 3.2).

Cooke and Cooch (1968) postulated that geese select mates of a plumage colour similar to that of their parents. This hypothesis was tested and supported in a series of papers by Cooke and associates (Cooke *et al.* 1972, 1976; Cooke and McNally 1975; Cooke 1978). In the wild, marked goslings of known family colour types returned to La Pérouse Bay in subsequent years with mates which overwhelmingly, although not perfectly, matched those of their natal family. Birds from mixed families returned with mates of either type, but a bias towards white (Fig. 3.6). Since birds choosing mates have access to mates of either plumage, though not necessarily in equal proportions in different parts of the wintering range (Geramita *et al.* 1982), the different types of mates obtained must reflect mate choice based on plumage colour, or some strongly correlated character. Non-random availability of mates, or colour-phase prevalence, cannot account for the general pattern, but it may account for the bias towards white among birds from mixed families and the apparently weaker colour preferences of birds from blue families (Cooke 1978).

The use of family plumage colour *per se* in subsequent mate choice was tested directly by experiments with young geese raised in captivity. Support for the hypothesis included a lack of assortative pairing when goslings were reared without parents in large mixed groups, which controls for pairing with respect to self-colour (Cooke 1978), and preferences for a novel family colour. A 'pink morph', made by dyeing white-phase adults, proved attractive to juveniles raised in pink families (Cooke *et al.* 1972)! This work confirmed that plumage colour was used as a character in active mate choice.

Plumage colour-phase differences are due to genetic differences, thus selection acting on colour preferences will affect the genetic structure of the population (Cooke and Cooch 1968; Chapter 3). Thus, the final question is: are fitness differences consistent with the pattern of mating preferences? The surprising answer is no. Mixed pairs' fitness components values are indistinguishable from matched pairs' (Chapter 9), including egg hatchability and gosling survival, which might detect genetic heterosis or decreased hybrid viability.

Sexual imprinting of young on parental type is a well documented, widespread mechanism producing inter-specific mate discrimination in waterfowl, and no doubt it continues to do so in Snow Geese. The component of fitness which presumably is selecting for the maintenance of plumage discrimination in mate choice is one not measured, namely the adverse consequences of mating with other species of geese. Nonetheless, one can argue that mate choice with respect to plumage type is out of evolutionary equilibrium during the period of secondary contact apparently occurring in this species (Cooke *et al.* 1988). If blue and white phases had differentiated genetically to the point where hybrids were less viable, the colour preference mechanism would be adaptive within the species. Since hybrids appear to be equally viable, assortative mating by colour phase constrains mate choice unnecessarily, an evolutionary cost, albeit a small one. Geese which continued to discriminate among species, but not between colour phases, would have a larger pool of potential mates from which to choose, and therefore potentially be at a selective advantage over geese which discriminate on the basis of colour.

5.1.2 Mate choice with respect to body size

Snow Geese are positively assortatively paired with respect to size. Pairs at La Pérouse Bay ($n \cong 350$) were positively assortatively paired with respect to mass, culmen, head length, and tarsus (Davies *et al.* 1988). Assortative pairing with respect to body weight may be due to common environment effects post-pairing, rather than mating preferences (Choudhury *et al.* 1992), but this cannot account for correlations with respect to structural size. Females in 44/48 pairs collected at McConnell River, NWT had males with larger culmens than their own, a higher proportion than would have been expected by chance, although there was not a significant correlation given the sample size ($r = 0.21$, $P > 0.05$; Ankney 1977). At La Pérouse Bay, with a larger sample, there was a significant correlation for culmen between pair members, despite a smaller correlation coefficient ($r = 0.11$, $P = 0.01$, $n = 436$).

At La Pérouse Bay, assortative pairing by size appears not to be due to active mate choice. There are significant differences in the mean body sizes of adult birds hatched in different years, due to annual differences in growth rates attributed to environmental variation (Chapter 13; Cooch 1990; Cooch *et al.* 1991*b*). Birds pairing for the first time tend to be the same age and sex-specific relative size simply because they have hatched in the same year. Since La Pérouse Bay females are nearly all pairing with males from other colonies, this interpretation requires a degree of common annual environmental effect on growth rates among the colonies contributing potential mates to the central wintering population. Widespread weather effects, especially variation in rainfall and post-hatch temperature, might produce such an effect.

We cannot assess directly the size distribution in the pool of available males, but we can test directly for assortative mating within cohorts. Assuming that a cohort of females will be competing for the same pool of males, be they large or small, this provides a quite powerful test for evidence of active assortative mating. Within cohorts, the correlations between male and female structural sizes are weaker than the pooled correlations reported by Davies *et al.* (1988). Using a sample of 187 females of known age, the correlations are positive in 9/13 cohorts, statistically significant in only 1, and have a non-significant overall correlation coefficient of about 0.06 (P = 0.46) ($PC1$ values based on culmen, tarsus, and head length; E. G. Cooch *et al.* unpublished data). This weak pattern may reflect differences in other variables weakly correlated with body size, rather than active assortative mating *per se*.

Body size has both significant genetic and environmental variance components (Chapters 8 and 13). In terms of fitness, larger female body size is not associated with higher fecundity in general (Davies *et al.* 1988; Cooch *et al.* 1992; Chapter 13). Cooke and Davies (1983) reported a trend towards higher clutch sizes for larger pairs, but this result was not confirmed with subsequent analyses and larger sample sizes (Davies *et al.* 1988). Thus, with respect to body size, Snow Geese show a pattern of non-random pairing which largely reflects environmental covariation in the prevalence of mates of different sizes, and which should have little if any effect on the genetic structure of the population.

Ridley (1983) argued that natural selection would favour assortative mating if (1) pairs formed for a significant fraction of a mating period, (2) male characteristics correlated with mate acquisition ability, and (3) female characteristics correlated with fecundity. Condition (1) obviously applies to Snow Geese and (2) is plausible, but there are no compelling data with regard to differential female fecundity as a function of colour or size (Chapters 9 and 13). This may be the primary reason why active adaptive mate choice is lacking with respect these characters. In summary, assortative mating with respect to colour is present, but there is no evidence for a fitness consequence of it. Assortative mating with respect to body size is weak at best, and should have little or no effect on the fitness measures calculated and analyzed.

5.2 *Mate fidelity*

Counting the number of eggs or young produced from a nest is a noisy, at best, and potentially biased, at worst, measure of the reproductive success of the male or female attending that nest. Here we consider how variation in pair and extra-pair copulation behaviour may affect individual fitness and our measures of it. The origins of eggs in nests will be considered in the next section.

5.2.1 Pair copulations

Early observers rarely if ever observed copulation in arctic nesting geese, including Snow Geese. Cooch (1958) believed Snow Geese to be most active sexually in the James Bay area, and noted that the only definite records of copulation in spring were from North Dakota and Wisconsin. Barry (1967) similarly believed that Snow Geese had already copulated when they arrived from their staging area to the Anderson River colony. Ryder (1967) saw no copulations among Ross' geese in the Perry River region and suggested that copulation probably occurs prior to arrival on the nesting grounds. Since fertilization occurs within 24 h of egg laying in birds (Howarth 1974), the above observations led Raveling (1978) to conclude that sperm storage was the rule among arctic nesting geese, when as many as 17 days might elapse between the arrival of a goose on the nesting area and the laying of the final egg.

During nest initiation and laying in 1976–78, Pierre Mineau and Ken Abraham watched geese from a live-in tower overlooking a dense portion of the La Pérouse Bay colony. Gurtovaya (1990) made similar observations on Wrangel Island between 1980 and 1985. Copulation of mated pairs was commonly observed at both sites. At La Pérouse Bay, all pair copulations occurred in water however, whereas pair copulations on land were the rule at Wrangel Island. At La Pérouse Bay, the male usually initiates the sequence by positioning himself near his mate high in the water and with tail cocked vertically. He then carries out a series of head dips of increasing frequency. If the female is receptive, she usually adopts a bowed posture with the lower neck in the water, often accompanied by bill dipping or head dipping. The transition from pre-copulatory display to mounting is sudden. The male grasps the neck feathers and mounts the female, who crouches if she is in shallow water (Fig. 5.1). Treading occurs for about 5 s and the female lifts her tail to one side as cloacal contact is made. Male Snow Geese, have intromittant organs, which are rare in birds in general but common in waterfowl. A post-copulatory display consists of the birds rising up with vertically extended head and neck. The birds usually vocalize and flap their wings. The sequence terminates with extensive preening and bathing.

Behavioural observations were made in the pre-laying, laying, and early incubation stages of the season. Ninety copulations were observed in 321 h of observations in 1977–1979. Most mating between pair members was seen in the pre-laying or laying period (Fig. 5.2). Observed copulation frequency increased gradually during laying and peaked just before the onset of incubation (Mineau 1978;

Fig. 5.1 A within-pair copulation of Snow Geese.

Mineau and Cooke 1979*b*). Once incubation began, copulation showed a dramatic drop in frequency and females often refused male copulation attempts. A roughly similar pattern was reported by Collias and Jahn (1959) on a captive flock of Canada Geese. Among 45 individually marked pairs which nested close to the observation tower in 1977, 29 were not seen to mate, 10 were seen copulating once, five twice, and one pair three times. These copulation frequencies are underestimates since birds were not under continuous observation and copulation takes place rapidly. Since individual pairs were seen to copulate on two (and in one case three) different days, we think it likely that most pairs copulate more than once and possibly every day during laying. Most copulations occurred early in the morning, and no pair was seen to copulate twice in the same day. Gurtovaya (1990) noted a similar temporal pattern on Wrangel Island.

5.2.2 Extra pair-bond copulation

Matings outside of pair bonds (EPCs) contribute significantly to the reproductive success of males, and perhaps also of females, in a number of species of birds (Birkhead and Møller 1992), including Snow Geese (Lank *et al.* 1989*b*). Mineau and Cooke (1979*a*) described 'rape' in Snow Geese based on observations of 116 EPC

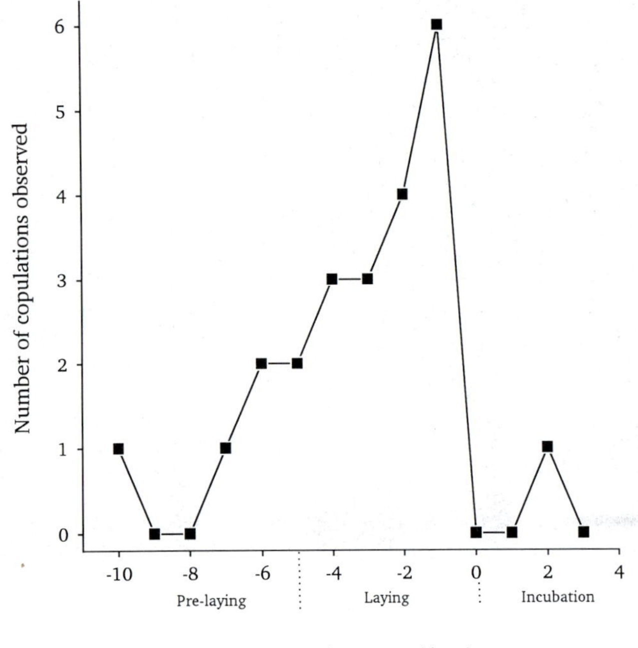

Fig. 5.2 The frequency of copulations seen, relative to nesting stage, among pairs observed intensively throughout the early nesting season in 1977 (data provided by P. Mineau and K. F. Abraham).

attempts, among the first papers to consider the potential significance of this behaviour for the reproductive success of male birds. The EPCs observed in Lesser Snow Geese differ significantly in character from matings between pair members, and did not involve solicitation by females. Unlike intra-pair copulations, EPCs never occurred in water. Males often approached females on the nest, but if not, females were forced to the ground and mounted from any direction. Treading is lengthier than the usual 4–5 s during pair copulation, and the male moves his tail from side to side to slip it under the female's. In nine cases, the female eventually definitely raised her tail and allowed cloacal contact. In general, however, it was impossible to see if the copulation was successful. No post-copulatory display was seen, and neither males nor females had the normal bouts of post-copulatory preening or bathing. B. Ganter (personal communication) observed an incubating female solicit a mating with her own mate 5 min after a forced EPC with a neighbour. In contrast to pairing, EPC attempts occurred without regard to male and female colour morph (Lank *et al.* 1989*b*). Mineau (1978) estimates that on

Fig. 5.3 A pair of Lesser Snow Geese, with the female on the nest, during laying.

average a female was subject to 0.2 EPC attempts per day in 1976 and 0.06 in 1977. The lower frequency in 1977 may be due to a lower density of nesting pairs in the observation area in that year.

The male Snow Goose breeding strategy is to mate-guard his female until she lays the final egg (Fig. 5.3), after which he may pursue EPCs (Mineau and Cooke 1979*a*,*b*). Most EPC attempts observed involved paired, nesting males attempting EPCs with females nesting in neighbouring territories. Most EPCs occurred when the resident male was absent from the nest, often (73 per cent) while the absent male was himself pursuing an EPC. Although there is no risk to a male's paternity at this point, there is some potential risk of damage to his undefended mate and/or eggs. Egg damage has occurred during EPCs among birds nesting in captivity (M. A. Bousfield, personal communication), but we have no data from the wild. The most curious aspect of this male strategy, however, is that it apparently results in most EPCs (*c*. 80 per cent, Mineau and Cooke 1979*b*) occurring outside of female fertile periods, in contrast to the patterns found in other birds (Birkhead and Møller 1992). As one component of a male's reproductive strategy, extra-pair copulation appears to be an inefficient one, with males usually unable to detect whether or not a female is fertilizable. EPCs are even seen well into the incubation period (Mineau and Cooke 1979*a*). Nonetheless, EPCs can and do result in fertilizations (Quinn *et al.* 1987; Lank *et al.* 1989*b*).

What are the fitness consequences of EPCs? Extra-pair copulations affect the paternity of *c.* 1.3 per cent of goslings hatched at La Pérouse Bay (Lank *et al.* 1989*b*). This rate was calculated by comparing the rates of mis-matched coloured goslings hatching in nests of homozygous white versus homozygous blue pairs. In homozygous blue nests, white goslings will only be produced by nest parasitism, since the dominant allele from the blue female must produce a blue gosling. In white nests, however, the rate of blue goslings hatched is a measure of nest parasitism plus extra-pair fertilizations by blue males, since any blue allele contributed by EPC males will be dominant to the homozygous female's white allele. By adjusting observed mismatch rates by the colour-morph frequencies of goslings produced in general and the estimated frequency of blue alleles in males, Lank *et al.* (1989*b*) calculated that mismatch rates were in fact higher than expected in white versus blue nests, and attributed this difference to extra-pair fertilizations. If the costs to males of pursuing EPCs are trivial, there is a fitness pay-off for the behaviour, even though the frequency of successful EPCs is so low.

We cannot assess the potential genetic consequences of EPCs for a female's young. Given the form of EPCs, with no female solicitation, it seems doubtful that females obtain any general benefit. However, if a pair of birds, or a single female, is attempting to lay parasitically, a female may have an easier time approaching a nest if it mates with the male. While this may seem far fetched, one apparent example has been reported in Barnacle Geese (*Branta leucopsis*) (Choudhury *et al.* 1993).

The male Snow Goose strategy of mate-guarding and then pursuing EPCs, combined with a short, synchronous breeding season, should produce a strong temporal asymmetry with regard to male reproductive pay-offs. Males breeding early in the season will be those most likely to obtain EPCs which result in fertilizations. The extent to which EPCs are costly to females is unknown, but this pattern of EPCs also favours females which lay early, since they are less likely to be approached by males (Birkhead and Biggins 1987). While we cannot assign the *c.* 1.3 per cent of goslings attributed EPCs to particular individuals, early breeding males may father *c.* 2–3 per cent more goslings than our raw data indicate. This provides a small, but non-trivial selective advantage for males of pairs which breed early in the season. Since older females lay earlier in the season (Chapter 7), and pairs are assorted by age, EPCs may slightly increase the fitness of older males.

In summary, patterns of pair and extra-pair mate guarding and copulation will only bias our measures of fitness components slightly, and only in those components involving goslings calculated with respect to timing in the season and age. This mainly affects males, and most of our major age and timing analyses refer predominantly to females, whose fitness is apparently negligibly affected. We have not tried to adjust for variation in copulation patterns in our analyses.

5.2.3 Maintaining pair bonds

Snow Goose pairs are indeed long-lasting, even if both members are not entirely genetically faithful during their tenure. There is no evidence that remating is more likely after reproductive failure than after success (Cooke *et al.* 1981), as expected if mate change were part of the reproductive strategy of the species (Diamond 1987). Separation and remating while both members of the pair are still alive appears to be extremely rare in our population, with just a handful of documented cases. Remating after the death of a partner occurs regularly, since 12–20 per cent of adults die each year (Francis *et al.* 1992*b*). In several species of birds with long term pair bonds, newly paired individuals have decreased reproductive performance relative to that of intact pairs (Rowley 1983). At La Pérouse Bay, re-paired birds had a lower average clutch size than intact pairs, however this decrease occurred only in the youngest age class of females and was attributed to an age effect, rather than partner change *per se* (Cooke *et al.* 1981). In Barnacle Geese, Forslund and Larsson (1991) argued strongly that re-pairing was likely to be disadvantageous because an individual was likely to obtain a younger mate. Pair status has little effect on our estimates of fitness components and has not been considered in our analyses.

5.3 *Intra-specific nest parasitism*

Intra-specific nest parasitism (ISNP) has recently been recognized as being widespread and of considerable reproductive significance in birds, and especially in waterfowl (Yom-Tov 1980; Andersson 1984; Rohwer and Freeman 1989). Intra-specific nest parasitism rates were first quantified in Snow Geese in association with testing genetic models of the control of the plumage polymorphism (Cooke and Mirsky 1972). The simple allelic dominance model proposed could not account for the low levels of blue goslings (*c.* 3 per cent) hatching in nests of white pairs. Cooke and Mirsky (1972) showed that such blue goslings occurred disproportionately in nests with larger clutch sizes, and suggested that nest parasitism accounted for their presence. They presented a simple method for estimating the overall frequency of ISNP among hatching goslings using the proportions of blue goslings in nests of white pairs, under the assumption that nest parasitism was random with respect to colour. If the frequency of blue goslings in W × W nests were x, then the probability of a blue gosling being a parasitic gosling is also x. The probability that a gosling of either colour was not produced by both nest attendants is simply x/p, where p is the overall proportion of blue goslings produced on the colony. This calculation actually includes the effects of EPCs and nest parasitism, but the two can be teased apart by comparing mismatch rates from both homozygous pair types (see above; Lank *et al.* 1989*b*).

The behaviour of nest parasites was described by Mineau (Mineau 1978; Lank *et al.* 1989*b*). Potential parasitic females, usually accompanied by mates, typically approach occupied nest sites rather than searching unattended sites for undefended nests. This prevents females from laying in abandoned nests (Lank *et al.* 1990), but also may simply save search time. Nesting pairs defend their sites against intruders. Resident and intruding males typically square off a short distance away from the nest. Thus nest parasitism is a reproductive tactic of a pair, as well as of a female. The resident female usually stays tight on the nest, while the intruding female approaches and attempts to dislodge her by pushing with her breast. If the parasite succeeds, she might lay in the nest. If not, intruding females may lay adjacent to the nest. Eggs laid just outside the nest may be rolled in by the attendant female. This apparently altruistic behaviour benefits the nesting female by decreasing the conspicuousness of the nest site to predators during the laying period, usually with little measurable cost to her own nesting success (Lank *et al.* 1990, 1991).

Parasitic eggs are added to clutches throughout laying and early incubation. Eggs added during incubation have reduced hatching probabilities, since they may be abandoned when the female leaves the nest when the rest of the brood hatches (Lank *et al.* 1990). However, eggs added to nests up to 4 days after the beginning of incubation can hatch successfully, albeit at a reduced rate (Davies and Cooke 1983*b*). Late-hatching goslings apparently both accelerate their own hatching and manipulate the parents into remaining longer with the clutch.

One female's eggs may also occur in another's nest due to nest 'takeovers'. Up to *c.* 25 per cent of the nests we find each year are abandoned with only one or two eggs in them (Chapter 4). While many abandoned eggs are removed by scavengers, some remain in nests which are subsequently used by a different female. If we use mismatching parent and gosling colour rates for first eggs in nests as a criterion for scoring a nest takeover, we conclude that at 6–8 per cent of nests a second pair of geese has adopted an existing egg and a nest site (M. Collins, personal communication). A second kind of takeover occurs when two females begin laying eggs in almost the same location, for example, on adjacent sides of a willow newly emerging from melting snow. Often such nests are both found at the one egg stage and when visited the next day, each contains two eggs. By day 3 or 4, all of the eggs may be combined into a single nest which is subsequently incubated by a single female. Clutches of 8 or more eggs account for *c.* 1.7 per cent of nests at La Pérouse Bay.

Sometimes two or more females may lay complete clutches in the same nest cup. In one unusual and thoroughly examined case, two females shared incubation at an eight-egg nest (Quinn *et al.* 1989). Analysis of nuclear DNA with RFLP markers showed that both females had contributed eggs to the clutch and that at least three males had fertilized the eggs. Biparental female care is extremely rare.

Parasitism rates calculated from hatching gosling colour ratios include all forms of multiple laying. Parasitism accounts for 2–9 per cent of the goslings hatched annually at La Pérouse Bay with higher rates usually occurring in years in which nest sites slowly become available as snow melts at the start of the laying season (Lank *et al.* 1989*a*). Birds unable to establish nests apparently become parasites at least for part of their clutch. Consistent with this, nests initiated early in the season are more likely to be parasitized than those begun later (Lank *et al.* 1991). One year deviated strongly from this pattern. In 1977, parasitism rates were as high as ever observed at La Pérouse Bay despite a total lack of snow cover when birds arrived. This lack of snow cover reflected a wide-spread drought in central North America that spring, and high parasitism rates may have been due to a greater than normal number of females in poor physiological conditions adopting the parasitic laying tactic (Davies and Cooke 1983*a*; Lank *et al.* 1989*a*).

Nest parasitism is usually a less effective method for hatching eggs than nesting (Lank *et al.* 1990), but the relative success of the two nesting strategies varies with parasitism frequency. At low levels of ISNP, eggs laid parasitically have similar survival probabilities to eggs laid by nesting females. Once parasitism accounts for more than *c.* 6 per cent of the eggs laid annually, however, the hatching probability of parasitic eggs is significantly lower, dropping on average

6.5 per cent with each 1 per cent increase in annual parasitism rate. Parasitic laying will not usually result in higher fitness for a female than nesting, unless parasitic females lay more eggs than they would if they nested or if the survivorship of parasitic females were sufficiently higher than that of nesting birds to offset the decreased hatching success. We have no data on this matter. However, the large amount of annual variation suggests that most parasitism occurs as a female's best option if she is precluded from nesting, or if she has only one or two eggs to lay, as might occur following partial clutch predation. In these situations, parasitism is a preferable tactic to simply abandoning eggs (Lank *et al.* 1990).

Nest parasitism affects our components of fitness model in three major ways. First, ISNP results in observed clutch sizes (TCL) overestimating the reproductive investment of nesting females. The magnitude of this discrepancy varies with annual ISNP level and systematically within seasons. Second, measures of hatching success are lower than those reflecting only eggs laid by the attendant female, due to the hatch synchronization problem. Third, the sample of nesting birds is probably a biased sample of breeding birds. Limited direct observations (Lank *et al.* 1989*b*), plus the fact that the mean size of parasitic eggs is smaller than that of normally laid eggs (unpublished data), suggest that younger birds (whose eggs are smaller, see Chapter 12), have higher probabilities of being parasites, for example. Thus analyses of nesting success with respect to age (Chapter 7) are likely over-estimating the success of younger birds as a whole, which are more sensitive to environmental conditions in general (Rockwell *et al.* 1985*a*).

We have dealt with the complications of ISNP in different ways in different analyses. In most of the work relating to components of fitness, clutches larger than seven, which are almost certainly the result of ISNP have been eliminated from our samples. In analyses of changes in annual mean clutch size, we have corrected for annual parasitism rate (Cooch *et al.* 1989). In analyses of patterns of egg size, individual nests were screened for obvious indicators of parasitism, such as two eggs in 1 day, or eggs laid after incubation had begun. In many analyses, parasitism will simply be an unbiased source of noise. In summary, the nest-based reproductive success analyses presented in this book are based on data which include contributions from non-nesting females. Where possible, we have taken parasitic eggs into consideration in our analyses. Where this was not possible, we have simply incorporated the fact of nest parasitism into our interpretation of the results.

5.4 *Biparental, uniparental, and alloparental care*

Biparental care by both a male and female occurs at over 99 per cent of our nests. However, each year five or six nests seem to be occupied by only a single female. The females at such a nest have fertile eggs and seemed more attentive on the approach of an observer. It is easy to assume that if two parents attend the nest, this means that both parents are essential at this stage. Martin *et al.* (1985; unpublished data) tested whether or not males play an essential role in nest and gosling success. In one set of experiments, conducted over three seasons, males were removed at the beginning of incubation. Widowed females ($n = 46$) were just as successful at completing incubation as control pairs at adjacent nests, but may have been more vulnerable to nest parasitism (four nests versus none for controls). Goslings from these nests survived and were encountered at banding at similar rates to goslings from control nests. In a second series of experiments, either the male, the female, or both parents were removed at hatch. Single parent goslings were encountered at similar rates as those from control nests, and two orphaned goslings were subsequently encountered. While the sample sizes and seasons tested are limited, the results challenge the notion that the benefit of biparental care is a major factor promoting long term pair bonds in Snow Geese. The no-parent results dramatically make the point that measures of brood success may be misleading because goslings may be adopted by other pairs.

Williams (1994*a*) estimated that least 13 per cent of Snow Goose pairs at La Pérouse Bay adopted one or more young. This estimate was based on the proportion of broods of individually identified adults that had an increased number of goslings on a second observation at foraging sites during the brood rearing period. The average increase among these broods was 1.3 young. Assuming that all family members were present and counted on the first sighting, and given an average brood size of approximately three, about one gosling in 20 is raised by alloparents. Larger broods have advantages in social encounters during the winter (Gregoire and Ankney 1990), and young in larger broods grow faster (Cooch *et al.* 1991*a*). These results could occur because parents with larger broods are better parents or because the addition of extra young is itself advantageous. If the latter, adoption may actually be advantageous for at least certain parents, as well as potentially for the gosling. Williams (1994*a*) found no demographic correlates of parents which did or did not adopt young.

Adoption will lower the values of brood success computed in our analyses below their true level, since goslings not captured in banding drives will not be attributed to their natal nest. Brood success values are based on re-encountering individual goslings tagged at nests, rather than direct counts of brood sizes of parents, thus all identified goslings captured will be attributed to their natal nest, whether or not they remain with the parents. The occurrence of adoption, however, means that our measures of brood size may not reflect the number of young attended by a pair, and are somewhat lower than the true values.

5.5 *Summary*

1. Variation in aspects of reproductive behaviour, including mate choice, pair bond maintenance, extra-pair copulation, intra-specific nest parasitism, and adoption can influence or bias our measures of fitness.
2. Snow Goose pairs are positively assorted by plumage colour and size.
3. Assortative mating by colour is a result of birds preferring a mate of the same plumage as that of members of the family with which they were raised. There appears to be no fitness advantage of such a choice within the population.
4. Assortative mating by size does not necessarily imply choice, since it can be explained by random mating within size-specific cohorts. There is no detectable fitness consequence of choosing a mate of a particular size.
5. Pair bond copulation occurs most frequently immediately prior to and during the laying period, when mate guarding is strongest. Some copulation occurs through early incubation.
6. Extra-pair copulation occurs during egg laying and early incubation and occasionally results in successful fertilization and gosling production. Of all goslings, 1.3 per cent are estimated to have arisen from extra-pair copulation.
7. Intra-specific nest parasitism occurs with varying frequency in different breeding seasons. From 2 to 9 per cent of all goslings hatch from eggs laid in the nests of other females. Parasites are thought to be females who have not found a nest site of their own and parasitism appears to be a less productive reproductive tactic.
8. Snow Geese pair for life, but re-pair after the loss of a mate. Divorces occur but are extremely rare.
9. Although biparental care is the rule, exceptions are found and

experimental evidence suggests that during the breeding season the absence of a male does not reduce fitness.

10. Adoptions of goslings to other families occurred in perhaps 13 per cent of the families.
11. The consequences of extra-pair copulation, intra-specific nest parasitism, and adoption to the accuracy of the fitness values calculated in our study are discussed. In general the values obtained are unbiased approximations to those pertaining to the parents attending the nest or goslings.

6 Annual variation in fitness components

Long term studies allow examination of a population's response to a range of naturally occurring environmental conditions. Fluctuation in selective regime, on an appropriate time scale, is a potential mechanism for maintaining genetic variation in a population. The work at La Pérouse Bay began in 1968, providing up to 25 breeding seasons' data with which to assess the selective regime and the birds' values for various components of fitness. We can determine whether patterns are relatively robust or highly sensitive to environmental conditions, and ascribe some of the total variation in characteristics to environmental or other variables acting on the population. We first outline the major factors which vary among seasons, then present data on the annual variability of fitness components and attempt to correlate this variation with one or more of these factors.

6.1 Processes affecting annual values

Three ideas should be kept in mind when interpreting annual variation in reproductive performance. First, different individuals comprise the population sampled in different years. Annual variation could reflect changes in the composition of the population, rather than responses of a constant population to environmental variation. Age structure provides the simplest hypothetical example. First-time breeders have a smaller clutch size than older birds (Chapter 7). Two years after a season with high juvenile survivorship, first-time breeders might make up a higher-than-usual proportion of the breeding population. The clutch size of the population would drop for purely demographic reasons, independently of environmental variation. Similarly, changes in the genetic structure of the population may change demographic parameters. If there is strong directional selection favouring birds producing larger clutches, for example, we might expect an increase in the mean clutch size of the population independently of changes in environmental factors.

Second, the La Pérouse Bay Snow Goose colony has not been at a population equilibrium during the study. It has grown from approxi-

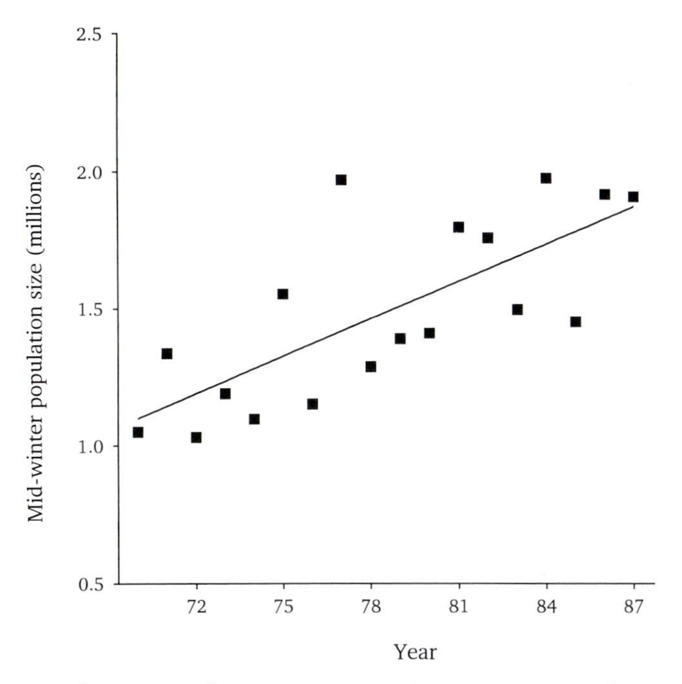

Fig. 6.1 Mid-winter mid-continent Lesser Snow Goose population size estimates. (Adapted with permission from Cooch *et al.* 1989).

mately 2000 pairs to 22 500 pairs from 1968 to 1990 (Fig 2.3). The total Hudson Bay–Foxe Basin population has had a concomitant increase (Fig. 6.1), and many of these birds stop over at La Pérouse Bay and feed prior to migration to colonies further north and west. We thus consider whether annual variation in fitness components may be attributed to the large change in breeding population size and the intensity of use of the area by residents and migrants.

Third, Snow Goose survival and breeding performance at La Pérouse Bay may be affected by environmental factors throughout the year. Hunting is a major source of adult mortality, and clutch size and laying date depend not only on weather conditions in the Churchill area but also on climatic factors in the US and Canadian prairies (Davies and Cooke 1983a). Overwinter survival rates also may change with changing land use practices (Chapter 2).

One complication in studies of the natural selection in the wild is that many features of the environment change with time. It is often difficult to isolate the key factors which are responsible for the variability detected in life history parameters. We have an advantage in this regard because we can correlate changes with either or both of two categories of variation: those which have changed systematically during the course of the study, plus the usual more random annual fluctuations. In this chapter, we describe patterns of annual variation, emphasizing those which have shown long term changes. We first describe major variables which may influence the reproductive performance and survival of the geese. We categorize these factors as extrinsic or intrinsic, although it is not always possible to fit the factors into such a neat classification system. We then document the patterns of variation among fitness components and investigate potential genetic, demographic, or environmental causes for the variation.

6.1.1 Extrinsic variables

Food availability on spring staging areas

La Pérouse Bay Snow Geese acquire most of the nutrients which carry them through nest initiation, laying and early incubation while feeding on newly growing vegetation and grain at staging areas in the prairie regions of the Dakotas and southern Manitoba in April and May (Campbell 1979; Wypkema and Ankney 1979; Alisauskas 1988). Additional nutrients, especially protein, may be acquired in salt marshes along the James Bay coastline (Alisauskas 1988). This nutrient acquisition pattern reflects modern agricultural land use which provides geese with a rich carbohydrate source in what was once grasslands. Food availability on the northern prairies is apparently related to the rate of snow disappearance and soil moisture content (Davies and Cooke 1983*a*). In years of low winter precipitation, soil water content is low and new plant growth is probably reduced. Food availability is reduced or at least delayed by late snow disappearance which is mainly a function of April temperatures.

Snow Geese also feed upon arrival at the colony itself (Mineau 1978; Gauthier and Tardif 1991; B. Ganter, personal communication). Although new vegetation is not available, geese grub for plant stolons, rhizomes and other overwinter storage organs. The relative contributions of endogenous nutrients acquired on migration versus exogenous local food to the total resources available for breeding are unclear.

Nest site availability at La Pérouse Bay

Snow Geese do not nest on snow, thus sites first become available at patches of snow-free ground. Additional sites become available after the melt water recedes following the spring flood. The environmental factors which influence snow disappearance are temperature and precipitation, mainly in the months of April and May. Within years, there is a strong correlation between Winnipeg and Churchill April temperatures ($r^2 = 0.61$), but little between precipitation values at the two locations ($r^2 = 0.19$). This suggests that geese cannot accurately predict northern snow melt phenology from that occurring in the prairies. Since the mean nest initiation date strongly correlates with other annual fecundity measures, including clutch size, nest parasitism rates, and gosling growth and survival rates, the timing of snow melt has a large effect on annual reproductive performance.

Weather conditions

Females spend most of their time on the nest during incubation, and nest abandonment is rare, thus a wide range of climatic conditions can be tolerated during incubation. However, on the rare occasions when severe snow storms occur late in the incubation period, some females abandon their nests, and we have found a few females dead on the nest. Variation in weather conditions during hatch itself could be crucial to whether goslings can successfully leave the nest.

After hatch, climatic factors affect goslings and adults both directly, by adjusting metabolic demand, and indirectly, by affecting the rate of plant growth. Most of the annual variation in weather during this period can be summarized along an axis of cold and wet versus hot and dry weather (Cooch *et al.* 1991*b*). Cold weather places goslings under more metabolic stress, but is actually more favorable to plant growth than hot dry weather.

Predators, diseases, and parasites

Predator levels at the colony vary from year to year, but we have not systematically indexed their annual abundance. Avian predators such as Northern Harriers (*Circus cyaneus*) and Short-eared Owls (*Asio flammeus*) fluctuate in numbers with fluctuations in the levels of microtine rodents, but rarely if ever prey on the geese. Caribou have increased in numbers during the study. Ravens and perhaps gulls also have increased, perhaps in response to the growth of the goose colony. Jaegers and foxes seem to have stayed relatively constant in numbers, although Red Foxes increased at the expense of Arctic Foxes in the early 1980s and subsequently declined again, and there have also been years when Arctic Foxes were abnormally frequent.

We know little about the effects of disease or parasitic organisms on annual fitness. We have not observed epidemics resulting in widespread obvious sickness or death. Clinchy and Barker (1990) did not find convincing correlations between the clutch sizes of adult females and either the prevalence or intensity of a variety of blood and organ parasites. Individual goslings have shown pathological damage induced by high levels of renal coccidia (Rainnie 1983), which are acquired by feeding goslings, and while the levels of infection vary among brood rearing localities, the consequences for gosling growth rate or survival of such variation is unknown.

Hunting pressure

Hunters are a major source of mortality in Snow Geese, and hunting intensity varies from year to year. Data from US Fish and Wildlife Service and the Canadian Wildlife Service show that there has been an overall decline in the numbers of hunters during the period of our study. This decline has occurred at the same time as the increase in overall numbers of mid-continent Snow Geese, so that the proportion of Snow Geese killed by hunters has declined steadily during the course of the study. This is reflected in a 2- to 3-fold linear decline in recovery rates (Francis *et al.* 1992*b*), defined as the rate at which banded birds are shot and reported to the US Fish and Wildlife Service Bird Banding Office. In 1969, band recoveries were obtained from nearly 10 per cent of the goslings banded, compared with less than 3 per cent in 1985.

6.1.2 Intrinsic factors affecting fitness

Population size and local food resources

The 10-fold increase in local colony size may well have affected fitness component values. Although the nesting area of the colony has expanded considerably, the original nesting areas are far less densely occupied than they previously were, suggesting changes in some aspect of habitat quality. There has been a decline in the distribution and availability of the main Snow Goose food plants in the La Pérouse Bay region, documented by R. L. Jefferies and his students from the University of Toronto. These changes apply both to underground storage structures 'grubbed' or pulled out by birds in early spring, which destroys individual plants, and to plant species grazed by adults and goslings after nests hatch. Although moderate grazing intensity can actually stimulate net primary production (Cargill and Jefferies 1984; Jefferies 1988*a*,*b*), over-exploitation at La Pérouse Bay has resulted in a substantial decrease in both the extent and productivity of the salt marsh vegetation during the course of our study (Chapter 2). Similar large-scale destruction of salt marshes by Snow Geese has been reported elsewhere along the Hudson Bay coastline (Kerbes *et al.* 1990). The geese apparently cope with this change in part by utilizing new nesting areas, and by dispersing to brood rearing areas much further from the nesting grounds than they did in the early years of the study. Whereas few post-hatching families of Snow Geese were observed by researchers working at Cape Churchill (Fig. 2.2) during the 1970s, thousands of geese now travel east past the Cape and continue south at least 30–40 km along the coast (Cooch *et al.* 1993).

The Snow Goose population in the entire central wintering region has increased, albeit less dramatically, and effects of food limitation on the wintering grounds and/or at staging areas may also occur.

Age structure

More young survive in some years than in others, which changes the recruitment rate of breeders 2–4 years later, and affects the age structure of the population. The proportion of birds of a particular age class which enter the breeding population also varies with cohort (Cooke and Rockwell 1988). There has been a long term decline in immature survival and an increase in adult survival during the course of the study, leading to an increase in age of the portion of the population under intensive study. For all of these reasons, the age distributions of our samples vary across years. Since age classes differ in several components of fitness (Chapter 7), different mean values in the population may simply reflect a different age structure, rather than responses to environmental conditions. We take this into consideration by controlling for year and age where sample sizes allow.

Breeding tactics

The proportion of females which lay in the nests of others varies among years, apparently largely due to changes in patterns of nest-site availability. The hatching success of both parasitic and non-parasitic eggs declines with increasing parasitism frequency, with parasites more strongly affected than hosts (Lank *et al.* 1989*a*, 1990).

Changes in genetic structure of the population

Gene frequencies may have changed during our study, due to selection and/or chance events. The massive genetic influx from immigrant males may have also changed gene frequencies (Chapter 3). However, we have seen only small changes in the one character for which we have a long term measure directly related to gene frequencies, namely the plumage colour locus (Fig 3.5).

6.2 *Annual variation in fitness components*

This section documents annual variation in our components of fitness and presents our best explanation for the variation in terms of

Table 6.1 Annual variation in fitness components. Fecundity components: 1973–1991, except 1979. Survival components: 1970–1987. *n* = number of nests or broods. Significant: yes = $P < 0.001$; no = $P > 0.10$; preincubation nest failure, total nest failure, and total brood loss tested by contingency analyses, other fecundity components tested by ANOVA and linear regression, survival estimates tested in Francis *et al.* (1992*b*). *Proportional survival adjusted for clutch or brood size (see Chapter 4; Rockwell *et al.* 1987)

Fitness component	Annual range	Significant annual variation	Long-term linear trend	
			Slope/year	Significant?
Fecundity components				
Preincubation nest failure	17–45%	yes	—	no
Total clutch size	3.6–4.8 eggs	yes	−0.05	yes
Total nest failure	1–25%	yes	—	no
Relative egg survival*	(−0.17)–(+0.08)	yes	—	no
Clutch size at hatch	3.5–4.5 eggs	yes	−0.04	yes
Relative hatchability*	(−0.14)–(+0.25)	yes	−0.01	yes
Goslings leaving nest	3.1–4.3	yes	−0.05	yes
Total brood loss	1–62%	yes	+0.02	yes
Relative gosling survival*	(−1.23)–(+0.25)	yes	−0.02	yes
Brood size at fledging	1.6–3.5 goslings	yes	−0.06	yes
Production index	0.5–3.3 goslings	yes	−0.11	yes
Survival components				
First year female survival	0.20–0.68	yes	−0.018	yes
First year male survival	0.15–0.60	yes	−0.011	yes
adult survival	0.68–1.0	???	+0.006	yes

the potential factors listed above. The patterns of annual variation are summarized in Table 6.1, and individual components are graphed in Fig. 6.2. All components show highly significant annual variation, except perhaps adult survival (Francis *et al.* 1992b), and most also changed systematically during the study. Our goal is to understand the causes of the variation in terms of the environmental, demographic, or genetic factors which influence the population. Our results are summarized in Table 6.2.

6.2.1 Pre-incubation failure

Pre-incubation failure rates vary annually around a mean value of 24 per cent, but have not changed systematically during the study (Fig 6.2(a)). Variation in predation, patterns of snow and flood conditions no doubt contribute to annual variation, as may female body condition. Failure results from both predation loss and nest abandonment. Nests are most vulnerable at the one- or two-egg stage, and birds failing at this stage probably re-lay or become nest parasites

Table 6.2 Patterns and causes of annual variation in fitness components of Snow Geese nesting at La Pérouse Bay

Component	Pattern of variation	Possible causes
Clutch size	long-term decline	food limitation at the colony or at migratory staging areas
	annual variation	related to timing of breeding
Nesting success	annual variation	predation levels?
Gosling survival to fledging	annual variation	weather-driven growth rate
	long-term decline	resource-base-driven growth rate
Post-fledging juvenile survival	annual variation	timing of breeding
	long-term decline	lower growth rate
Adult survival	long-term increase	proportional decrease in human hunter kill
Adult structural size	long-term decline	growth rate as goslings due to food availability
Adult body condition	long-term decline	feeding conditions on migration or at the colony

(Chapters 2 and 5). Birds starting their nests early or late in the initiation period are more vulnerable to nest loss than synchronous layers, which Collins (1993) showed to be consistent with predator swamping.

6.2.2 Total clutch size

Mean annual clutch sizes have ranged from 4.85 to 3.58 eggs between 1973 and 1991, with a significant long term decline (Fig. 6.2(b)). What accounts for annual variation? Real clutch sizes (that is the number of eggs the bird actually laid, as determined by counting the number of post-ovulatory follicles) vary with mean laying date, pre-laying female body condition, and some aspect of the environment which has changed systematically during the period of study. Our observed clutch sizes also vary with changes in the proportions of birds laying continuation clutches following pre-incubation failure, and in the level of nest parasitism. These factors are considered below.

Pre-incubation failure and intra-specific nest parasitism

Although nests failing prior to incubation are not included in the clutch size sample (Chapter 4), pre-incubation failure rate indirectly affects measured clutch size by changing the numbers of birds laying small continuation clutches and those becoming nest parasites. Annual pre-incubation failure rates have varied randomly over the course of the study (Fig. 6.2(a)). There is no significant correlation

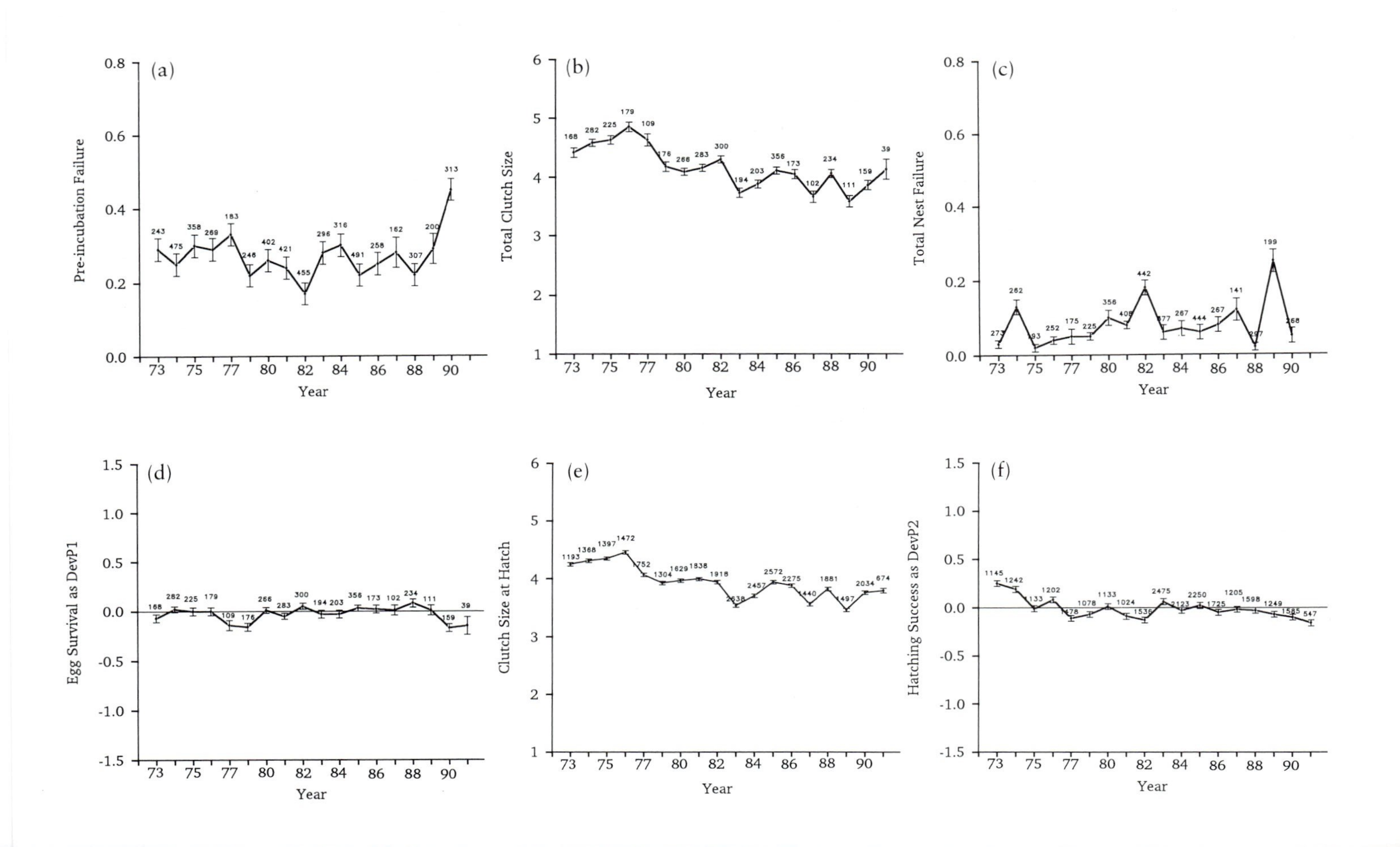
(a)
Pre-incubation Failure
Year
(b)
Total Clutch Size
Year
(c)
Total Nest Failure
Year
(d)
Egg Survival as DevP1
Year
(e)
Clutch Size at Hatch
Year
(f)
Hatching Success as DevP2
Year

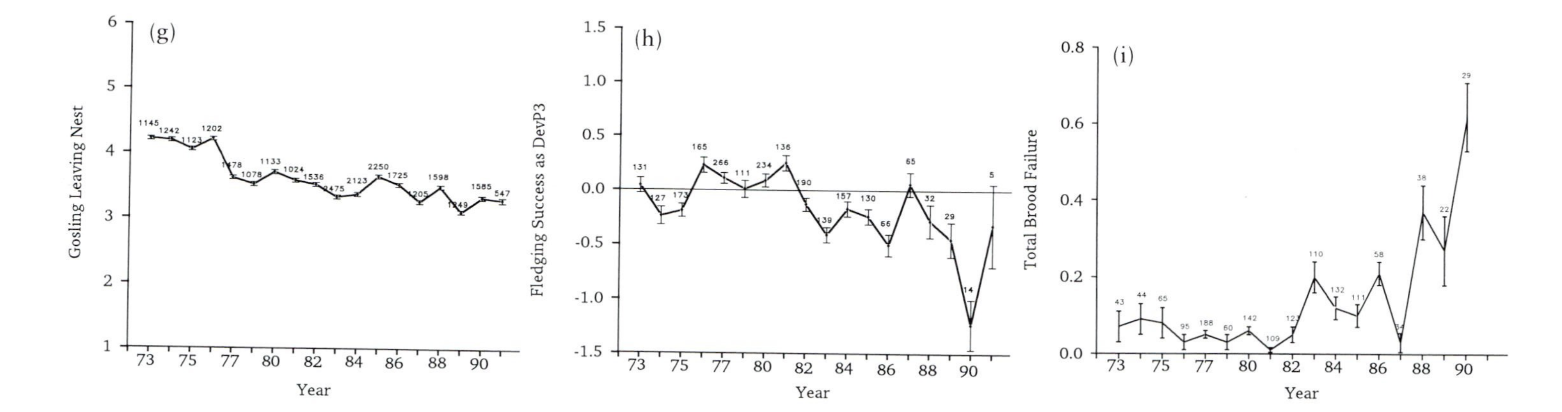

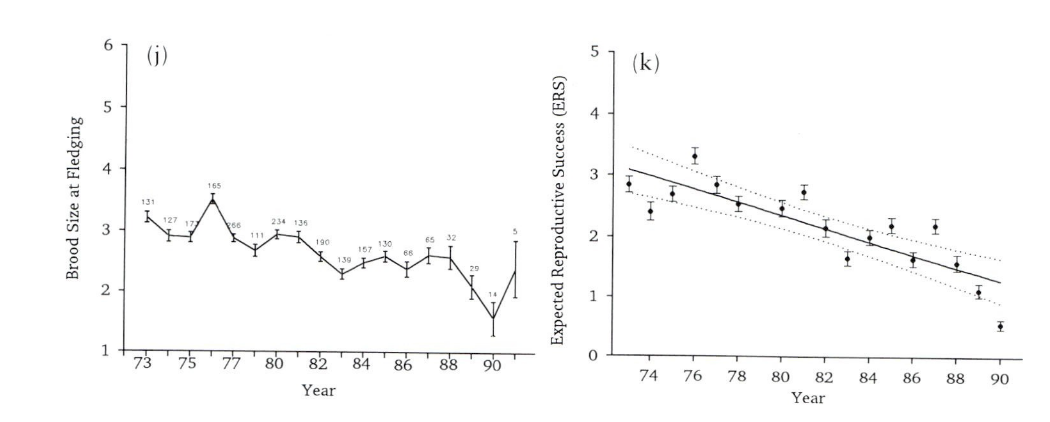

Fig. 6.2 Annual variation in fitness components of Lesser Snow Geese breeding at La Pérouse Bay. These values are not adjusted for covariates. (a) Pre-incubation failure. (b) Total clutch size. (c) Total nest failure. (d) Relative egg survival. (e) Clutch size at hatch. (f) Relative hatching success. (g) Brood size leaving nest. (h) Fledging success. (i) Total brood failure. (j) Brood size at fledging. (k) Expected annual reproductive success through fledging.

between pre-incubation failure rates and mean clutch size, perhaps because large clutches which include parasitic eggs compensate for smaller incubated continuation clutches resulting in little net change. Annual pre-incubation failure covaried positively with parasitism rate, thus the addition of small continuation clutches may to some degree cancel out a higher proportion of larger clutches. Also, unparasitized clutch sizes apparently averaged somewhat lower in years with high parasitism rate, as might be expected in relatively stressful years (Lank *et al.* 1989*a*).

Annual nest parasitism rates varied from 4.1 to 11.6 per cent of eggs laid from 1973 to 1987 (Lank *et al.* 1990), with no systematic trend over the study period. Years with higher parasitism rates had unusually large proportions of larger clutch sizes, but mean clutch size did not correlate with parasitism rate (Lank *et al.* 1989*a*).

The effects of pre-incubation failure and parasitism on observed clutch size have been documented within a season. In a sample of birds whose physiological clutch size was determined by ovary examination after laying was complete, birds which initiated nests prior to peak laying averaged fewer eggs laid than they were incubating, reflecting nest parasitism or takeovers, whereas those initiating after peak laying averaged more eggs laid than they were incubating, suggesting that some were sitting on continuation clutches (J. Hamann, personal communication).

Laying date

Mean laying date has varied from 18 May to 11 June between 1973 and 1991. This 24-day range is enormous relative to the length of the arctic summer. There is a strong negative correlation between annual mean clutch size and mean laying date (Fig. 6.3; $r = -0.55$, $P = 0.008$, 1971–1992), but date-specific clutch sizes appear to have dropped dramatically after 1980, and stronger correlations are produced if separate regressions are calculated for the two time periods (1971–1979: $r = -0.80$; 1980–92: $r = -0.76$). In general, delayed breeding seasons have smaller mean clutch sizes. However, the 2 years with the earliest laying dates appear to differ radically from this pattern and weaken the correlations. In 1977 and 1980, clutch sizes were well below the value expected from a linear relationship between laying date and clutch size (Fig. 6.3). We suggest two possible explanations for this. 1977 and 1980 were years of drought at prairie staging areas and the ground was bare at La Pérouse Bay when the geese arrived. The small clutch size, which was accompanied by a decline in the total number of breeding birds, an increase in nonbreeders, and a decrease in the proportion of 2-year-old breeders,

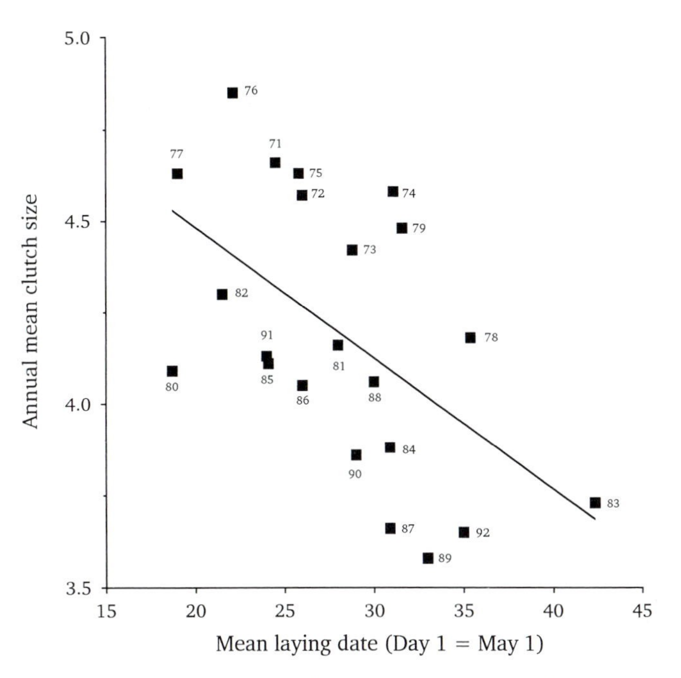

Fig. 6.3 Correlation between annual mean clutch size and laying date of Snow Geese at La Pérouse Bay, annotated by year.

may have resulted from a poor availability of nutrients in the dry spring staging areas (Davies and Cooke 1983*a*). An alternative explanation is that earliness per se accounts for the reduction in clutch size, and a steeply curvilinear relationship is in fact a better description of the relationship. This could occur if birds which nest immediately upon arrival do not 'top up' the flight cost of the last leg of their journey by feeding locally, trading off early breeding against and increased clutch size (Chapter 11).

Why is mean clutch size generally lower in years when mean laying date is delayed? Since clutch size depends primarily on the amount of nutrient acquired prior to egg laying, delayed seasons could provide more, not less, time to acquire nutrients. At a proximate level, therefore, it is not self-evident that a delayed season should result in a smaller than average clutch size. Nevertheless, the negative relationship between clutch size and laying date occurs within individuals and within seasons (Chapter 10), as well as among seasons. When an individual female's laying date was delayed relative to other birds in a particular season, there was also on average a decrease in her clutch size (Hamann and Cooke 1989). Within females, therefore, there appears to be a close physiological relation between how many eggs are laid and when they are laid. Clutch size declines systematically within seasons as well, even within age groups (Hamann and Cooke 1989; Chapter 10). We suggest that birds differ in their success at acquiring nutrients which affects both clutch size and laying date.

Long term changes

Mean annual clutch size has decreased over the course of the study (Fig. 6.2(b), 6.4; Cooch *et al.* 1989), approximately half an egg in 22 years, about an eighth of an average female's annual production. This dramatic change in a natural population could reflect changes in population genetics, age distribution, or the environment in which the birds are living.

Natural selection favours Snow Geese which lay larger clutches (Rockwell *et al.* 1987; Chapter 10). If there is a genetic response to selection, we would expect an increase, rather than a decrease in mean clutch size in the population. Therefore, local selection alone cannot explain the decline in mean clutch size during the course of the study. An alternative genetic explanation is that gene flow is altering the genetic composition of the La Pérouse Bay population,

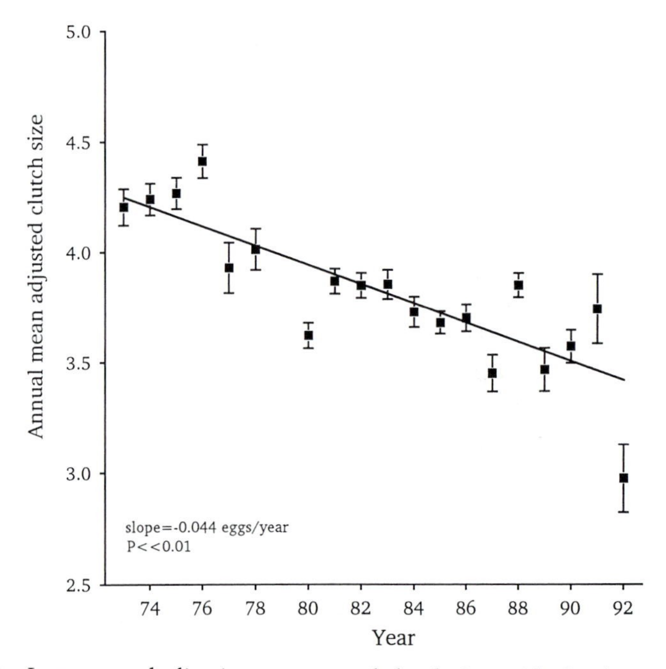

Fig. 6.4 Long-term decline in mean annual clutch size, with clutch size adjusted for the effects of annual mean laying date. (Adapted with permission from Cooch *et al.* 1989).

through the immigration of males from other colonies with smaller clutch size genotypes. This seems unlikely since we would expect selection at these colonies too to favour birds laying larger clutches, and, as outlined in Chapter 3, the large amount of gene exchange between colonies makes it unlikely that major genotypic differences between colonies exist, with the exception of genes related to plumage colour.

Changes in the age distribution of the population also lead us to expect a pattern opposite to that observed (Cooch *et al.* 1989). The increased survival rates of adults and decline in survival of rates of immature birds suggest that proportionally fewer young birds now breed at the colony. Since older birds lay larger clutches (Chapter 7), we thus expect an increase in mean clutch size. When we examine 2- through 5-year-olds directly, we find that clutch size has declined within each age group. Finally, our most telling analysis, despite relatively small samples, shows that the clutch sizes of individual adult geese have declined (Fig. 6.5). This analysis controls for potential changes in genetic structure and body size, which has also declined (Chapter 13). Since clutch size in general remains constant

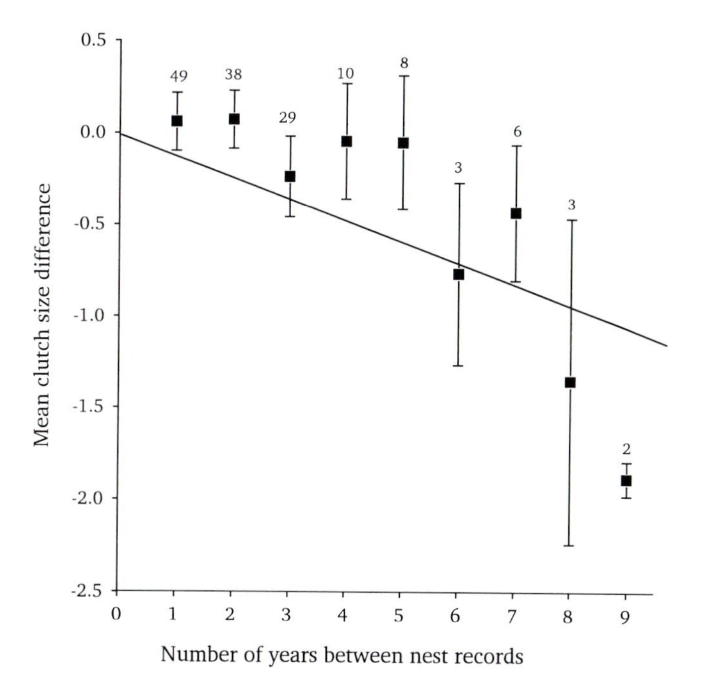

Fig. 6.5 Regression of mean difference in individual clutch sizes against the number of years between nest records of adult females. (Adapted with permission from Cooch *et al.* 1989).

between the ages of 5 years and at least 12 years, this decrease cannot be ascribed to age itself. It appears that some aspect of the birds' environment has changed such that individual females are no longer able to produce as many eggs as they could when the study began.

Clutch size in Snow Geese is widely believed to be limited by the size of a female's pre-laying nutrient stores. There is a close relationship between the nutrient reserves carried by individuals prior to egg-laying and the number of eggs which they then laid (Ankney and MacInnes 1978; Hamann *et al.* 1986). Egg size itself has also declined about 2 per cent during the study (Chapter 12). We are confident that breeding females at La Pérouse Bay are not as well fed as previously. We directly examined changes in female body condition by analysing differences in individual females' body weights when captured in banding drives in different years. Mean body weight when recaptured declined at a rate of −4.5 g/year ($n = 576$, $P < 0.001$, 1976–1993). Whether this change reflects events prior to arrival in the spring, during nesting, or post-hatching is unknown. The decline is small, given a *c.* 1900 g bird, but the analysis shows clearly that females are in somewhat poorer condition during their post-breeding moult, and may have been earlier.

The decline in mean clutch size over the period of the study suggests that there has also been a decline in *per capita* food availability, but it is not clear where this change has occurred. Have we documented a local or a more widespread phenomenon? The geese might face increased competition for food with the entire central wintering population in the northern prairies or along the Hudson and James Bay coastlines, and/or be obtaining nutrients less well after arrival at La Pérouse Bay itself. The decline in clutch size correlates strongly and negatively both with estimates of La Pérouse Bay colony population size ($r = -0.701$, d.f. $= 14$, $P = 0.003$) and with the combined Central and Mississippi flyway wintering population ($r = -0.554$, d.f. $= 14$, $P = 0.032$). Unfortunately, there are no data from other colonies in which to look for comparable declines.

It would seem most parsimonious to attribute the decline to the local habitat degradation described earlier. The partial abandonment of formerly dense areas of nesting habitat is also consistent with depletion of local food sources. This explanation conflicts with the broader view of arctic goose breeding biology, which stresses the transport of nutrients to the north for use in breeding (Newton 1977), but it may account for some or all of the trend. The use of local pre-laying food sources has been documented for Lesser Snow Geese (Mineau 1978), Brant (*Branta bernicla*) (Ankney 1984), Greater Snow Geese (Gauthier and Tardif 1991) and for Dusky

Canada Geese (Bromley and Jarvis 1993). On the other hand, increasing numbers of geese are also using the prairie staging areas and Hudson Bay coastal marshes, and *per capita* food availability could have decreased at the sites. Effects may occur at both levels, and detailed feeding and energetic studies are needed to resolve this question.

Summary

Four major factors influence annual clutch size variation. Pre-incubation failure and nest parasitism influence observed clutch sizes, and appear to largely compensate for one another in their net effect on annual mean. Real clutch sizes will be affected only if pre-incubation failure increases the rate of follicle absorption. Annual clutch size is usually lower in seasons with later mean laying date. In our two earliest seasons, however, clutch size was lower than predicted, whether due to inadequate food at the spring staging areas or an onset of nesting prior to 'topping off' with local food. Clutch sizes at La Pérouse Bay have decreased during the course of the study as the population has increased, both at the local colony and wintering population level. Food availability prior to nesting is the prime proximate mechanism influencing clutch size. Food availability can be influenced by seasonal differences in weather, numbers of birds competing for the food, and as we will show in the next chapter, individual experience in acquiring food.

6.2.3 Nest, egg, and hatchling survival

Total nest failure during incubation, eggs lost from the nest during incubation, eggs failing to hatch, or goslings being abandoned in the nest when the brood leaves lead to fewer goslings leaving the nest than eggs initially laid. In most years, losses at these stages are low, as per the global fitness values given in Chapter 4. Nonetheless, annual variation in the patterns of this loss could lead to variation in gosling production, and even outweigh the annual variation in initial clutch size documented in the previous section. Table 6.1 and Fig. 6.2 summarize annual variation in these fitness components. All varied significantly annually, but only hatching success showed a long term, downward trend.

Nest failure during incubation was usually under 15 per cent, and often less than 5 per cent (Fig. 6.2(c)). In 1982 and 1989 concentrations of Caribou on the colony destroyed unusually large numbers of nests. Despite an overall increase in Caribou in the Cape Churchill area during the study, this depredation has not been repeated in every recent year.

There were significant differences in rates of egg survival among years, with 1977, 1978, and 1990 unusually low, and 1988 unusually high. Nevertheless, the magnitude of the differences is small and variation at this stage contributes little to variation in overall annual success. Annual variation in clutch size at hatch, including the long term decrease (Fig. 6.2(e)), thus largely follows the decrease in the number of eggs laid, although the slope of the decrease is slightly less (Table 6.1).

The hatching success of surviving eggs was somewhat more variable than losses during incubation, with a small but statistically significant downward trend, largely attributed to two unusually high values at the start of the study and two low ones in the last 2 years. The high values at the start of the study may be methodological artifacts, but the trend in recent years seems real, if small. One mechanism which may account for part of the trend is poorer performance at this stage by females 9 years or older (Rockwell *et al.* 1993; Chapter 7), which make up a larger part of our nesting sample in more recent years. It is also possible that poorer female body condition at the end of hatch leads to hastier departures with broods and lower hatching success. The most alarming possibility is that environmental contaminants are affecting the hatchability of eggs. We are currently examining embryos from unhatched eggs collected for the past 8 years for symptoms of such effects.

6.2.4 Gosling survival

The numbers of goslings leaving the nest declined during the study at essentially the same rate as that of eggs laid, despite the small decrease in hatching success (Fig. 6.2(f), Table 6.1). In contrast, both total and partial brood survival components decreased dramatically in the 1980s relative to levels in the 1970s. Brood size at fledging shows a steeper decline than than that of goslings leaving the nest (Table 6.1, Fig. 6.2(j), Williams *et al.* 1993*b*). This measure of gosling survival, based on brood sizes in banding drives relative to the number of goslings leaving particular nests, is consistent with the drop in daily gosling survival probabilities in the fifth week after hatch presented earlier (Fig. 2.11). Our measure of total brood failure, which is a conservative index (Chapter 4), shows the largest proportional change of all the fecundity components, increasing from under 10 per cent during the early 1970s to 20–40 per cent during the late 1980s (Fig. 6.2(i)).

We attribute the long term decrease in pre-fledging survival primarily to the deterioration of gosling feeding conditions at La Pérouse

Bay, as reflected in slower growth rates in more recent years (Chapter 13; Cooch *et al.* 1991*b*, 1993). Gosling growth rate, which may affect survival, varies with post-hatching weather. However, there have been no systematic changes in post-hatch weather, or in numbers of potential predators, as far as we can assess, which parallel the decline. Some food plants used by goslings differ from those which could be contributing to the decrease in clutch size (Chapter 2, see above), thus the specific mechanism responsible for the decline in survival of goslings is separate from the clutch size effect. If both sets of plant resources have been over-exploited by the local birds however, one can view both effects as resulting from the local population increase.

As with hatching success, the age, or more accurately, the breeding experience of geese in our nesting sample may interact with the environmental change to produce part of our results. Geese which rear their young at 'traditional' feeding areas at La Pérouse Bay have significantly smaller goslings and tend towards smaller brood sizes at fledging than those rearing young at sites more distant from the breeding colony (Cooch *et al.* 1993). These adults may be paying a cost of philopatry which contributes to the decline in gosling survival.

6.2.5 Post-fledging survival

The period immediately following fledging is the most vulnerable time of life for many species of birds, and one during which survival rates are notoriously difficult for ornithologists to estimate. Such estimates are possible in Snow Geese because of the large numbers of young birds banded and recovered. We estimated annual survival rates of specific gosling cohorts, from the time the birds were banded through the next year's banding, using band recovery analyses, and estimated first-year 'local survival' rates of banded female goslings, using capture-recapture techniques (Chapter 4; Brownie *et al.* 1985; Francis *et al.* 1992*b*; Francis and Cooke 1993; Lebreton *et al.* 1992). Since there is some mortality between banding and fledging, it is not strictly correct to refer to either of these as post-fledging survival, but we do so for convenience. Figure 6.6 shows the first-year survival rates and 95 per cent confidence limits, based on band recoveries, for female and male goslings hatched from 1970 to 1987. Both sexes show large, significant, annual variation, with estimated values ranging from 14 to 68 per cent, and significant declines during the period (females: $t = -2.8$, $d.f. = 17$, $P = 0.01$; males: $t = -2.3$, $d.f. = 17$, $P = 0.04$.). Local survival estimates for females based on recapture data are shown in Fig. 6.7, on which we have also re-plotted the

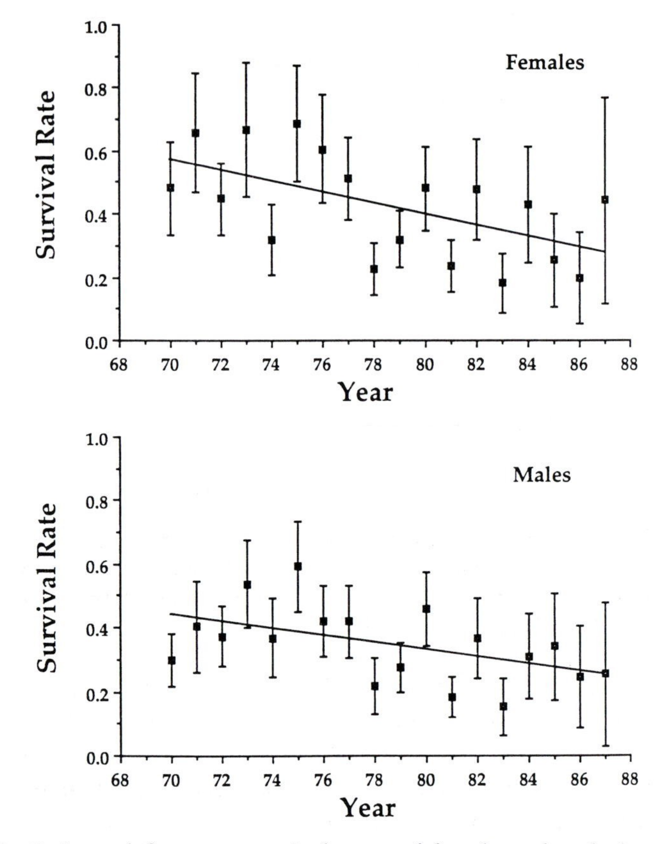

Fig. 6.6 Estimated first year survival rates of female and male Snow Goose goslings banded at La Pérouse Bay, based on band recoveries, showing 95 per cent confidence limits and the regression line calculated from the point estimates. (Reproduced with permission from Francis *et al.* 1992*b*).

female survival estimates based on recovery data for comparison. The year-to-year variation in the two measures is perfectly concordant, despite being based on completely different re-encounter data bases. Our interpretation of the recapture analyses assumes that the cohort-specific return rate to the colony is proportional to the immature survival rate. Most surviving females return to their natal colony (Chapter 2), and there is relatively little annual variation in the survival rates of birds older than one year (see below). Thus, variation in cohort-specific recruitment primarily reflects variation in immature survival.

We attribute the decline in post-fledging survival rates to the same changes in forage quality and abundance to which we attributed decreased pre-fledging survival. Much of the residual annual varia-

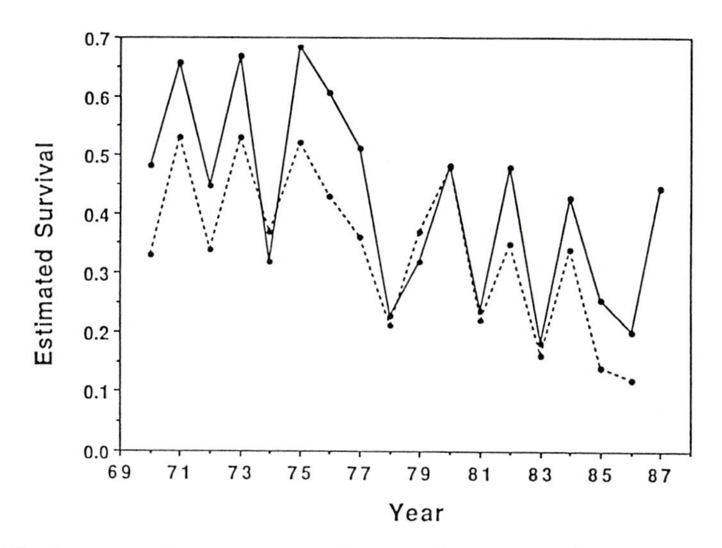

Fig 6.7 Apparent first-year survival rates of immature female Snow Geese banded at La Pérouse Bay between 1970 and 1987, estimated from recapture data (dashed line) and recovery data (solid line). (Reproduced with permission from Francis and Cooke 1993).

tion around the decline can be explained by variation in mean hatch date and/or mean banding date (Francis *et al.* 1992*b*). Both terms are significant (partial correlation $P < 0.05$) in multiple regressions which incorporated a linear year term. Survival was highest when breeding was early and when banding was latest. For logistical reasons, we tend to band geese at a relatively constant date each year, but hatch date varies considerably, so that in years with an early hatch, the goslings are usually older when banded. Thus we cannot statistically disentangle these two effects. Older goslings may be better able to survive the stress of the banding activities (Williams *et al.*, 1993*d*), but in years of early hatch, goslings also grow faster (Cooch *et al.* 1991*a*,*b*), and post-fledging survival is higher, even when corrected for the effects of the long term decline (Francis *et al.* 1992b).

Effects of hunting?

What causes such high immature mortality and inter-year variation? An obvious source of mortality is hunting. Ninety-seven per cent of Snow Goose band recoveries are birds killed by hunters (Table 2.2). This sample is strongly biased, since hunter-killed banded birds have a far higher probability of being reported to the US Fish and Wildlife Service than those dying of other causes. The rate at which banded geese taken by hunters are reported is thought to be approximately 33 per cent (Martinson and McCann 1966; Henny 1967; Boyd *et al.*

1982), but may be higher for La Pérouse Bay birds which also carry coloured leg bands. In contrast, a bird which dies of natural causes somewhere on the Hudson Bay Coast is unlikely to be found and reported by anyone. Human causes of mortality are thus massively exaggerated in the raw recovery data.

Two relationships suggest that variation in hunting is not the major factor determining variation in first-year survival rates. First, as survival rates have declined, the proportion of geese shot has also declined, exactly opposite to what we would expect if hunting was the major influence on mortality rates. There has been a significant decline in recovery rates of immatures, and adults, between 1970 and 1987, during which time the total mid-winter population of central region Snow Geese doubled, and there was a small decline in the total harvest. An estimated harvest index thus declined about 30 per cent during our study (Francis *et al.* 1992*b*). A decline in band reporting rates over this period could also produce these results, a potential alternative interpretation. While there have been no studies from geese, reward band studies in Mallards (*Anas platyrhynchos*) failed to find changes in reporting rates between the early 1970 and late 1980s (Nichols *et al.* 1991). Thus, hunter kill has not kept up with the growth in the central region Snow Goose population which has occurred coincident with the decline in gosling survival rates, and contributes proportionately less to total mortality as survival rates continue to decline.

This lack of relationship is directly shown in our second analysis, which compares the annual proportions of goslings to adults reported shot by hunters each year with the first-year survival estimates made from recapture data from geese surviving and returning to La Pérouse Bay (Fig. 6.8). Instead of the negative relationship we would expect if hunting accounted for a large fraction of the total mortality, we have a positive relationship. Young birds survive better in years when hunters kill proportionately more immatures.

These results suggest that the bulk of variation in annual first-year survival rates occurs before the birds arrive in locations where hunters reporting bands are numerous. The bulk of the hunting occurs in settled parts of North America, southern Manitoba, the Dakotas, Iowa, Missouri, Texas, and Louisiana, and these hunters can be viewed as consistently sampling the populations of geese which encounter them. There is some hunting along the Hudson Bay and James Bay coasts but this is unlikely to account for more than 10 per cent of the total harvest. Since people are infrequent along the arctic coasts and in the boreal forests over which the geese must fly, it seems likely that much immature mortality goes undetected by band recoveries.

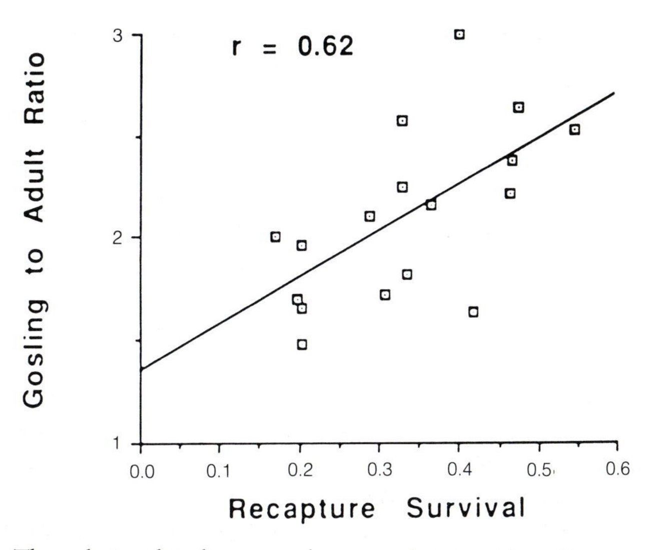

Fig 6.8 The relationship between the annual ratio of goslings to adults in recoveries of shot geese, and annual local survival rate of immatures, estimated from recapture data, 1970–1986. (Reproduced with permission from Cooke and Francis 1993).

We have attempted to estimate the proportion of young birds killed by hunters versus those dying from other causes. We must make some simplifying assumptions about hunter reporting rate to do so. First, we assume that 90 per cent of geese shot by hunters are actually retrieved by them. Second, based on reward band studies, wildlife agencies believe that hunters report about a third of the bands on birds they have shot. Finally, we assume that the above values have not changed annually or over the long term. Under these assumptions, a band recovery rate of 10 per cent translates into a hunter harvest rate of 33.6 per cent of banded goslings. Since our juvenile recovery rates vary from 3.0 per cent to 10.5 per cent, we estimate that death from hunting varies annually from 10.8 per cent to 35.4 per cent of the young birds, compared against total first-year mortality rates ranging from 32 per cent to 86 per cent (Fig. 6.6). A year-specific comparison of survival and recovery estimates suggests that hunter kill comprises as much as 85 per cent of immature mortality in years with the lowest mortality, but as little as 14 per cent in years with the highest mortality. Variation in natural mortality thus appears to be a far larger contributor to annual differences in immature mortality than variation in hunter kill. Changes in natural mortality are the most probable cause of the long term decline in first-year survivorship which has occurred during the course of the study while the proportion of geese shot has declined.

Effects of body size

It is likely that variation in natural mortality reflects variation in the body condition of the goslings after fledging (Chapter 13), which ultimately depends on food availability in brood-rearing areas. The most direct analysis shows that mean annual gosling growth rate, standardized for gosling age when measured, correlates positively with annual immature survival rate (Fig. 6.9, $r^2 = 0.85$). If adequate nutrients exist, goslings mature in good condition and are able to migrate south. If not, they may die from starvation or increased vulnerability to predation and disease before becoming part of the population sampled by hunters.

While the growth rate and survival rate correlation is strong, both seasonal timing and gosling age when measured are tight covariates of growth rate, and both correlate with immature survival rates in a larger data set (see above). Changes in food availability, indirectly measured as growth rate, could explain the lower post-fledging survival of goslings in delayed seasons. In years when laying date has been delayed by late snow melt, the timing of plant growth may also be affected. A late season results in both a delayed goose hatch and, although direct evidence on this point is lacking, could result in lower overall plant productivity on the salt marsh at the time when goslings are actively growing. Lower survival in late years may also reflect the

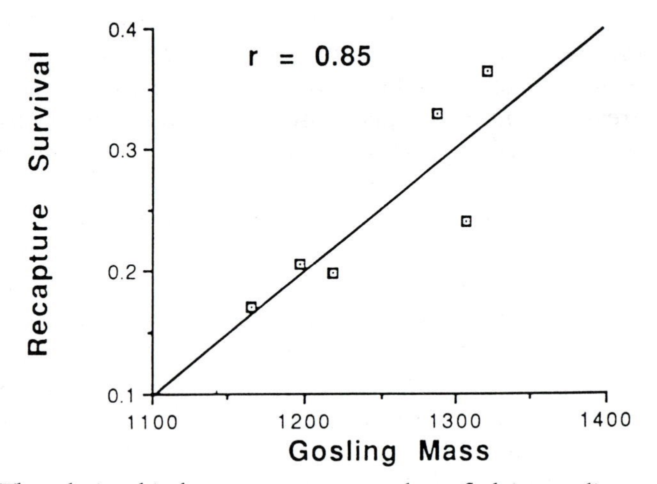

Fig 6.9 The relationship between mean annual pre-fledging gosling mass and apparent local first-year survival rates, estimated from recaptures, 1981–1986. (Reproduced with permission from Cooke and Francis 1993).

younger age of goslings at the time they are banded. Juvenile geese are stressed in mass banding drives, and if younger or smaller goslings are less able to cope with this stress, post-banding mortality may be higher, thus an artifact of our activity. Few goslings die during banding, but more do so soon after. In recent years, banded families have been observed to have on average fewer goslings than unbanded families (Williams *et al.*, 1993*d*).

This brings out the important point that survival rates in this and all other banding studies are the survival rates of the banded sample and are not necessarily reflective of the population as a whole. Survival rates so calculated will underestimate those of the population if there is enhanced mortality due to the banding operation itself. There is some evidence that banding-induced gosling mortality has increased during the course of the study, despite improvements in our methods of handling the geese (Williams *et al.*, 1993*d*). Banding mortality may simply hasten the demise of goslings with already poor survival prospects. The recent decline in pre-banding gosling survivorship suggests that banding aggravates problems which goslings are already experiencing. Even in recent years, the survival rates of goslings banded in richer feeding areas are higher than those banded in degraded sites (Cooch *et al.* 1993). Thus, the effect of banding per se on variation in survival rates may be smaller than it appears to be.

Changes in food availability are also consistent with the long term decline in immature survival rate. Most of the birds we band are caught in La Pérouse Bay itself, rather than the surrounding areas. The Bay usually contains the highest concentration of geese and is close to our camp. The disappearance of the salt marsh vegetation in the Bay referred to earlier is dramatic in its effect. Large areas of former marsh have been converted by the geese into vegetation-free areas of mud. Fewer geese use the area now, and those which do have shown a striking decline in growth rate during the study (Chapter 13; Cooch *et al.* 1991*b*, 1993). Increasing numbers of geese have moved to other feeding areas, often many kilometres from the nesting areas. These geese do not show the declines in growth rates typical of birds in the Bay itself (Cooch *et al.* 1993). This suggests that the long term decline in first-year pre-fledging and post-fledging survival may be partly a local phenomenon within the colony itself, and not necessarily typical of the breeding population as a whole. Nevertheless the decline in survival of the banded sample is real and dramatic and can be most reasonably be attributed to a decline in the availability of salt marsh vegetation in La Pérouse Bay.

The salt marsh vegetation along the shores of La Pérouse Bay has been the major nesting and brood rearing area of the colony. As the colony expanded, this vegetation has been degraded in extent and forage quality. Birds which have remained have shown a slower gosling growth rate, fewer goslings surviving to fledging, and a higher post-fledging mortality. Other geese have increasingly moved to other brood-rearing areas often at a considerable distance from their nesting areas. Changes in food availability seem to be the main proximate factor causing these changes in the various components of fitness.

6.2.6 Index of annual production

Large variation in annual productivity is well known in arctic geese. At high arctic nesting areas, no young at all may be produced in certain years. E. V. Syroechkovsky (personal communication) notes that at the Wrangel Island colony no goslings were produced in 2 of 22 years, and F. G. Cooch reports 3 years of total failure on at Boas River on Southampton Island (personal communication). In the low arctic colony of La Pérouse Bay, the variation is less extreme. By combining all the components of fitness through fledging and first-year survival, we can estimate the average number of goslings surviving to fledging, or to 1 year of age per nesting pair at the beginning of egg laying (Chapter 4). The index does not include variation due to pre-incubation failure, may exaggerate the effects of total brood loss, as discussed earlier, ignores variation in breeding propensity in particular years, and is not corrected for age effects. Nevertheless, this index summarizes the annual reproductive performance through fledgling of our sampled population of La Pérouse Bay Snow Geese over time. Figure 6.2(j) shows significant annual variation in expected annual fledging success, but also the clear steep linear decline of *c.* −0.11 fledglings per year. Since first-year female survival rates have also decreased, with a slope of *c.* 0.018 per year (Table 6.1), the decline in the production rate of yearling females would be even steeper.

6.2.7 Adult survival

Estimates of annual survival rates of adult geese, based on band recoveries (Chapter 4; Brownie *et al.* (1985); Francis *et al.* 1992*b*), present a striking contrast to the pattern for juveniles (Fig 6.10). There are high variances in the values from the individual years, mainly due to high sampling variances and strong negative sampling covariances inherent in the method. Nevertheless, based on likelihood ratio tests, a model which assumed constant survival with time was rejected in favour of models in which survival was assumed to

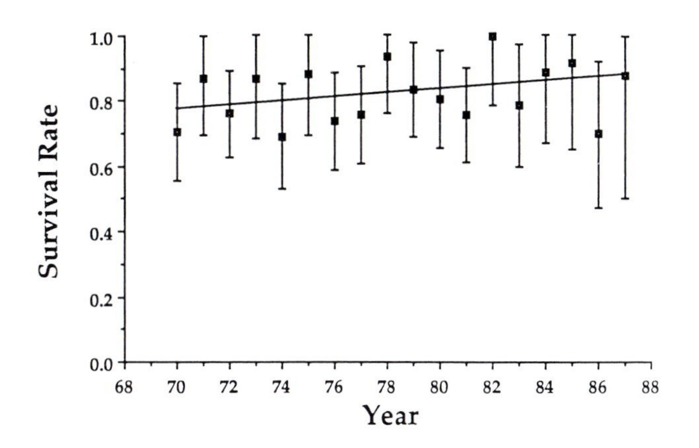

Fig. 6.10 Estimated survival rates of adult Lesser Snow Geese banded at La Pérouse Bay, with 95 per cent confidence limits, and the regression line calculated from the raw data. (Reproduced with permission from Francis *et al.* 1992*b*).

change with time. We tested whether survival increased or decreased over time by modeling survival (S) as a linear function of year (t), and produced the equation:

$$S_t = 0.78 + 0.006t,$$

where $t = 0$ in 1970. This model fits the data significantly better than one assuming constant survival. Assuming a linear relationship, annual adult survival has increased from 78 per cent in 1970 to 88 per cent in 1987. Assuming that survival rate does not change with the age of the adults (Rockwell *et al.* 1993; Chapter 7), this change in adult survival rate is equivalent to a doubling of the expected mean adult life span from 4.0 to 7.8 years.

This increased survival of adults contrasts dramatically with the decreased survival over time of the immature birds estimated with the same methods, suggesting that different causes are involved. In contrast to immatures, most adult mortality of Snow Geese can be attributed to hunters. As an example from our own data, nearly all of the birds marked in the first years of the study must now be dead. We received hunter-kill recovery records of 22.4 per cent of the 1680 birds marked between 1969 and 1971. Assuming reporting rates and retrieval rates of 33 per cent and 90 per cent respectively, as done previously for juveniles, ca. 75 per cent of these banded adults died as a result of hunting. This is a relatively crude approach, but it is consistent with the findings in other arctic goose populations (for example Owen 1980; Ebbinge 1991).

If hunters account for 75 per cent of adult mortality, changes in the proportion of geese shot may affect overall adult mortality rates. This is referred to by waterfowl biologists as additive mortality, meaning the mortality due to hunting is to a large extent additional to natural mortality, as opposed to the possibility of compensatory mortality, in which fewer geese die of natural causes when more are shot by hunters. Under additive mortality, we expect a negative correlation between annual survival rates and annual recovery rates. As fewer birds are shot, survival rates should increase. We examined the relationship between adult survival rates (S) and adult recovery rates (F) in year t by modeling survival as a linear function of recovery rates producing the equation:

$$S_t = 0.92 - 2.9\ F_t.$$

This model provided a significantly better fit than one assuming constant survival, indicating that survival rates were significantly higher in years of low recovery rates. We showed earlier that recovery rates for both adults and immatures have declined 2–3-fold during the course of the study, and attributed this primarily to a dilution of a slightly declining hunter kill over the expanding central flyway Snow Goose population. While this negative correlation between adult survival and recovery rates is not proof of a causal relationship, it seems likely that the decline in the proportion of geese shot has led to increased survival among the adults (Francis *et al.* 1992*b*).

The evidence for a major effect of hunting on adult survival rates is consistent with observations that survival rates on several European species of geese increased when hunter mortality was reduced (Cabot and West 1983; Owen 1984; Ebbinge 1985, 1991). If we assume total additivity and that non-hunter mortality has remained roughly constant with time, annual adult survival in the absence of hunting can be estimated at about 92 per cent from the intercept of the regression of survival on recovery rate. This value is comparable to the survival rates observed in the non-hunted population of Barnacle Geese breeding in Svalbard and wintering in Scotland (Owen 1984).

In addition to simple dilution of hunter kill in a larger population over time, the decline in immature survival could indirectly increase adult survival at our colony. Adult geese with young are more likely to be shot than those which fail to breed successfully (Francis *et al.* 1992*a*); in some way the presence of young increase adult vulnerability to hunters (Giroux and Bedard 1986). As total brood loss has increased at La Pérouse Bay, an increased proportion of adults will have no young with them when they reach hunted areas.

6.3 *Summary*

1. Nearly all fecundity and survivorship components of fitness vary annually. The most important contributions to variation in annual recruitment come from the total clutch size and pre- and post-fledging gosling mortality.
2. Environmental variation in weather conditions, predator frequency and food availability contribute to the annual variation, as do changes in the age structure of the study population. However, we have concentrated on the long term changes in the fitness components in this chapter since they provided the best opportunities for isolating the environmental factors responsible for the variation. The major long term environmental changes are a 10-fold increase in breeding population size and a concomitant decrease in local food availability to laying females and both adults and goslings after hatch.
3. Clutch size decreased by about 25 per cent between 1973 and 1991, probably due to decreased food availability for laying females. This may be a local phenomenon, but changes in *per capita* food availability at staging areas also may contribute.
4. Variation in clutch size unrelated to the long term decline can be attributed to variation in (1) the frequency of intra-specific nest parasitism, (2) mean initiation date, which is mainly determined by rates of annual snow disappearance, and (3) prairie droughts.
5. Mortality due to nest, egg or loss at hatch contributes little to the annual variation in recruitment, suggesting that variability in predation is of minor importance to annual nest success.
6. Pre-fledging gosling mortality is important, with approximately 25 per cent of the goslings disappearing at this stage, mainly during the first few days after hatch. In recent years there has been a considerable increase in pre-fledging mortality towards the end of the brood-rearing period, coincident with a decline in food availability in the major salt marsh feeding areas.
7. Immature mortality varies massively from year to year, but has increased during the study. Hunters contribute relatively little to the annual variation and in fact fewer juvenile geese are killed by hunters in years of high gosling mortality, suggesting that most juvenile mortality occurs before the young birds reach areas heavily populated by hunters. High immature mortality occurs in late seasons.
8. Gosling growth rate prior to fledging, which largely reflects

food availability, appears to be the most important proximate factor affecting juvenile mortality rate. Food supply in the post-hatch feeding areas determines the condition of goslings at the time of fledging and their ability to survive fall migration.

9. The overall production of young from the colony declined at a rate of 0.11 goslings per nest per year between 1973 and 1991.
10. Adult mortality, in contrast to immature mortality, has declined over the period of the study. This results from a smaller proportion of the geese being shot by hunters, because of a slight reduction in hunter numbers, a considerable increase in the global population of Snow Geese, and possibly because more adult geese travel south in the fall with no goslings and are less vulnerable to hunters than those with families.

7 *Age effects*

A population's potential ecological and evolutionary dynamics are determined in large measure by its life history strategy. By strategy, we mean those 'coadapted traits designed, by natural selection, to solve particular ecological problems' (Stearns 1976). A life-history strategy includes age-specific schedules of breeding propensity, reproductive success and survivorship. Combined, these parameters determine the lifetime reproductive success of individuals. This chapter describes the age-specific patterns of the fitness components for the Lesser Snow Goose, and considers the implications of these patterns on the ecology of the species.

Four major methodological problems should be outlined before presenting and interpreting the data. First, our analyses depend upon having marked birds of known age in our sample. Not all birds are seen every year and so statements about age-specific effects rely on different samples in different age classes. Secondly, the effects of age per se and the long term changes are not always easy to separate. As the study has proceeded, the frequency of older birds in our marked sample has increased simply because we cannot know that a bird is, say, 10 years of age unless it was marked as a gosling 10 years earlier. Thus older marked birds are more prevalent in our more recent samples, which are the years when environmental conditions are worse. The statistical methods for dealing with this are described in detail by Rockwell *et al.* (1993). Thirdly, the marked sample of known-aged birds does not necessarily represent a random sample of the age distribution of birds in the colony. In recent years, the banded sample of known-aged birds may be representative of the population as a whole, but the nesting sample is certainly not. Since nesting geese are somewhat philopatric, and we sample similar areas of the colony each year, we under-represent young birds which settle disproportionately in newer parts of the colony which we do not visit. Much of the fecundity data in this chapter comes from marked birds in the nesting sample, whereas survival and breeding propensity data come from the banding sample. Fourthly, our analyses were based on mean performances of individuals in various age classes. This could confound changes in yearly performance of aging individuals with progressive appearance or disappearance of individuals with different constant performances (Newton 1989*a*). For example, if birds laying larger clutches lived longer, the mean performance of advancing age classes would increase. Such problems are expected if any component of reproductive success is correlated with survival, and can be examined by analysing changes in the performance of individual birds, as was done for long term changes in clutch size in the previous chapter.

7.1 *Breeding propensity*

Breeding propensity, as defined in Chapter 4, has two major components, (1) age of first breeding and (2) the probability of nesting in years after first entry into the breeding population, referred to as 'adult breeding propensity'. The age of first reproduction is at least as important a determinant of the evolutionary and ecological dynamics of a species as is the total number of offspring produced, particularly for long lived species (Cole 1954), and in this section we put most emphasis on this aspect of age-specific breeding propensity.

7.1.1 Age of first breeding

Population growth rate is quite sensitive to changes in the age of first reproduction (Caswell 1989). In principle, the proportion of females that begin breeding at each age can be determined by estimating the numbers of a given cohort that could be breeding and comparing them to the numbers that actually are breeding. The former is obtained by depreciating the size of a given cohort by age-specific survival estimates for each year. The latter should be obtainable from the age distribution of nesting females. Using this approach, approximately 50 per cent of the surviving 2-year-olds and 86 per cent of the surviving 3-year-olds bred relative to the breeding propensity of older birds (Cooke and Rockwell 1988). No yearlings have been found nesting.

Another approach to assessing age-specific recruitment is to adapt Jolly–Seber models for calculating age-specific recapture probabilities, assuming known survival rates taken from our recovery analyses. Recapture rates reflect the probability that a bird is detected in the banding sample, given that it is alive. Since nearly all adult plumaged birds captured in the banding drives are successful breeders, recapture rates estimate the fraction of successful breeders re-encountered. The recapture probabilities for birds of ages 1 through 4 years and older were: 0.02, 0.10, 0.19, and 0.22. These translate into relative detection rates of 0.09, 0.45, 0.86 and 1.0 relative to the probabilities of birds older than 4 years. Two-year-olds are thus 45 per cent and 3-year-olds 86 per cent as likely to breed successfully as were adults. The few yearlings caught at banding are non-breeders which may have remained with their parents throughout the following nesting season. We had no evidence from either method that 4-year-olds had lower breeding propensity than older birds.

Although the values using the two methods give similar results, there are several problems. Neither considers cohort-specific recruitment rates, nor do they calculate absolute, as opposed to relative, age-specific recruitment rates. Additionally, since they are based on birds caught at the end of the nesting period, the comparison is among successful nesters rather than birds which attempt to nest. As we show later, younger females have lower nesting success than older ones, and so the values above underestimate the proportion of the younger birds which attempt to breed. Nor are nest parasites likely to be included, and these are probably biased towards younger birds. Finally, birds are often unsuccessful one year after their first successful breeding season (see below), suggesting that many of the 3-year-old breeders had not nested as 2-year-olds.

7.1.2 Adult breeding propensity

Without accurate numbers of the total number of adult birds alive at the beginning of the breeding season, we cannot assess the breeding propensity of birds once they have entered the breeding population. We have no data on whether adult breeding propensity changes with old age, cohort or year.

We have discovered one interesting age-specific effect on breeding propensity, however. Female are less likely to breed successfully the year after their first breeding. Specifically, the probability that a female nests successfully the year following her first successful nesting is much lower than after subsequent nesting attempts (Viallefont *et al.* in press). This effect is strong for 2-year-olds, which are only 55 per cent as likely to be recaptured as 3-year-olds than they are in the year following later successful nestings. Birds first detected breeding at ages 3 years or older show a smaller, but not significantly lower recapture rate relative to that after subsequent breeding. These results suggest that first-time breeding, especially at the age of two, is more costly than subsequent breeding.

7.2 *Fecundity components*

Rockwell *et al.* (1993) calculated age-specific values for the fecundity components of reproductive success that span the period from egg laying through fledging of the goslings (Chapter 4), and produced a composite estimate of brood size at fledging. Known-aged, locally hatched females, which carry a year-specific colour band, provided most of the data for these analyses. However, birds banded as adults each year formed an additional sample of individuals for which a minimum age is known, and we used a portion of these data to augment our analyses of older females. Such adults were considered to be at least 2 years old when banded, and thus a minimum age in subsequent years could be calculated. The reproductive success data spanned the period from 1973 to 1988. The analyses standardized for annual variation by including year as a covariate for all components except total nest and brood loss, where samples sizes were too small to allow for meaningful corrections. These components were pooled over years.

7.2.1 Clutch size

Mean clutch sizes ranged from 3.0, for 2-year-olds, to 4.4, for 9-year-olds (Fig. 7.1, Table 7.1). Two-year-olds laid a smaller clutch

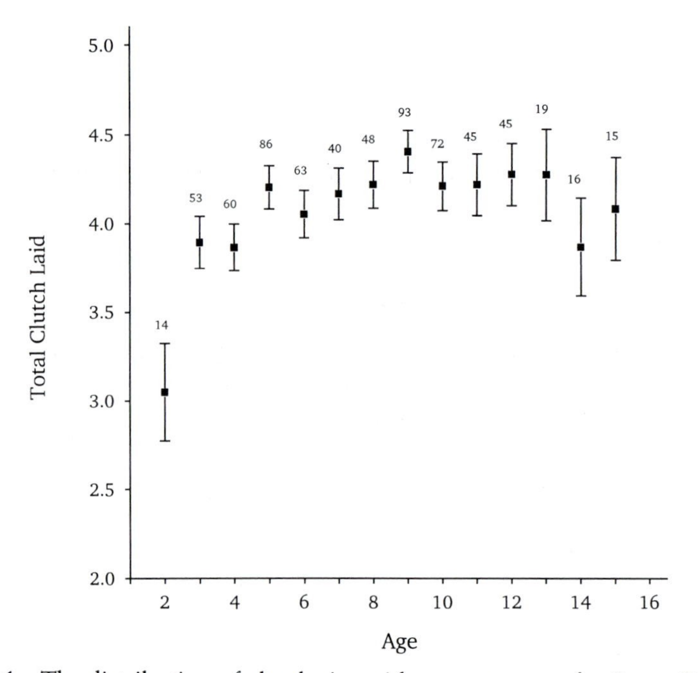

Fig. 7.1 The distribution of clutch size with respect to age for Snow Geese at La Pérouse Bay. The estimates of clutch size were adjusted for annual variation using a covariance procedure. Standard errors and sample sizes are indicated for each estimate. Age classes 9 and above include both known and minimum age individuals. Age class 15 is a composite of individuals 15 and older. (Reproduced with permission from Rockwell *et al.* 1993).

Table 7.1 Age-specific fitness components for breeding birds. Reproductive components from Rockwell *et al.* (1993), survival components from Francis *et al.* (1992*a*)

Fitness component	Age-specific range	Significant variation?	Pattern of variation
Total clutch laid	3.0–4.4 eggs	yes	2 < 3 < 4 ≤ 5–14
Total nest failure	3–14%	yes	2,4 > 3,5–10
Relative egg survival*	(−0.18)–(+0.05)	no	2–10 no significant difference
Relative hatchability*	(−0.26)–(+0.06)	yes	2–9 > 10–16
Relative gosling survival*	(−0.42)–(+0.20)	yes	2, 3 < 4–10, 14 > 11–13
Total brood loss	3–21%	yes	2, 3 > 8–10 > 4–7
Expected brood size at fledging	1.4–2.8 goslings	yes	2 < 3 < 4 < 5 < 6,7 > 8–10
Post-fledging juvenile survival (relative)	0.06–0.15	no	2–10 no significant difference
Breeding propensity (relative)	0.50–1.0	yes	2 < 3 < 4-older; lower year following first success as 2-year-old
Survivorship	0.83	no	2–16 no significant difference

* Adjusted for clutch or brood size.

than 3- and 4-year-olds, which laid a smaller clutch than 5- and 6-year-olds, changing an average of 14 per cent per year between 2 and 6 years. There was no significant change in clutch size in the 5–10 or 9–15 and greater age ranges. Thus the initial clutch size of Lesser Snow Geese increases through ages 5–6 and then does not significantly change to at least age 15 years.

7.2.2 Egg survival and hatchability

Egg size increases slightly from ages 2 thorough 4 years, but remains constant thereafter (Robertson *et al.* 1994; Fig. 12.2). Both egg survival and hatchability are independent of egg size (Chapter 12, Williams *et al.* 1993*c*). Consistent with these general findings, there was no significant effect of age on egg survival from ages 2 through 15 years (Table 7.1, Fig. 7.2). In contrast, hatchability declined in older birds. Pair-wise contrasts indicated a significant decline from age 9 to age 16 years, and a linear comparison suggested a rate of −4.2 per cent (±0.9 per cent) per age class. This decline was not related to increased nest abandonment by older females, but rather a decreased hatchability of the eggs themselves.

7.2.3 Total nest failure

There were insufficient data to assess annual variation in total nest failure, so these data were pooled over years. We combined data from known aged and minimum aged females for birds at least 10 years old. Total nest failure rates varied with age, with mean values ranging from 3 per cent for 6- and 7-year-olds to 14 per cent for 2-year-olds (Table 7.1). Age classes 2 and 4 had unusually high rates, with the remaining classes forming one homogeneous subset. Age class 3 was a member of that subset, but also was not significantly different from ages 2 and 4. We conclude that the relative rate of total nest failure was greater for younger females.

7.2.4 Gosling survival

Gosling survival within successful broods was highest for middle-aged birds (Fig. 7.2). There was a significant linear increase of 9.4 per cent (±4.0 per cent) per age class in the age range 2–6 years, and an apparent, but not significant decline, perhaps due to the high value for 14-year-olds. Total brood failure varied with age in a similar fashion, with survival highest for middle aged mothers (Table 7.1). Females 4–7 years of age formed a homogeneous subset, losing under 10 per cent, while 2-year-olds lost over 20 per cent, and the oldest birds had intermediate values. These estimates of total brood failure are minima (Chapter 4) and depend on the assumption that the probability that a bird will be captured in the banding drives after brood failure is independent of age.

7.2.5 Composite age-specific brood size at fledging

We derived an average measure of reproductive success to fledging for each age class by combining the age-specific clutch sizes and transition variables, as presented earlier (Fig. 7.3, Chapter 4). A second-order equation provided the best fit to the data. The expected brood size at fledging thus increases through age 6 years and then declines with age. While additional data on older age classes are needed to establish the extent and precise pattern of this decline past age 10 years, the significant decrease in hatchability from age 9 to age 16 years suggests that the decay continues. Age-specific pre-incubation failure could not be estimated and incorporated into this analysis. Since such failure is biased towards early nests (Chapter 11), adding it to the analysis would lower somewhat the relative success of middle-aged and older birds.

7.2.6 Post-fledging offspring survival

Ratcliffe *et al.* (1988) concluded that the recruitment rate of hatchlings varied with maternal age, with highest values for 'middle aged'

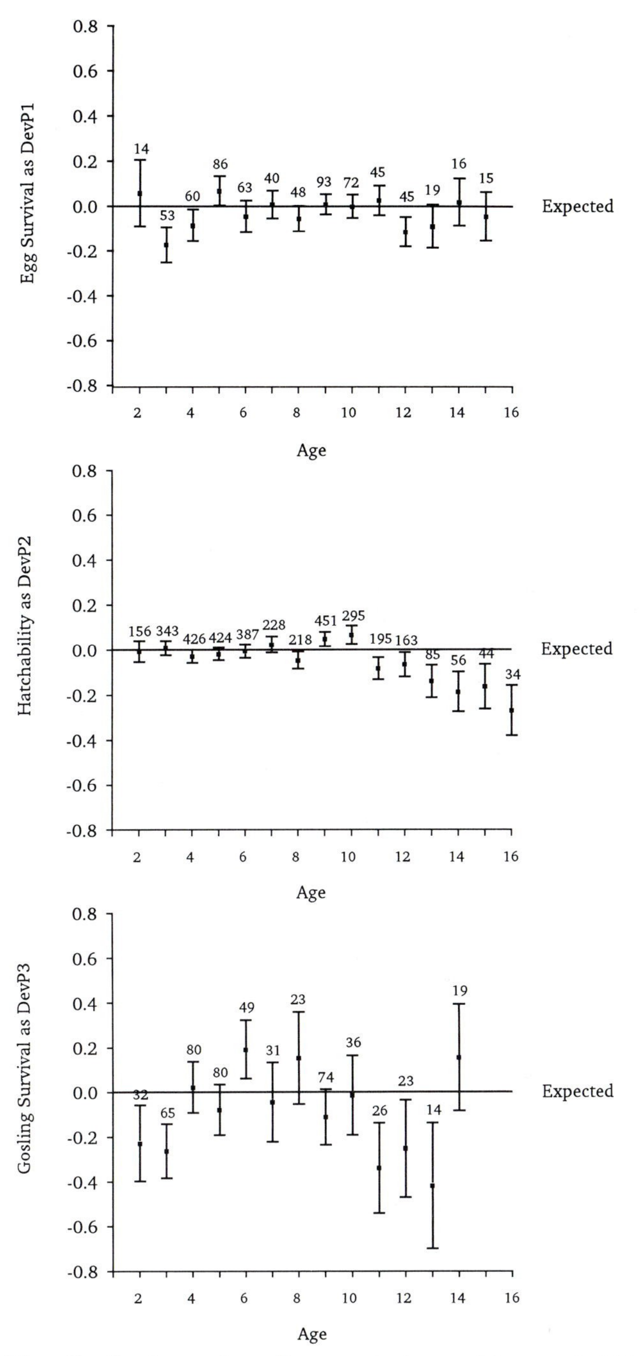

Fig. 7.2 The distributions of gosling survival transition probabilities with respect to age for Snow Geese at La Pérouse Bay. Deviations of actual survival to that expected under a model assuming no age differences (solid line) are presented, with standard errors and sample sizes for each estimate. (Reproduced with permission from Rockwell *et al.* 1993).

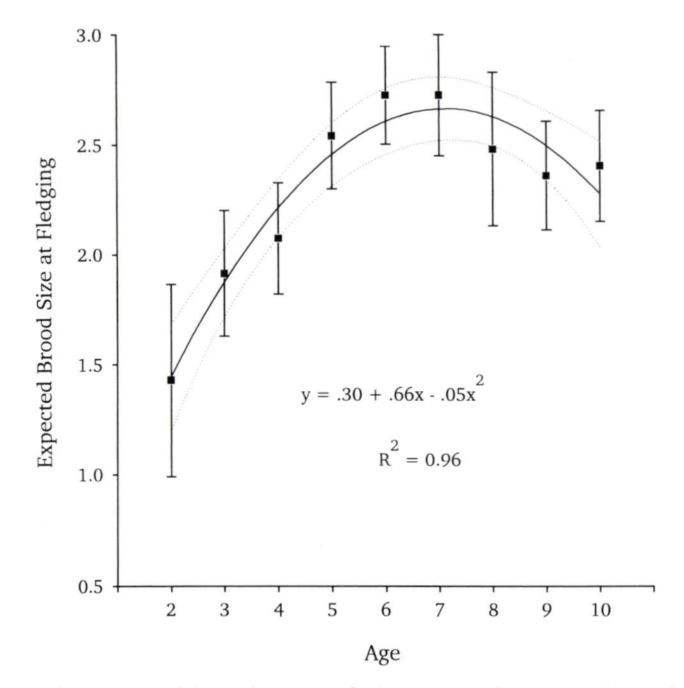

Fig. 7.3 The expected brood size at fledging as a function of age for Snow Geese at La Pérouse Bay. Points are the age-specific expectations with Monte Carlo estimated 95 per cent confidence intervals. The solid lines correspond to the regression equation given. Upper and lower 95 per cent confidence limits for the regressions are indicated by dotted lines.

mothers. This pattern could be produced entirely by the pattern of pre-fledging brood loss documented above, or could also include post-fledging survival effects. A direct analysis of recruitment rates per banded gosling failed to find differences in re-encounter rates of banded goslings with respect to maternal age (Rockwell *et al.* 1993). This suggests that maternal age effects, and perhaps parental effects in general, are smaller after fledging, consistent with our general view of a transition from fecundity to survival components of fitness at this stage.

7.3 *Survival*

Age-specific survival analyses complement the reproductive success analyses to provide a complete picture of Snow Goose life history. Francis *et al.* (1992*a*) evaluated the effects of age on the survival of Lesser Snow Geese banded primarily at La Pérouse Bay, plus supplementary information from Snow Geese banded at other colonies of the Hudson Bay/Foxe Basin population, and those banded along the flyway used by these birds.

7.3.1 Immature survival

Females banded as goslings at La Pérouse Bay have an average immature (first-year) survival rate of 42.4 per cent, estimated from band recovery data (Brownie *et al.* 1985). The immature rate varied between 10 per cent and 70 per cent of the year-specific adult rate, averaging 49 per cent, and declined significantly over time (Fig. 6.6). Young birds are more vulnerable to hunters throughout the season, with band recovery rates averaging 1.8 times higher than those of adults. Nonetheless, we argued in Chapter 6 that much of the annual variation in immature mortality occurs prior to migration to areas sampled by hunters. Consistent with this, survival estimates from first-year birds banded at fall migration stopovers are substantially higher than those based on birds banded at the colony (Fig. 7.4).

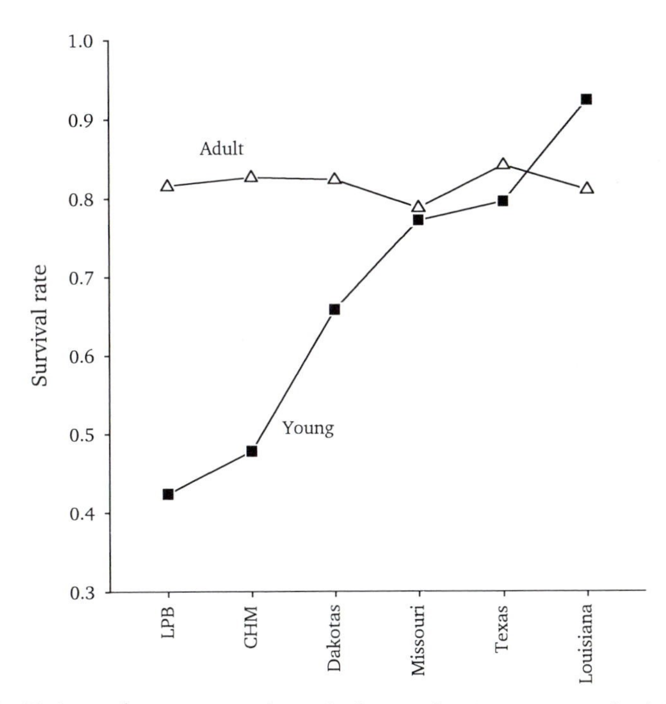

Fig. 7.4 Estimated mean annual survival rates for immature and adult Lesser Snow Geese banded on breeding grounds, at migratory staging areas, or on the wintering grounds (Francis *et al.* 1992*a*). LPB, La Pérouse Bay; CHM, Cape Henrietta Maria.

Even at the more northerly stopover sites, the survival rate of the immatures is considerably higher than it is when they leave the breeding grounds. By the time the immature birds reach Louisiana and Texas, their expected survival rate is similar to that of adults. The differences in survival rate estimates suggest that less than half of the immature mortality of the La Pérouse Bay birds is due to hunting.

7.3.2 Yearling survival

Estimating the survival of yearlings is complicated by the fact that few of them are banded (as yearlings) at La Pérouse Bay. Without large sample sizes, estimates based on recovery data are unreliable. Large numbers of birds of all ages and breeding status were banded at the McConnell River colony (Fig. 3.2), 750 km to the west and north of La Pérouse Bay, by the Canadian Wildlife Service in 1977 and 1978. With only 2 years' data, the recovery analysis methodology does not allow us to distinguish between differences in recovery rate, differences in survival or some combination of the two (Brownie *et al.* 1985). However, yearlings are slightly more likely to be shot than adults in the hunting season following banding, with 'direct' recovery rates of 3.13 per cent versus 2.71 per cent ($P = 0.03$), and survival rates 90.1 per cent of the rates of adult breeders. This suggests absolute average yearling survival rates around 75.6 per cent.

We looked for additional evidence for differential survival rates of yearlings and adults by comparing the patterns of recovery of females banded as goslings with those banded as adults. If yearlings have a higher recovery or lower survival rates than adults, the proportion of goslings recovered in their yearling year (that is in the second hunting season after banding) versus the number recovered in later years should be higher than the proportions of recoveries of newly banded or re-encountered adult birds over the same time intervals. At La Pérouse Bay, the recovery proportions of yearlings were higher than those of adults in 12 of 17 years, and significantly so in 3 years and in an overall analysis. Similar results were obtained from data from three other breeding colonies (Francis *et al.* 1992*a*). Since we don't know the number of birds banded as goslings which entered their yearling year, we cannot distinguish whether yearlings have a higher recovery rate (for example greater vulnerability to hunters), a lower survival rate in general, or both, relative to breeding adults. However if hunters are a major cause of mortality in non-juvenile geese (Chapter 6; Francis *et al.* 1992*b*), this result strongly suggests that yearlings have a lower survival rate than breeding adults.

7.3.3 Adult survival

Adult survival has increased during the course of the study, largely due to a dilution of hunting pressure in an expanding population (Chapter 6; Francis *et al.* 1992*b*). Ignoring this trend, the average adult survival rate is 83 per cent, with no significant differences between the sexes (Francis and Cooke 1992*b*). This estimate assumes that survival rate remains constant once birds enter the breeding population. Francis *et al.* (1992*a*) tested for changes in survival as birds age by estimating survival rates relative to the year of initial capture, when adults were assigned a minimum age of 2 years (Fig. 7.5). This method provided larger samples than those available from known-aged females, and should detect linear changes in survivorship. There is no evidence of a change in survival rate at least up to a minimum age of 10 years, although statistical power decreases with advancing age. In contrast to the reproductive success analyses, we have no evidence of senescence affecting survival rates in our population.

Age-related breeding status could affect adult survival probabilities. We can calculate both the recovery and survival rates of non-breeders relative to breeders because a large sample of non-breeding birds were banded at McConnell River in 1977 and 1978. Non-breeders had a recovery rate approximately 65 per cent (1.75 per cent versus 2.71 per cent) that of breeding adults (Francis *et al.*

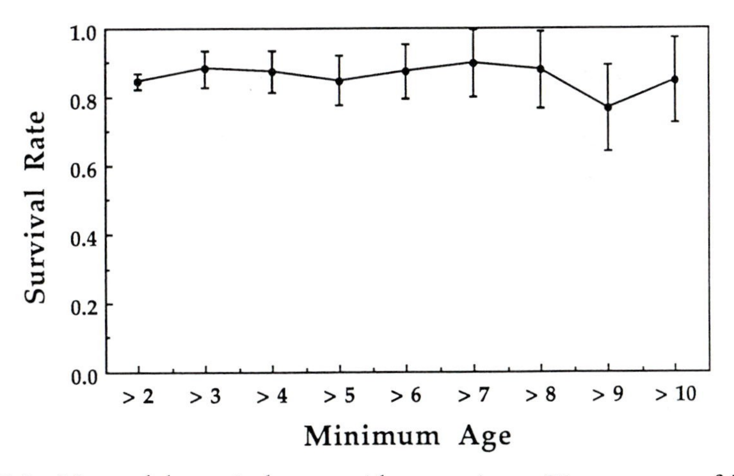

Fig. 7.5 Mean adult survival rates, with approximate 95 per cent confidence limits, for female Lesser Snow Geese of known minimum age captured and banded at La Pérouse Bay. (Reproduced with permission from Francis *et al.* 1992*a*).

1992*a*). Adults thus appear to have a reduced probability of being shot when unaccompanied by young. Despite this, there is no evidence of a higher survival rate of the non-breeders. Non-breeders had a survival rate 96 per cent of that of the breeding adults, but this difference was not significant. The discrepancy between the differences in band recovery and survival rates suggests that despite a lower probability of being shot, non-breeders have a higher likelihood of succumbing to other causes of mortality.

7.4 *Lifetime reproductive success*

An individual's contribution to the next generation is measured by its lifetime reproductive success (LRS), which combines reproductive and survivorship components of fitness. Annual or age-specific reproductive success will be a poor indicator of an individual's contribution if there are trade-offs between, for example, reproductive success and survival or breeding propensity and survival. In such cases, evaluation of the lifetime reproductive success of individuals or phenotypic segments of the population are required. The variance in LRS among members of a population is an important evolutionary parameter. In one sense it sets an upper limit on the magnitude of selection (Arnold 1986), while in another it is viewed as an indicator of at least phenotypic selection within a population (for example Sutherland 1987).

Ideally, LRS is estimated by following the annual reproductive success and fate of individuals from several cohorts. Unfortunately, few studies have such complete data on individuals. Barrowclough and Rockwell (1993) developed several simulation methods for estimating the expected mean (ELRS) and variance (VLRS) of LRS from annual demographic data. Their model follows the members of the cohort through their lives, sampling performance with respect to age-specific breeding propensities, reproductive success, and survival rates. Each female is monitored until she dies, and her total production of fledglings defines her LRS. Once the entire cohort has been sampled, ELRS and VLRS are calculated, using fledging as a point of reference. We chose fledging since it is the age of demographic independence for Snow Geese and is used by numerous other authors (for example Newton 1989*b*).

The model treats each component of fitness as a random variable, and includes no covariation—positive or negative—among components. Positive covariation ('good birds' versus 'bad birds', or 'enhancement') would increase the estimated VLRS, while negative covariation ('trade-offs') would decrease VLRS. There are no large covariances among the reproductive components (Rockwell 1988) nor have we yet found convincing reproductive enhancement or trade-offs with respect to female survival (Rockwell *et al.* 1987, see above). The only potential phenotypic trade-off we have documented is that between age of first breeding and subsequent breeding propensity (Viallefont *et al.* in press).

The output for one cycle using a cohort of 500 fledglings is summarized in Fig. 7.6, where the percentage of the cohort having a given LRS is plotted against LRS. Seventy-seven per cent of fledglings have LRS = 0, including those that died before their first breeding attempt as well as those that survived to age 2 years and tried to breed at least once, but failed at any stage of the reproductive cycle. Among birds breeding successfully, the distribution declines gradually, as expected from the relatively high adult survival and breeding propensity probabilities and a randomization of fecundity and survival variables. In this particular simulation, the luckiest female produced over 40 fledglings.

Figure 7.7 presents estimates of ELRS and VLRS with confidence

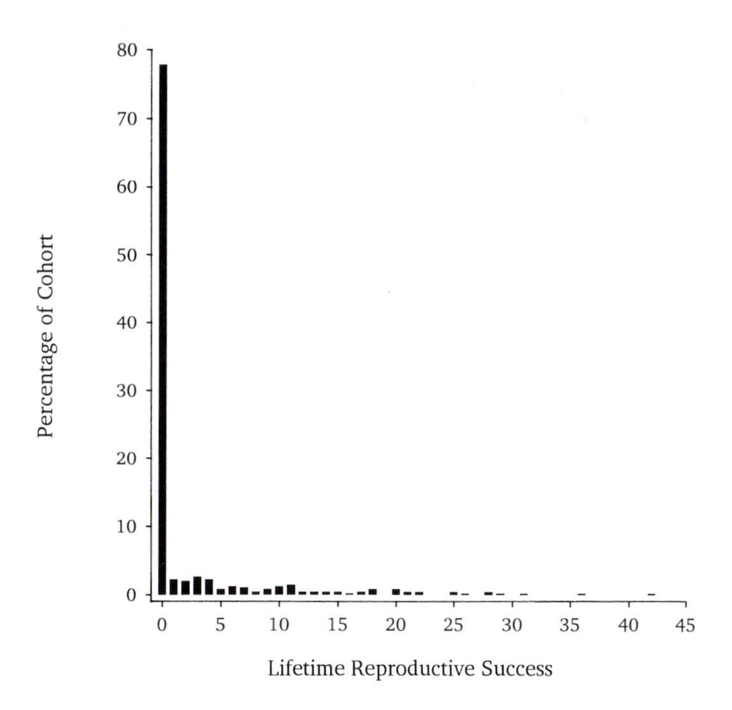

Fig. 7.6 The distribution of estimated lifetime reproductive success in a cohort of 500 fledglings.

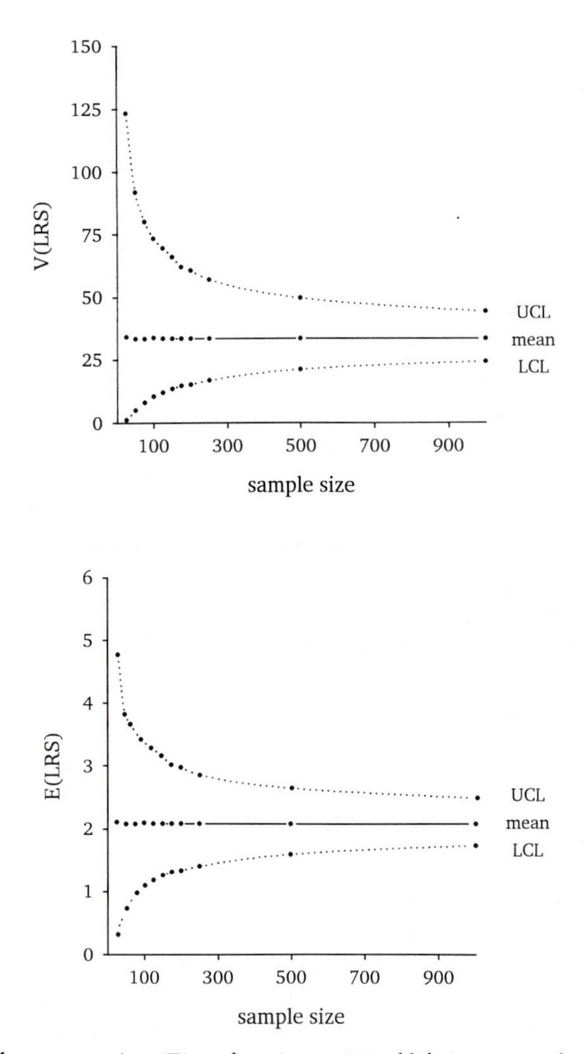

Fig. 7.7 The expectation (E) and variance (V) of lifetime reproductive success (LRS) of Lesser Snow Geese. (Adapted with permission from Barrowclough and Rockwell 1993). UCL, upper confidence limit; LCC, lower confidence limit.

limits based on 10 000 Monte Carlo trials. The mean ELRS value is 2.1, slightly above the replacement value of 2 expected from a stable population. The 95 per cent UCL and LCL correspond to the upper and lower 2.5 percentiles of the distributions, but it is not obvious what sample sizes to use to estimate the 95 per cent confidence limits. Sample size in the figure refers to the size of the cohort used in the simulations. With our data, it would be best to use the same individuals to estimate the magnitude of each model component used in the model, however we have opted instead for maximizing the sample size for each (see Rockwell *et al.* 1987). The most statistically reasonable single sample size (and hence cohort size) for our estimates would be the harmonic mean of the various components' sample sizes, but we present the confidence limits over the range of sample sizes to provide a better feel for the behaviour of the system.

Our estimate of VLRS is approximately 32, far higher than that indicated graphically for the species summarized by Newton (1989*a*). The values are not comparable however since the data he presents includes only the LRS of female fledglings that lived long enough to make one breeding attempt. Thus, his LRS = 0 individuals are those that lived long enough to try once but just did not succeed. Needless to say, such omission deflates VLRS. Similarly, Owen and Black's (1989) information on Barnacle Geese is also not comparable, since they do not estimate LRS over the entire cycle. They examined LRS of those birds of a cohort that survived about 15 months (yearlings in the fall or winter). LRS of those birds is defined in terms of fledglings 'in tow' in the fall and has been estimated from incomplete data. One might estimate the VLRS of hatched goslings, or even eggs, which would produce even higher variances. We chose to estimate from the fledgling stage because the age of demographic independence (Chapter 4) is an appropriate place to start.

7.5 *Lesser Snow Goose life history strategy*

Snow Goose females are long-lived iteroparous birds which do not nest as yearlings, and some of which may not breed until they are 4 years old. Adult survival rates are high, even in hunted populations. Breeding propensity, clutch size, and other components of reproductive performance increase for the first 5 or 6 years of life, and several components decrease in old age. Our discussion centres on understanding the selective regime and proximate mechanisms responsible for the increases and decreases in fitness components.

7.5.1 Individual maturation or changes in phenotype distributions?

Does age-specific variation reflect changes within individuals, populations, or both? Our evolutionary interpretation depends first on answering this question. For clutch size at hatch, we have sufficient longitudinal data to show that aging individuals mirror the population pattern of age classes (Hamann and Cooke 1987), as we also found with the long term change in clutch size. Changes in populations would imply a relationship between particular phenotypes and survivorship, but we have not found such relationships. We also have not found survivorship differences as a function of adult female fitness component values (Chapters 10–13). It seems reasonable to assume that the mean age-specific patterns reflect the performance of aging individuals in this species. Age-specific patterns for several components of reproductive success were similar in both longitudinal and cross-sectional studies of Barnacle Geese (Forslund and Larsson 1992).

The changes in survival rates reflect both population and individual processes. The population changes, by definition, when some birds die while others survive, but we would expect a more gradual stochastic change in age-specific survivorship schedules if individuals did not also improve their annual prospects. We conclude that we are examining primarily aspects of the life history strategies of individuals.

7.5.2 Causes of age-specific variation

Two general hypotheses have been proposed to account for increases in parental effort and/or success of individuals beyond the age of physical maturity. The restraint hypothesis posits that young birds limit their reproductive effort to avoid future costs associated with reproduction early in life (Williams 1966; Curio 1983), reflecting what Stearns (1992) refers to as 'intra-individual microevolutionary trade offs', which may then result in 'intergenerational microevolutionary trade-offs'. The constraint hypothesis asserts that developmental processes involving morphology, physiology or behaviour have not advanced to a level that allows full reproductive output. Curio (1983), Newton (1989a), and Lessells and Krebs (1989) discuss the difficulties in distinguishing among these hypotheses. Newton concluded that there is little evidence in support of the restraint hypothesis, especially for species such as Lesser Snow Geese, where survival for all adult age classes is high. We consider our data with respect to these hypotheses below.

Age of first breeding and breeding propensity

The variable age of first breeding, and the age-specific cost of first breeding, have substantial fitness consequences for individuals and the population. The variable age of first breeding increases simulated VLRS by ca. 30 per cent, and slightly increases ELRS, relative to a simulated population in which all birds started to breed at age 2 years (Barrowclough and Rockwell 1993). Age-specific breeding propensities after first breeding, not incorporated into this particular simulation, would lower somewhat both the variance contributed by a variable age of first breeding and the total estimated variance.

Birds which first breed as 2-year-olds have on average the highest ELRS, assuming equivalent survival as non-breeding 2-year-olds to age 4 years, despite their lower breeding propensity as 3-year-olds (Viallefont *et al.* in press). Although the ELRS advantage is lower than that which would accrue without this effect, this life history strategy is still the most productive in the population, producing 5.13 (±0.58) versus 4.05 (±0.45) goslings, on average, during the first 4 years of life.

Variation in breeding propensity is the first apparent 'intra-individual microevolutionary trade-off' we have documented in this study (sensu Stearns 1992). Viallefont *et al.* (in press) suggested that the lower breeding propensity of birds first breeding at 2 years also supported physiological constraint, rather than restraint, as the reason why only half the females in the population did not breed as 2-year-olds. They argued that breeding as a 2-year-old was a condition-dependent trait, since the most productive strategy was pursued by only by the fraction of the population able to do so. However, the situation is more complex, and restraint may indeed affect a 2-year-old's reproductive decisions by influencing their hypothetical 'condition-dependent breeding threshold'. We have within our population roughly equal proportions of females in three categories, rank ordered with respect to ELRS: those which breed as 2- and 3-year-olds (2,3), those which breed as 3-year-olds only (0,3), and those which breed as 2-year-olds only (2,0). These categories result from the 'decisions' to breed at ages 2 and 3 years, plus stochastic events. Viallefont *et al.* (in press) suggested that the higher ELRS of birds which breed successfully at age 2 rather than 3 years provided evidence for constraint as the major determinant of age of first breeding, but we need to consider the ELRS of both categories of the 2-year-olds to understand the evolutionary dynamics. The 2,0 females are likely to do more poorly than 0,3 females, whereas 2,3s will do much better. It is at this point that restraint should play a role. Suppose that a female's decision whether to nest as a 2-year-old, rather than be a nest parasite or not breed, depends in general on her body condition. The body condition threshold at which a 2-year-old nests should be affected by the conditional probability that doing so will put it into the more (2,3) or less (2,0) productive category. The risk of becoming a 2,0 rather than a 2,3 raises the threshold at which nesting as a 2-year-old is favoured. Any genetic variation impinging on this threshold would be available to respond to selection, resulting in an 'intergenerational microevolutionary trade-off' (Stearns 1992). Thus, while constraint on a females' resources or abilities is likely to be the overriding factor determining breeding propensity, evolutionary feedback from the costs of breeding at a younger age may result in a degree of restraint on the part of younger females.

Annual reproductive effort

The number of eggs laid by a female Snow Goose is her investment in reproduction for a breeding season. The expected brood size at fledging is the return on that investment for an average female during the critical period in which her offspring reach an age of demographic independence, the point at which their survival depends as much on their own age as that of their parents (Cooke *et al.* 1985). Expected brood size increased up to age 6 or 7 years, and declined thereafter. Age-specific increases in clutch size were the primary source of the increase among maturing birds, but lower rates of total nest failure, and partial and total brood loss also contribute.

The initial increase in clutch size with age is consistent with earlier population analyses of Snow Geese (Finney and Cooke 1978; Rockwell *et al.* 1983), and corresponds to the general pattern observed for most species of geese (Kear 1973; Cooper 1978; Hardy and Tacha 1989; Forslund and Larsson 1992). While some of the increase may be due to genuine physiological maturation, it is likely that much of it results from increased experience in acquiring the nutrients needed to produce a larger clutch of eggs (Aldrich and Raveling 1983; Lamprecht 1986; Cooch *et al.* 1989; Saether 1990; Forslund and Larsson 1992).

Performance during both the incubation and brood rearing periods also improved over ages 2–6 years. As with clutch size, we feel that this is mainly due to increased experience (Aldrich and Raveling 1983; Lamprecht 1986; Saether 1990). Experience and age-dependent social status could certainly improve parents' ability to find and defend good forage areas and to warn and defend against predators. Increased reproductive performance of Barnacle Geese has been associated with age and behavioural complementarity of partners (Black and Owen 1988).

Williams (1966) argued that as individuals age, their reproductive effort should increase since, having lived most of their life span, they have little to lose from any cost of such effort. Pugesek (1981) used this hypothesis to explain the increased breeding success of California gulls (*Larus californicus*). This version of the restraint hypothesis may apply at the start of a Snow Goose's life, but it does not appear to carry over after maturity. For Snow Geese, there is certainly no increase in 'realized or observed investment', measured as clutch size, other components, or estimated brood size, past age 6 years. This does not mean that aging individuals might not be 'trying harder', but if they are, it is both without any return (Fig. 7.3) and without apparent cost (Fig. 7.5), since there is no evidence of an age-specific decline in survival.

7.5.3 Senescence in Snow Geese?

Patterns of age-specific effects differed among fecundity measures. For some, an initial improvement in performance was followed either by no further change (clutch size, total nest failure, and gosling survival) or by a decrease in performance during old age (total brood failure). Egg survival did not change over the entire range of ages examined, while hatchability was constant until age 9 years and then declined. Thus, there is no single age-specific phenomenon affecting reproductive success in general. Rather, there were several phenomena that, taken together, gave us the composite age-specific pattern of reproductive success depicted in Figure 7.3 as an inverted 'U'.

Ratcliffe *et al.* (1988) reported that the rate of recruitment of hatchling females varied with maternal age, increasing to age 7 years and then declining to pooled age class 10–13. Recruitment rate was defined as the probability that female hatchlings from mothers of a known age class were recruited into the breeding population during the succeeding 4 years. Our fitness component approach showed that the age-dependent pattern of hatchling recruitment resulted primarily from the combined age-specific effects of total and partial brood success rather than maternal effects on fledgling survival (Rockwell *et al.* 1993). This is in interesting contrast to the finding that maternal age in Barnacle Geese appeared to correlate with post-fledgling survival (Owen and Black 1989).

The reproductive performance of female Snow Geese declined past age 8 years for both hatchability and brood loss. We suggest two possible explanations for these old age declines in reproductive success. The declining performance of older females, especially during brood rearing, may reflect physiological senescence and/or habitat degradation and philopatry.

If senescence is the explanation, it must be more specific than the '. . . general wear and tear, which reduces efficiency and social status' as defined by Newton (1989*a*). There was not a decline in all components of reproductive success, nor was there a decline in survival for older geese. Age-specific declines in hatchability have been reported for domestic chickens (summarized by Lindauer 1967), where both infertility and embryonic death increased with age and were attributed to senescence. The change in this component may truly reflect physiological senescence.

The alternate explanation relates to both habitat degradation at La Pérouse Bay and philopatry in this colonial species. The decline in habitat quality over the past 20 years may have made it more difficult for adults using those areas to find enough forage during the brood rearing period to successfully raise their young (Chapter 6). Female Snow Geese display philopatry not only to their natal colony but also to nesting and brood rearing areas within the colony (Cooke and Abraham 1980; Cooke *et al.* 1983). If females remain philopatric despite declining food supplies, then a declining success with age may reflect a deteriorating environment rather than a deteriorating bird (Cooch *et al.* 1993). A similar situation has been described in Barnacle Geese, where females that continued to use traditional and degraded spring-staging areas displayed lower reproductive success than individuals that shifted to new habitat (Black *et al.* 1991).

The potential relation of reduced reproductive success to continued use of degrading habitat is particularly interesting since it is a consequence of female philopatry—a nearly universal feature of migratory waterfowl, thought to be intimately connected with the pattern of parental care in this group (Rohwer and Anderson 1988). By displaying breeding philopatry, our sample of older females may experience reduced reproductive performance and fitness. In effect, this maladaptive response would act as phenotypic selection against philopatry in this population and one might predict a decline in the general tendency of females to return to nesting or brood rearing areas. However, even if there were heritable variation for philopatry, such a decline is not likely to occur rapidly since aging females make an ever decreasing contribution to the population's gene pool (Charlesworth 1980).

7.5.4 The effects of hunting on life history parameters

Lesser Snow Goose survival rates increase during the first two years of life, after which they remain constant for at least 8 years. The initial increase is similar to that seen in other species of migratory waterfowl. What has been reported less frequently is the fact that much of the immature mortality occurs either before the birds leave the breeding colony or, at least, early in the migration period. A similar finding has been documented by Owen and Black (1989) in Barnacle Geese, which make a trans-oceanic migration. Much of the immature mortality is independent of hunting. If one-third of the geese killed by hunters are reported, then hunting mortality of immatures is approximately 20 per cent. Average annual immature mortality is estimated as 59 per cent, so hunting accounts for less than half of the immature mortality (Chapter 6). A similar calculation for adults indicates that approximately 11 per cent are shot while annual mortality is only 17 per cent. Francis *et al.* (1992*b*) showed that adult hunting mortality was largely additive (i.e. over and above the natural mortality rate) suggesting that in the absence of hunting, the natural mortality rate would be around 6 per cent, similar to estimated mortality rates calculated from non-hunted populations of geese (Owen 1984), and to the 8 per cent estimated by a different method in Chapter 6.

The current hunting regime has been in effect for perhaps 20–40 generations of Snow Geese. We cannot assess how human activity may have affected other components of the species' life history; although hunting mortality appears additive under current environmental conditions, those conditions have also changed, perhaps in ways which favour over-winter survival of geese. Nonetheless, if hunting has increased adult mortality, it should select for geese which take somewhat greater risks to breed earlier in life and/or lay larger clutches. Variation in the age of onset of first breeding, and the apparent costs to some of doing so, may in part be one life history consequence of hunting.

7.6 *Summary*

1. Snow Geese have a variable age of first breeding, with no birds breeding as yearlings and nearly all entering the breeding population by the age of 4 years. Variation in age of first reproduction increases variation in lifetime reproductive success.
2. After the first successful nesting attempt, birds have a lower probability of success in the following year than after later breeding attempts. This effect is strongest among 2-year-old breeders.
3. The number of eggs laid increases up to the age of 5 years. There is no evidence for a decline among older females.
4. Nest failure is higher among young female breeders.
5. Total brood loss is higher among young female breeders and also increases in old age. The latter effect may be cultural and local, reflecting old females' attachment to degraded habitat, rather than reflecting physiological senescence.

6. Egg hatchability declines with the age of the female parent, whereas gosling survival increases as the female parents get older, at least up to the age of 6 years.
7. Annual reproductive output of La Pérouse Bay females increased up to the age of 6 years and a declined thereafter.
8. The calculated variance in lifetime reproductive success in this population is considerably higher than in other well studied species, but this is probably due to our calculating variance starting with fledglings, rather than restricting the population to birds attempting to breed.
9. Immature survival averages 41 per cent, with most mortality occurring soon after fledging. Immatures are almost twice as vulnerable to hunters as are adults, but despite this, hunters account for a lower proportion of the immature than adult mortality.
10. Second-year (yearling) survival is around 76 per cent, significantly lower than adult survival.
11. Annual adult survival is around 83 per cent with no evidence of senescence, at least up to the age of 10 years.
12. Non-breeding adults have a lower vulnerability to hunters than breeders, but despite this, have a slightly lower survival rate.

8 *Heritability of quantitative traits*

In the most traditional view, heritability is the proportion of total phenotypic variance attributable to additive genetic variance. The latter results from segregation of alleles with additive effects. Other contributors to total variance include dominance, epistatic and environmental effects and covariances and interactions among all these factors (Falconer 1989). Additive genetic variance is the basis of Fisher's Fundamental Theorem—'the rate of increase in fitness of any organism at any time is equal to its [additive] genetic variance in fitness at that time'. When factors other than additive genetic variance contribute to phenotypic variance, the theorem should be stated in terms of heritability rather than genetic variance. In this sense, heritability sets an upper limit on the instantaneous rate of evolution of a trait in response to a selection gradient.

Many studies have compared heritabilities among traits and tried to reach historical conclusions and generalities in relation to the fundamental theorem. The rationale is that under Fisher's formulation, traits that are closely related to fitness should have been under strong directional selection and should thus retain less additive variance and display lower heritabilities than traits less associated with fitness, all else being equal. Under this view, for example, it has been inferred that traits displaying low heritabilities are likely to be fitness characters. Mousseau and Roff (1987) and Stearns (1992) provide extensive summaries of heritabilities for traits that are actual fitness components (for example clutch size) and traits that are less obviously related to fitness. While there is a general inverse relation between heritability and 'obvious association to fitness', Stearns notes that there are exceptions, particularly since all else is often not equal (Price and Schluter 1991).

Even granted such broad agreement, there are limitations to comparisons and to historical inference. For example, the additive genetic variance of fitness components themselves will seldom be 0. Numerous models for maintaining some level of additive genetic variance have been proposed. Balances between selection and either recurrent mutation or gene flow as well as epistasis, pleiotropy and the interaction of genotypes and environments are among the more common mechanisms suggested (see Charlesworth 1987; Barton and Turelli 1989; and Stearns 1992 for readable and detailed reviews.) Unlike additive genetic variance, heritability is a proportion whose value is determined by the relative magnitudes of all the factors contributing to phenotypic variance. As such, differences between heritabilities estimated for several traits or between a single estimate and its expectation could result from differences in factors other than additive genetic variance. In short, comparative and retrospective inference from heritabilities must be done cautiously lest, as in much of evolutionary biology, the wrong process be inferred from an observed pattern.

Alternatively, one can examine heritabilities in a more 'predictive' vein. For example, by assessing the current heritability of a character, one knows the maximum rate at which the trait would respond to selection over a short period of time. This view is consistent with Endler's admonition that the study of evolution in natural populations must consider both whether there is a selection gradient associated with observed phenotypic variance and if there is a heritable basis to that variance.

The rate of response to selection need not be as high as the heritability. Indeed, under some situations involving gene frequency dependencies or negative covariances, selection will not proceed even when heritability is substantially greater than 0 (for example Cooke *et al.* 1990). It remains true, however, that if heritability is 0, there can be no response to selection as long as conditions do not change. It is partly in this predictive vein that we evaluated the heritability of the traits we have examined in this work. We also believe that resolution of the more general issues related to the interpretation of heritability will benefit from additional estimates from natural populations.

Central to both the concept and estimation of heritability is the transfer of genetic information from one generation to the next. Since there are various means by which this can be achieved, we must expand our view of heritability from a traditional focus on nuclear genes to one that includes mitochondrial genes, chloroplast DNA (in plants) and other forms of non-nuclear DNA that are passed from parent to offspring. We must also at least consider mechanisms such as cultural transmission that allow the reliable transfer of non-genetic information between generations (Cavelli-Svorza and Feldman 1981). For example, if parents that produce more offspring provide those offspring with information critical for the offsprings' own enhanced productivity, then the evolutionary consequences may not differ from those driven by the transmission of additive genes. Such mechanisms, however, are often confounded by the complexities associated with common environmental effects, a topic to which we return later.

In the following sections we explain our general approach to estimating heritability in this species and present the results for several sets of traits. We discuss the limitations and implications of the estimates for the various types of traits and make some comparisons and generalizations across the traits.

8.1 *Methodology*

Heritability is generally estimated from the phenotypic similarity of individuals of known genetic relatedness. Owing to female natal philopatry, our primary relation is that of daughters and their mothers. In estimating heritability from the regression of daughters' scores on mothers', several factors could contribute to phenotypic similarity in addition to nuclear alleles with additive effects. These include maternal effects, such as mitochondrial genes and cytoplasmic DNA, and common environmental effects. As noted above, however, since most of these would contribute to a response to a selection gradient, this is not necessarily a problem given our goals (see below and Schluter and Gustafsson 1993).

We do not have data on the same numbers of daughters for each mother. Treating each mother/daughter pair as an independent sample violates the assumption of independence of data points underlying the probability calculations of the methodology. Using family means over unequal family sizes violates equality of error assumptions. Weighting those means by within family variance is problematic since many of our families have one daughter and, hence, no variance. Using family size as a weight yields robust estimates of the slope but unstable estimates of the confidence limits of that slope (H. Levene, personal communication). We used family size as a weight in our regressions but estimated the mean and confidence limits of heritability following a bootstrap technique (B. Riska, personal communication).

The bootstrap method re-samples pairs of mother/daughter-means using mother as a focus. For example, if you have 17 such pairs, you randomly select 17 of them with replacement and regress daughter mean value on mother value using family size as a weight. The slope is doubled to estimate heritability since mothers and daughters share 50 per cent of their genes. The process is repeated 10 000 times and the mean of those estimates is used as the estimate of heritability. The upper and lower 2.5 percentiles are used as the estimates of the upper and lower 95 per cent confidence limits. Because re-sampling is based on mothers and done with replacement, variation in family size correctly contributes to sampling error and these limits. Confidence limits bootstrapped in this fashion are stable but tend to be wide (B. Riska, personal communication).

Finally, heritability estimates apply to particular populations in particular environments. Our population has expanded, males and females have immigrated and emigrated, and the breeding environment and fitness values have changed substantially during the study (Chapter 6). As a proportion of total phenotypic variance, heritability values obviously will change in different seasons which vary in food abundance for breeding females or growth conditions for offspring, resulting in changes in phenotypic values (for example Gebhardt-Henrich and van Noordwijk 1991; Larsson 1993). Since heritability values change, and may be driven by 'maternal effects', there has been much confusion and controversy about the interpretation and utility of measuring heritabilities in natural, changing environments (for example Turelli 1988; Schluter and Gustafsson 1993). The global values we are presenting are calculated from data taken over a range of conditions. Although annual variation has been minimized through covariance procedures, the estimates reflect the maximum potential for an evolutionary response over the range of environments in which Lesser Snow Geese breed, and should be considered in this context.

8.2 *Heritability estimates*

8.2.1 Fitness components

We measure the reproductive output of female Snow Geese at four points during the breeding season (Chapter 4). The heritabilities of those measures are summarized in Table 8.1. In one sense, we could argue that none of the estimates is significant since the lower 95 per cent confidence limits all overlap zero. However, these limits are wide, and such evaluations are, thus, conservative. This is particularly true for total clutch size, where the overlap is minimal. Even allowing that this initial measure of heritability of reproductive success differs from zero, the heritabilities are in general low and in broad accordance with Fisher's theorem.

The decline in heritabilities at successive state variables (Fig. 8.1) is perhaps more interesting than the absolute magnitude of any of them. One could argue that this means there is less additive genetic variance for components of reproductive success measured later in the breeding season. However, these components follow a natural time progression and as such are part/whole correlated in that each

Table 8.1 Heritability estimates for female Lesser Snow Geese at La Pérouse Bay

Trait	Lower confidence limit	Heritability	Upper confidence limit	Sample size
Fitness components				
Total clutch size	−0.083	0.294	0.636	14
Clutch size at hatch	−0.074	0.083	0.259	414
Goslings leaving nest	−0.161	0.033	0.240	321
Brood size at fledging	−0.915	−0.239	0.341	47
Egg survival	0.480	1.087	1.976	11
Hatching success	−0.224	−0.008	0.200	274
Gosling survival	−0.942	−0.347	0.215	47
Laying date	−0.939	0.334	1.572	24
Egg mass	−0.483	0.513	1.668	20
Body size				
Tibiotarsus	0.284	0.663	1.000	144
Culmen	0.148	0.408	0.710	144
Mass	0.329	0.586	0.869	144

Estimates are based on bootstrapped, family-sized-weighted regressions of mean daughter scores on their mother's scores. All scores were adjusted for annual variation with covariance procedures.

sets the upper bound for the succeeding one. This extends to the genetic domain as well since all of the loci affecting the total clutch laid must affect the clutch size at hatch and so on. While some loci may affect later measures but not their predecessors, it is reasonable to conclude that the four measures share a large, common genetic basis.

Clearly, the genetic constitutions of individuals do not erode during the breeding season. Moreover, the addition of more loci to the underlying basis of later measures of reproductive success is unlikely to reduce genetic variance within the population. Thus, the decline in heritability at successive stages is not likely due to a decline in additive genetic variance. It is more reasonable to assume that the level of environmental variance affecting the traits increases. In effect, when reproductive success is measured later in the breeding season, phenotypic differences among individuals are more likely to reflect random environmental variation rather than inherited differences. Predation events, for example, may be quite random with respect to clutch size. This interpretation conforms nicely to the model presented by Price and Schluter (1991).

This result can be usefully recast using a view of heritability borrowed from animal breeders. From their perspective, heritability

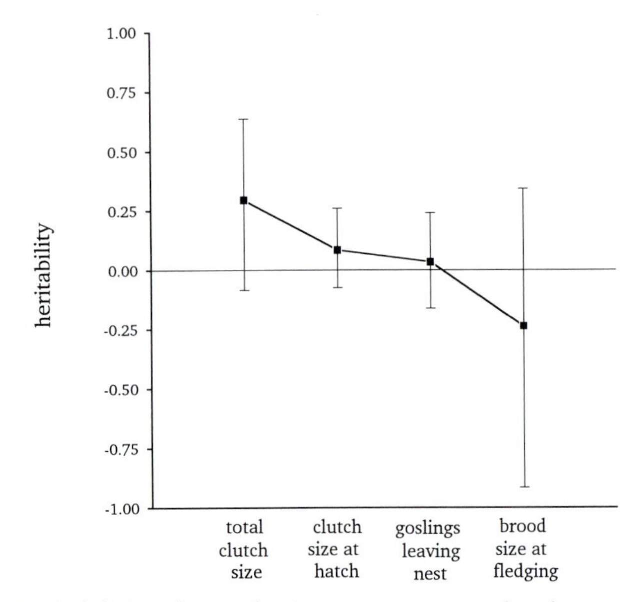

Fig. 8.1 Heritabilities of reproductive output measured at four points during the breeding season, based on family-size weighted regressions of mean daughter scores on mother scores. All scores were adjusted for annual variation with covariance procedures prior to analyses. The bootstrapped means and their 95 per cent confidence intervals were based on 10 000 resampling trials.

is the reliability with which an individual's phenotype reflects its breeding value. That is, heritability is a measure of how accurately one's inherited potential (which will also be passed on) has been translated into phenotype via functional interactions between genotype and environment. If we consider the measures as expressions of a largely shared set of underlying genotypes, then the decline in heritability over the breeding season may simply indicate increasing difficulty in realizing one's inherited potential over that 10–12 week period.

Heritability estimates for the three transition probabilities connecting the preceding measures are summarized in Table 8.1. The estimate for egg survival is the only one for which the lower 95 per cent confidence interval does not include 0. From that perspective, it is certainly greater than 0 and in fact is particularly large. Recall that we estimate heritability as twice the slope of the regression to allow for a daughter only receiving 50 per cent of her genes from her mother. As mentioned earlier, similarity could be based on more than just additive nuclear genes. If it were due to mitochondrial genes or use of a behaviour learned from the mother, the doubling is unwarranted and would lead to overestimation. The heritabilities of

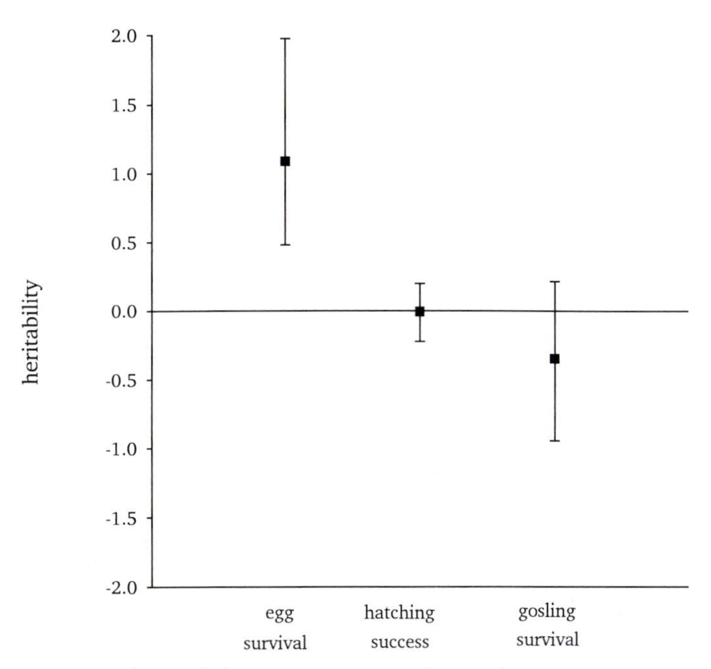

Fig. 8.2 Heritabilities of three components of reproductive success, on family-size weighted regressions of mean daughter scores on mother scores. All scores were transformed to a deviation scale similar to that described in Rockwell *et al.* (1993) prior to analyses. The scale removes any effects of annual variation and bias related to constraining these transitions to values greater than 0. The bootstrapped means and their 95 per cent confidence intervals were based on 10 000 resampling trials.

hatching success and gosling survival are, like the previous components of reproductive success, low and in broad agreement with Fisher's theorem. Egg survival, however, seems to be another of the exceptions noted by Stearns (Stearns 1992).

In comparing the heritability estimates over the three transitions (Fig. 8.2), we must keep in mind that unlike the measures of reproductive output, these components are not part/whole correlated in that they do not set upper limits on their successor. Genetically, of course, they could reflect expression of a common or at least shared set of genes (for example genes affecting 'nest tenacity' could be involved in all three transitions). In that fashion, the comparison is similar to that for state variables and leads to a similar interpretation. Phenotypic differences in components measured later in the season are more likely to reflect environmental than inherited differences. Again, it may become increasingly difficult to realize one's inherited potential over the breeding season.

8.2.2 Laying date

The heritability of laying date is 0.334 (Table 8.1). Unfortunately, the sample size is small and the broad 95 per cent confidence interval overlaps 0. Using estimates based on hatching date, Hamann (1983) calculated the repeatability of laying date which provides an upper limit on heritability of the trait (r = 0.22). Our point estimate of exact laying date is similar.

8.2.3 Egg size

The heritability of fresh egg weight is 0.513 (Table 8.1). While broad confidence intervals again reflect the small sample size, the value is close to that estimated by Lessells *et al.* (1989) using a slightly different approach. That estimate was also based on a small sample, had a large standard error and did not differ significantly from 0. Both values are similar to the heritability of estimated egg mass, an estimation based on the weight of hatching goslings and involving a much larger sample size (Lessells *et al.* 1989). That estimate was significantly greater than 0.

8.2.4 Body size

We used three measures of body size, tarsus, culmen and weight and estimated the heritabilities from a common set of mothers and daughters. The heritabilities are: tarsus = 0.69, culmen = 0.33, and weight = 0.47 (Table 8.1). Confidence intervals are narrow and all values depart 'significantly' from 0. The estimates are nearly identical to those of Davies *et al.* (1988) who used a variety of approaches. Body size is the one character for which we can compare estimates based on father–daughter and midparent–daughter, as well as mother–daughter regressions, which allows us to some degree to try to untangle the relative importance of 'maternal' versus additive genetic effects. We find in fact little difference in the estimates (Chapter 13), despite the fact that fathers lack mitochondrial or other extra-nuclear inheritance, and would be expected to have lower non-genetic parental effects than mothers, since they are immigrants from other colonies.

8.3 *General considerations*

A small portion of the phenotypic variation in initial clutch size is due to heritable variation in Lesser Snow Geese. For egg survival, another fitness component expressed early in the breeding season, the proportion of heritable variation is substantially higher. This is also the case for laying date, egg size and three measures of body size. Taken together, our results broadly conform to the general inverse pattern of a trait's heritability and its relation to fitness. We do, however, provide a few more of the types of exceptions mentioned by Stearns (1992).

Owing to our estimation procedure, there are several sources for heritable variation, including nuclear genes, mitochondrial genes, other cytoplasmic DNA and common environments. Common environmental effects result when the relatives used in estimation are exposed to conditions that increase their phenotypic similarity for the trait beyond that expected from environmental effects for randomly chosen individuals.

Two basic forms of common environmental effects are relevant to mother-daughter regressions in Snow Geese. The first is *transient* and affects only the mother and daughter. Consider an example situation in which a mother is exposed to a high quality level of an environmental factor like food that increases her initial reproductive output and performance. It is possible that her daughter could reap a benefit, perhaps in terms of size, that would similarly enhance her own reproductive performance. However, unless the daughter encountered the same high quality conditions as her mother, she would not necessarily transmit this advantage to her own daughters. In this scenario, we also expect that some mothers would encounter lower quality food and would have daughters similarly displaying poor performance. Overall, mothers and daughters would have more similarity than expected from their genes, but mothers and their granddaughters would not. The effect is transient, ceasing after two generations within a family lineage.

Suppose, however, that it was probable that the daughter would encounter the same environmental conditions that affected both her mother's and her own performance. The second form of common environmental effect is *persistent* and requires cultural transmission in its broadest sense. Continuing the example, we need only posit that obtaining the high or low quality food has to do with behaviours that are culturally transmitted from parent to offspring. This could involve, for example, learning where to eat, what to eat, or even how to eat from the mother. The key difference from the the transient form of common environmental effect is simply the enhanced probability that the daughter will encounter the environmental factor the mother encountered. The proximate consequence is that the increased similarity of mother and daughter will also exist for mother and granddaughter. The common environmental effect persists within a family lineage.

It is possible that both forms of common environmental effects as well as nuclear, mitochondrial and cytoplasmic genetic effects contribute to the heritable variation in early reproductive output, egg survival, and body size at La Pérouse Bay. Unfortunately, we do not have sufficient data to distinguish among them. Determining the relative importance and contribution of each will require not only additional information on family lineages but also a detailed understanding of the way genotypes and environments jointly influence the traits (Schluter and Gustafsson 1993). For egg and body size, studies from Barnacle Geese suggest substantial gene–environment interaction is common (Larsson and Forslund 1992; Larsson 1993).

Heritability estimated from mother–daughter regression most likely overestimates evolutionary response potential. Its poorest predictions result when transient common environmental effects are a major contributor to phenotypic similarity between mother and daughter. The degree of bias can be determined a priori only by ascertaining the relative importance of several contributors to that similarity. However, like all estimates of heritability, its predictive power can be tested by examining the response of a population to an observed selection gradient. Such testing should be the essence of evolutionary biology and an example using clutch size is presented in Chapter 10.

8.4 *Summary*

1. The importance and constraints associated with heritability in studies of evolutionary biology are considered.
2. The estimation procedure we used is presented.
3. Heritabilities of fitness components, for the most part, tend to be low and are certainly less than those of egg size and several body size characters. The heritability of laying date is intermediate.
4. Heritability estimates of measures of reproductive success made later in the breeding cycle decline. This is likely due to environmentally based variation (for example random predation events) having increasing influence on reproductive performance.
5. We discuss the potential importance of transient versus persistent environmental effects.

9 *Plumage colour*

Field studies of natural selection have been carried out on both polymorphic and quantitative characters. Although the latter are more common in nature, above the molecular level, population geneticists have until recently concentrated on polymorphic variation for both technical and theoretical reasons (Endler 1986). Mendel developed the fundamental models of heredity by interpreting phenotypes as genotypes, and characteristics for which population allele frequencies could be determined stimulated Fisher, Haldane, Wright and others to formulate the fundamental population genetics models of mutation rates, genetic drift, gene flow and selection. The empirical literature of ecological genetics similarly concentrated on tractable polymorphic traits (Ford 1975), including Kettlewell's (1973) studies of industrial melanism in the Peppered Moth (*Biston betularia*), which remains the paradigmatic textbook example documenting natural selection and evolutionary change in the wild.

Snow Geese have one highly conspicuous polymorphic trait, namely plumage colour. The frequency of the blue morph has slowly increased during our study (Fig. 3.6). We initially attributed this increase not to potentially higher local fitness of the blue morph (Cooke *et al.* 1985; Findlay *et al.* 1985), but rather to gene flow within the central breeding population of Snow Geese (Chapter 3). The slow rate of change is consistent with familial imprinting and assortative mating with respect to colour.

This chapter presents new analyses of field data which are consistent with a simple genetic polymorphism in plumage colour variation, considers potential differences in a variety of morphological and behavioural attributes of the two colour phases, and summarizes the selective regime operating on the plumage colour locus in Lesser Snow Geese at La Pérouse Bay. With our components of fitness approach, we assess not only the net reproductive success of the two morphs, but also look for any potential life history differences which might produce equivalent fitnesses in different ways. We compare the success of white versus blue females, and test for evidence of heterozygote advantages or disadvantages. We conclude by reexamining the breeding success of different pair types (Findlay *et al.* 1985). The strong positive assortative mating in Snow Geese would lead many 'adaptationists' to expect demonstrable selective advantages to matched versus mixed pairs. Since the two plumage phases appear to have have only recently merged, population geneticists might expect either selection against hybrid offspring or, potentially, a general heterozygote advantage among the young of mixed pairs.

9.1 *The genetics of plumage colour*

The first pedigree analysis of parental and gosling colours from nests of wild Lesser Snow Geese breeding in a mixed colour-phase colony showed that that the basic blue–white plumage colour polymorphism was controlled by two alleles at a single autosomal locus, with the blue allele (B) incompletely dominant to the white (b) (Cooke and Cooch 1968). Graham Cooch's data from the Boas River colony suggested blue phase birds were either homozygous dominant (BB) or heterozygous (Bb), while white phase birds were homozygous recessive (bb). There was no evidence of sex linkage. The first publication from the La Pérouse Bay study confirmed that this genetic model applied elsewhere, and showed that minor discrepancies in parent and gosling colours could be attributed in part to intraspecific nest parasitism (Cooke and Mirsky 1972). More complex threshold polygenetic or multiple allele models of genetic control were subsequently tested and rejected, and evidence accumulated to support the earlier suggestions that darker-bellied blue adults were predominantly, but not absolutely, homozygotes, and lighter-bellied blues heterozygotes (Rattray and Cooke 1984). In this section, we review the evidence that a single autosomal gene with two alleles is responsible for the basic plumage colour difference, and explore the validity of genotypic assignments among various phenotypes of blue birds.

A detailed genetic analysis of field data is complicated by three facts. First, not all goslings are the genetic progeny of the parents attending the nest, which adds noise to pedigree analyses and causes significant deviations from expected offspring colour ratios associated with matings of particular genotypes. While blue goslings in nests of white parents are the obvious case, gosling colour ratios from nests with blue parents will be generally slightly biased towards white goslings, since white birds predominate in the colony as a source of parasitic eggs (Lank *et al.* 1990). Second, although white phase birds vary little in their plumage colour and can be readily identified as bb genotypes in adult, immature, and gosling plumages, blue phase birds vary considerably, and cannot be assigned a genotype with certainty in any plumage. All blue adults have dark col-

Fig. 9.1 Plumage variation among blue phase Lesser Snow Geese. In the foreground, from left to right, these birds would receive plumage belly darkness scores of 4, 2, and 6, respectively.

oured backs and wings, but parts of the rump, neck, breast and belly, may be light or dark (Fig. 9.1). The plumage variation is related to genotype, but unlike the blue–white difference, the relationship is not absolute. Although goslings vary in down colour, they also cannot be reliably classified as BB or Bb genotypes (Cooke and Cooch 1968). Finally, since mating is highly non-random with respect to plumage colour, the simple rules of the Hardy–Weinberg law do not apply. Thus, allelic frequencies cannot be estimated simply from the frequencies of recessives, and we must be able to identify heterozygotes or homozygotes to do so.

The range of adult plumage variation can be seen from the photographs in Fig. 9.1. Although the variability within the blue phase is essentially continuous, we divided the plumages into five categories (2–6), with higher numbers indicating the increasing extent of the dark plumage (Rattray and Cooke 1984). Table 9.1 shows the relative frequency of the different categories of blues. Extreme white-bellied birds (categories 2 and 3) are relatively rare. In general, we refer to categories 2–4 as light-bellied, and categories 5 and 6 dark-bellied blues.

The numbers of blue and white goslings produced from nine combinations of parental plumages, whites and three forms of blues

Table 9.1 Frequency distribution of belly colour classes of blue phase Lesser Snow Geese (see Fig. 9.1)

Belly colour class	Proportions
2	0.013
3	0.166
4	0.247
5	0.235
6	0.338

Table 9.2 Gosling colour ratios observed in nests of different plumage phenotypes. To minimize effects of nest parasitism, only nests of clutch sizes at hatch of three and four eggs were used. 'exp' = expected proportions from genetic cross

	Proportion of blue goslings		
Pair type	Mean	SD	*n* pairs
White × White	0.015	0.079	9710
exp: bb × bb	0.000		
White × light bellied blue	0.491	0.304	1226
exp: bb × Bb	0.500		
White × medium bellied blue	0.608	0.326	774
exp: bb × BB	1.0		
White × darkest bellied blue	0.777	0.322	612
exp: bb × BB	1.0		
Light × light bellied blue	0.786	0.271	392
exp: Bb × Bb	0.750		
Light × medium bellied blue	0.874	0.219	501
exp: Bb × BB	1.0		
Light × darkest bellied blue	0.926	0.196	623
exp: Bb × BB	1.0		
Medium × darkest-bellied blue	0.967	0.121	521
exp: BB × BB	1.0		
Darkest × darkest bellied blue	0.973	0.109	609
exp: BB × BB	1.0		

(plumage classes 2–4, 5, 6), are summarized in Table 9.2, along with expected proportions under a one-gene hypothesis with hypothetical genotypes for different plumage morphs. The observed data fit the expected proportions well for white × white, white by light blue, and dark blue × dark blue crosses, strongly supporting the one-locus two-allele model. Under that model, the gosling ratios from other crosses suggest that medium and dark bellied blues are an ordered

Table 9.3 Frequency distributions of proportion blue goslings for selected pair types from nests with four hatching goslings

Pair type	Number of blue goslings 0	1	2	3	4	*n*
White × white	0.952	0.040	0.007	0.001	0.000	4764
exp: bb × bb	1.0	0.0	0.0	0.0	0.0	
White × light blue	0.101	0.250	0.333	0.220	0.097	577
exp: bb × Bb	0.063	0.250	0.375	0.250	0.063	
White × darkest blue	0.040	0.050	0.163	0.156	0.591	301
exp: bb × BB	0.0	0.0	0.0	0.0	1.0	
Light blue × light blue	0.016	0.054	0.217	0.223	0.489	184
exp: Bb × Bb	0.004	0.047	0.211	0.422	0.316	
Light blue × darkest blue	0.010	0.006	0.045	0.107	0.831	308
exp: Bb × BB	0.0	0.0	0.0	0.0	1.0	
Darkest blue × darkest blue	0.000	0.006	0.010	0.061	0.923	310
exp: BB × BB	0.0	0.0	0.0	0.0	1.0	

mixture of homozygotes and heterozygotes, with increasing proportions of homozygotes in darker colour classes. Blue plumage category thus is at best a partial predictor of genotype.

A more detailed look at offspring colour ratios among nests with four hatching young is consistent with the genotype assignments suggested above (Table 9.3). Given four hatching young, we can make exact predictions of the expected distributions of nests with 0–4 blue young for different genetic combinations. White × white, white × light blue, and dark blue × dark blue match well the expected proportions for their presumptive genotypes. The blues in the white × dark blue sample appear to be a little over half homozygotes, which would account for *c*. 50 per cent of families being all blue, with the remaining heterozygotes crosses producing families in approximately the same offspring proportions as expected for bb × Bb crosses. Blues in other crosses to blues also appear to be mixtures of Bb and BB genotypes.

Mate colour is also a good predictor of blue genotype. Blues mated to white geese are more likely to be heterozygotes, within a plumage category, than blues mated to blue geese. Dark blue birds with white mates, for example, are about half heterozygotic, as shown above. In contrast, dark blues mated to blue birds, even light-bellied ones, appear to be 80–90 per cent homozygous (Tables 9.2 and 9.3). This at first seems an odd relationship, but it makes perfect sense when one considers the mechanism of sexual imprinting (Chapter 5). Blue geese mated to white geese predominantly were raised in mixed families (Fig. 3.6), and are therefore nearly all heterozygotes. Blue geese mated to blue geese may have come from matched blue or mixed families. Thus, the observed relationship exists not because genotype influences mate choice, but rather because learned mate preferences covary with probable genotype. This is one case where a 'cultural' behavioral trait works in tandem with genetic transmission to affect population structure.

In summary, the genetic basis for the plumage polymorphism is well established. Blue is incompletely dominant to white, with heterozygotes stochastically distinguishable from homozygous blues on the basis of adult plumage and mate colour. The simplicity of the system makes it ideal for examining microevolution in the wild.

9.2 *Selection on correlated traits?*

It is difficult to exclude the possibility that selection patterns attributed to a particular trait in fact operate on phenotypically correlated traits, especially when using observational data. Fortunately, the two colour phases are virtually identical in the ancillary aspects of morphology, physiology, and behavioural traits we have measured (Table 9.4). This table, and the analyses of fitness components below, compares aspects of the colour phases in the same environment, rather than comparing snows from one colony against blues from another. We have restricted the analysis to females to compare attributes of birds known or likely to be native to the colony. Finally, we have contrasted attributes of whites against samples of likely homozygote blues, as determined by plumage colour class and pair status, where possible. The phases do not differ in nesting habitat or nest size, egg size, gosling size at hatch, hatch (and presumably laying) date, antipredator behaviour at hatch, gosling energetics or growth rate, and most measures of body size. The slightly smaller average culmen length of blue birds is the only significant result in the table. We conclude that the comparisons of fitness components with respect to colour presented in the following sections are not likely to be strongly biased by selection on other characteristics.

9.3 *Colour phases and fitness*

F. G. Cooch's (1958, 1961, 1963) studies suggested that white and blue Snow Geese differed in overall fitness, with blues having a higher overall fitness at that time. These observations were consistent

Table 9.4 Tests for relationships between plumage colour and other traits among geese at La Pérouse Bay. Analyses for most variables were restricted to contrasting whites versus dark-bellied blue females. For nest-related variables, the blue data set was further restricted to blue females with blue mates to maximize the likelihood of contrasting the two homozygote genotypes. Statistical probabilities are from *t*-tests, unless otherwise noted

Character	Mean + SE white	Mean + SE blue	*n* white, blue or statistics	*P*	References
Female body size[1]					this study
Mass (g)	955.8 + 1.3	956.8 + 1.7	1823, 579	0.49	
Tarsus (cm)	192.9 + 0.5	192.0 + 0.7	1823, 579	0.10	
Culmen (cm)	53.9 + 0.1	54.2 + 0.1	1823, 579	0.006	
Nest characteristics[2]					Jackson *et al.* (1988)
Nest cup size	—	—	$G = 0.5$,4 d.f.	0.98	
Nest area	—	—	$G = 0.8$, 4 d.f.	0.92	
Vegetation type	—	—	$G = 2.7$, 4 d.f.	0.66	
Mean egg mass (g)[3]	124.2 + 0.2	124.6 + 0.5	1349, 247	0.42	this study
Mean gosling mass at hatch (g)[3]	97.4 + 0.07	97.3 + 0.18	9578, 1815	0.67	this study
Hatch date[3,4]	23.8 June + 0.02	23.7 June + 0.11	12 741, 440	0.17	this study
Female distance from nest at hatch (m)[3,5]	14.5 + 0.14	14.9 + 0.31	8279, 1810	0.11	this study
Gosling energetics (ml CO_2/h)	522 + 12.4	506 + 10.0	6, 9	0.32	Beasley and Ankney (1988)
Female gosling mass at fledging (g)[6]	1329.5 + 7.3	1345.4 + 16.4	917, 164	0.96	this study

[1] Locally hatched known-aged females only; statistics are least-squared means and type 3 probabilities from 2-way ANOVA model including cohort and colour.
[2] Based on a categorical analyses of nest size classes.
[3] Data only from nests which produced three to six all white or all blue goslings.
[4] Least-squared means and type 3 probabilities from a 2-way ANOVA model including year and colour.
[5] Statistics based on log-transformed distances.
[6] Least-squared means and type 3 probabilities from a 3-way ANCOVA model including year and gosling age.

with the increasing frequencies of blue birds in the previously all-white western and southern colonies within the flyway. However, no comparable observations of the penetration of snows into the predominantly blue colonies on Baffin Island were available. Cooch (1963) further argued that these fitness advantages could account for the increase in frequency of blue phase birds at a number of arctic breeding grounds. From data gathered at the Boas River colony on Southampton Island, and at Eskimo Point, NWT, Cooch (1958, 1961) concluded that the phases did not differ in nest sites, construction or defense, or clutch size. In most seasons, however, white birds nested earlier and experienced higher nest and egg loss than blues. In mixed broods, the proportion of whites declined with time, suggesting differential predation or other mortality of whites. Finally, whites had higher band reporting rates, which were interpreted as indicating a greater susceptibility to hunters. There is no reason for the findings to be the same in the two studies. Boas River is a more northerly colony and the data were collected during a different time period than our current study. Cooch's studies did not consider the potential survival advantages for goslings fledging earlier in the season, and the statistical basis for claims of fitness difference between the phases was generally weak in this pioneer study. We have compared morph fitnesses over 20 years at La Pérouse Bay using the detailed fitness model presented in Chapter 4.

9.3.1 Reproductive success

Our analysis of fitness differences with respect to colour have been done without corrections for the age effects documented in the previous chapter because the relative age distributions of the colour morphs in the nesting samples do not significantly differ (Fig. 9.2; $G^2 = 19.44$, 12 d.f. $P = 0.25$; Cooke *et al.* 1985). Thus our analyses should not be biased by age effects.

The mean performances of females, classed according to the three genotypic colour phases, are summarized in Tables 9.5 and 9.6. Pre-incubation failure, total nest failure and total brood failure are discrete, categorical variables and are evaluated with multidimensional contingency analyses. We tested for colour class differences of the continuous variables total clutch size, egg survival, hatching success and gosling survival with analyses of variance. As in Chapter 8, we used partial sums of squares which remove any potential year biases from our colour comparisons, and the three transition probabilities were transformed to a deviation scale to remove any potential differences in initial clutch size. The results are clear: despite large samples,

Table 9.5 Components of reproductive success for the genotypic colour phase classes of female Lesser Snow Geese from La Pérouse Bay. Means and standard errors (SE) of clutch size, egg survival, hatching success, and gosling survival are estimated with a least-square means procedure that removes any annual variation. Means and standard errors of pre-incubation failure, total nest failure, and total brood failure are based on weighted annual estimates (Rockwell *et al.* 1993)

Component	White		Light bellied blue		Dark bellied blue	
	Mean	SE	Mean	SE	Mean	SE
Preincubation nest failure[1]	0.022	0.023	0.030	0.049	0.018	0.035
Clutch size	4.163	0.022	4.192	0.071	4.160	0.047
Total nest failure	0.056	0.040	0.043	0.061	0.042	0.034
Egg survival	0.945	0.002	0.934	0.008	0.942	0.005
Hatching success	0.933	0.001	0.941	0.003	0.937	0.002
Total brood failure	0.106	0.186	0.106	0.246	0.071	0.137
Gosling survival	0.726	0.008	0.765	0.022	0.709	0.015

[1] Pre-incubation failure is a minimum estimate since failure can occur before the individual's phenotype can be observed.

Table 9.6 Analyses of variance of potential colour phase differences in components of reproductive success of female Lesser Snow Geese from La Pérouse Bay. Sample sizes differed among the four components. The appropriate error degrees of freedom, from left to right, are: 3361, 3411, 21 410, and 2035. Mean squares are based on partial sums of squares (type 3) from which the effects of other sources are removed. Egg survival, hatching success and gosling survival were transformed to a deviation scale before analyses to eliminate any bias related to excluding total failure (see Rockwell *et al.* (1993) for details)

Source	d.f.	Mean squares			
		Clutch size	Egg survival	Hatching success	Gosling survival
Colour (C)	2	0.90	0.33	1.15	1.55
Year (Y)	16	10.25**	0.25	0.47	0.42
C × Y	32	0.69	0.38	0.31	0.63
Error	a[1]	1.11	0.29	0.33	0.73

** $P \leq 0.01$.

[1] a = component degrees of freedom given in table title.

there are no differences among the colour phases in any of these components of reproductive success.

Our estimates of pre-incubation failure are minimum estimates since failure during that period often occurs before we have been able to identify the phenotype of the attendant female. However, assuming that our inability to make that identification is independent of colour, the relative comparison is valid. We did not have sufficient data to include year as a term in the evaluation of either pre-incubation failure or total brood failure and data were pooled over time. In neither case was the effect of female colour significant (pre-incubation failure: $G^2 = 1.44$, $P = 0.49$; total brood failure: $G^2 = 2.05$, $P = 0.36$). We had sufficient data to include year as a variable in the analysis of total nest failure. There too, however, the effect of female colour was not significant ($G^2 = 3.17$, $P = 0.21$).

In sum, there are no significant differences in any component of reproductive success between the colour phases of female Snow Geese nesting at La Pérouse Bay, as found earlier with a smaller data base (Cooke *et al.* 1985).

We have failed to find differences in fitness components of different colour classes of females. What about pair types? Since there were no differences between the two genotypic classes of blue females, we merged them in our analysis. The four pair types are simply: white female × white male, white female × blue male, blue female × white male, and blue female × blue male. The analyses of the continuous variables are summarized in Table 9.7. Clearly, there are no differences among the pair types. Multidimensional contingency analyses of the three discrete components also failed to find differences (pre-incubation failure $G^2 = 1.10$, $P = 0.78$; total nest failure $G^2 = 3.72$, $P = 0.29$; total brood failure $G^2 = 0.84$, $P = 0.84$). The lack of difference in total nest failure contrasts with the findings of Findlay *et al.* (1985). The current analyses are based on a much larger sample size, and we conclude that there are no differences among the pair types in any component of reproductive success.

9.3.2 Viability measures

We began the reproductive success section by showing that there were no significant differences in the age structure of nesting females

Table 9.7 Analyses of variance of potential differences in components of reproductive success between the four pair-type classes of Lesser Snow Geese from La Pérouse Bay. There were different sample sizes for the four components. The appropriate error degrees of freedom, from left to right, are: 3326, 3326, 20 550, and 2035. See Table 9.6 for treatment of variables

Source	d.f.	Mean squares			
		Clutch size	Egg survival	Hatching success	Gosling survival
Pair type (P)	3	0.95	0.28	0.54	1.65
Year (Y)	16	10.83**	0.20	0.48	0.37
P × Y	48	1.15	0.28	0.30	0.86
Error	a	1.11	0.30	0.33	0.72

** $P \leq 0.01$.

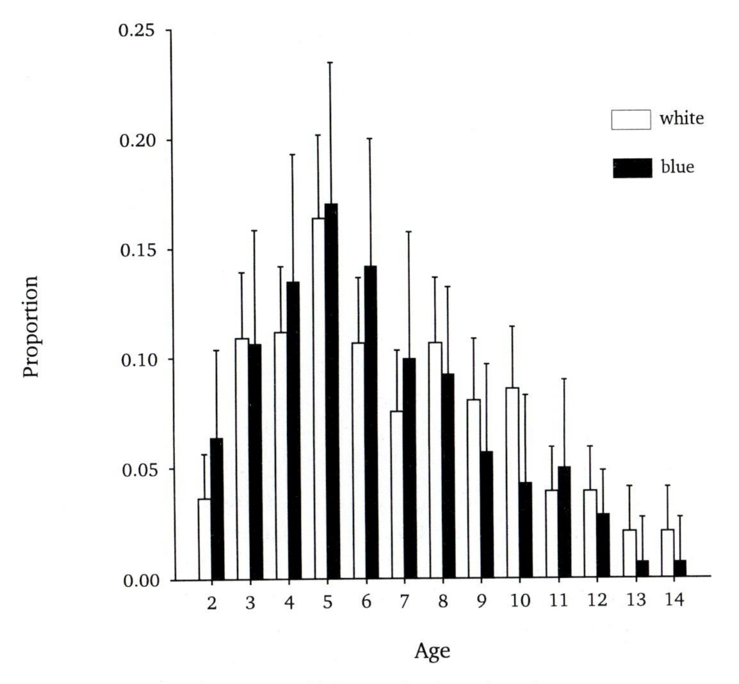

Fig. 9.2 The age distributions of blue and white female Lesser Snow Geese at La Pérouse Bay. Data were pooled over years 1973–1990 after we demonstrated no significant year term in a multidimensional contingency analysis. Data are presented as mean proportion and upper 95 per cent confidence limit. The distributions for the two colour phases do not differ.

(Fig. 9.2), which is consistent with survivorship of the two morphs being similar. We present here formal analyses of survivorship with respect to colour phase, based on both band recovery and recapture data (Chapter 4).

Immature and adult survival rates of the two colour phases, calculated from band recovery data (Brownie *et al.* 1985), show no differences for immatures or adults of either sex (Table 9.8). However, band recovery rates differed among males, with both adult and immature blue males significantly lower than white males, implying that they had a lower probability of being reported shot. This is probably due to the fact that males have lower philopatry than females and frequently enter other migration routes with lower probabilities of being reported shot (Francis and Cooke 1992*a*). Because of the patterns of assortative mating, blue birds are more likely to migrate back with females from more easterly colonies such as Baffin Island and Cape Henrietta Maria. Birds from these areas are either less likely to be shot or less likely to have their bands

Table 9.8 Annual survival and band recovery rates (means and 95 per cent confidence limits) of male and female Lesser Snow Geese banded at La Pérouse Bay 1969–1998, with respect to colour phase and sex. Survival rates were calculated assuming variable recovery rates and constant survival over time using model 2 of the program ESTIMATE (Brownie *et al.* 1985) to estimate adult survival and the program BROWNIE to calculate immature survival (C. M. Francis, personal communication)

Age	Sex	Colour	Survival rate (mean % ± 95% c.l.)	Recovery rate (mean % ± 95% c.l.)
Adult	female	white	81.5 ± 1.0	3.3. ± 0.2
		blue	82.1 ± 1.5	3.4 ± 0.2
Adult	male	white	82.5 ± 0.9	3.3 ± 0.1
		blue	82.9 ± 1.3	2.9 ± 0.2
Immature	female	white	43.0 ± 1.9	6.7 ± 0.2
		blue	41.0 ± 2.6	6.7 ± 0.3
Immature	male	white	34.5 ± 1.6	6.8 ± 0.2
		blue	34.7 ± 2.4	6.1 ± 0.3

reported; perhaps the latter since there is no evidence of a difference in survival.

Relative local survival of the phases, estimated from recapture data (Lebreton and Clobert 1986; Chapter 4), also failed to suggest differences between the colour phases. Blue phase adult females had an estimated survival rate of 77.2 per cent (0.753–0.790) compared with 77.0 per cent (0.757–0.783) for whites (R. Pradel, personal communication), using a model which assumed that recapture rate varied with time, but survival rate did not. Only female rates were calculated because the high rate of male emigration confounds interpretation. As suggested above (Chapter 4), calculating 'absolute' and local survival rates allows us to infer emigration rates. We can conclude from the lack of differences in both recovery and recapture-based survival estimates that the emigration rates of the morphs are also similar.

We detected no differences in absolute survival rates, based on band recoveries, of light-bellied versus dark-bellied blue birds. The two had identical values of 82.6 per cent, assuming constant survival over time. We found no annual variation in the phase specific patterns of survival, but our data are such that differences would be difficult to detect because of small sample sizes.

9.3.3 Breeding propensity

We examined patterns of age-specific recruitment into the breeding population of the two colour phases using the method described in

Chapter 7. If the two phases differ in the age of first breeding, this should be detectable by the relative changes in recapture rate with age. Simple recapture models showed no differences among the colour phases, but more elaborate models have not yet been developed or tested. Finally, we tested for differences in the relative breeding propensities of females of the two colour phases, but failed to find any (Rockwell *et al.* 1985*b*).

9.4 *Discussion*

The similarity of the colour phases in all components of fitness is striking. We failed to detect differences despite substantial sample sizes and powerful statistical techniques for most components. *A priori*, we might have expected to find some differences as a direct result of the physiology resulting in differing plumage pigmentation, as an ecological consequence of plumage differences, or as an indirect result of the hypothesized difference in evolutionary histories (Chapter 3). However, none of these seem relevant to the current fitness values of Snow Geese breeding at La Pérouse Bay. We know little about the direct physiological costs associated with the production or use of white versus grey-brown feathers, except for the work done on growth energetics of chicks (Beasley and Ankney 1988), which failed to find energetic differences. Despite suggestions that hunters (or other predators) might take birds differentially with respect to colour, the survival rates of different coloured birds breeding in a single colony are indistinguishable.

As documented in Chapter 3, historically the blue and white phases have had different breeding ranges, migration routes, and wintering grounds. Throughout the year, they would have been subject to a different range of predators, food supply, and seasonal phenologies which should have selected for different characteristics. If we are correct in assuming that there was little exchange of genes between the two morphs from at least as far back as the most recent glaciation, then random drift should have led to differentiation across the entire genome, and there is still residual evidence of a difference at the allozyme level (Cooke *et al.* 1988; Chapter 3). Although these differences may have lessened during the forty or fifty years since sympatry has been restored, we would still expect some differences in genetic composition of the two morphs which might be reflected in variation in demography at our study site. Yet, the two morphs, differing in major plumage colour and evolutionary history, appear to have virtually identical life histories.

One possible explanation is that the environmental variance in fitness components, or the gene × environmental variance, is large enough to swamp potential genetic variation. If the life history strategies of both morphs are sufficiently sensitive to environmental conditions, phenotypic differences attributable to genetic differences may be extremely small. Since the heritability estimates of fitness components were low (Chapter 8), we already have established that relatively little of the variance in these values is attributable to inheritance within our population.

Before accepting the conclusion that the two morphs have the same life history strategies however, we must be aware of the limitations of our data. Sample sizes for some of our measures are still small and if slight differences in fitness were present we may fail to detect them. If selective differences were present only under extreme conditions, we may not detect them despite many seasons of data. We have investigated selective factors at a single breeding colony for a limited number of years. The La Pérouse Bay colony is both a young and a southerly colony. There have been no total breeding failures during our 25 years of study, in contrast to the situation at Wrangel Island, Russia, for example, where there no young at all fledged during two of 22 years of study when complete data were collected between 1970 and 1993 (E. V. Syroechkovsky and colleagues, personal communication), or at Boas River, where total failure occurred in 3 years when studied (F. G. Cooch, personal communication). Production is lower in the more northerly colonies and perhaps there is more opportunity for selective differences to be detected. The contrast of our results with those of Cooch at the Boas River colony on Southampton Island, reviewed at the start of this chapter, suggests this may be the case. However, although phase-specific fecundity components may differ among colonies, this is less likely for viability components, since environments are shared for at least parts of the year.

Despite a lack of selective difference between the colour phases, there has been a gradual increase in the frequency of blue phase birds in the colony (Fig. 3.6). We attribute this increase to non-random gene flow (Cooke 1988), with the immigration rate of blue phase birds exceeding the emigration rate. Immigrants have had a significantly higher frequency of blue phase birds than do residents. Twenty-eight per cent of breeding birds known to be native to La Pérouse Bay were blue, approximating the proportion of blue goslings produced. This differs significantly from 31.7 per cent blue among males of unknown origin, nearly all of whom will be immigrants. Immigration, rather than local selection coefficients, appears to account for the changes in plumage allele frequencies at La Pérouse Bay.

We suggested that the net immigration of blues was due to the slow equilibration of formerly allopatric colour phases within the central wintering population (Chapter 3). One potential mechanism which could generate net immigration of blues without intra-colony fitness differences would be intra-colony productivity differences. If the overall fecundity at predominantly blue colonies, such as those on Baffin Island, were higher than that in the more southerly and westerly predominantly white colonies, net blue immigration could occur. Since the higher latitude colonies are in fact generally less productive, this explanation is unlikely.

9.5 *Summary*

1. There are no detectable differences in reproductive success between the colour phases.
2. Heterozygous and homozygous blue birds do not differ for any components of reproductive success.
3. The four pair types in terms of plumage colour do not differ for any components of reproductive success.
4. Global and local survival rates of immatures and adults are similar for both colour phases, and heterozygous and homozygous blues do not differ in survival rates.
5. Despite the lack of fitness differences, blue phase birds have increased in frequency during the course of the study. This increase is best explained by a proportionally higher immigration rate of blue phase birds into the colony.

10 *Clutch size*

This chapter considers clutch size, one of the most obvious and easily measured components of fitness. We determine the selection gradient on this univariate phenotypic trait (Arnold and Wade 1984*b*), and consider its ecological and evolutionary constraints and consequences. Thus, we assess whether and how other fitness components and net fitness vary with respect to clutch size.

10.1 *Clutch size limitation of reproductive effort*

Lack (1947) stimulated modern interest in clutch size variance from his study on Swifts (*Apus apus*). Most female Swifts lay two eggs, but one- and three-egg clutches also occur. If there is a heritable component to clutch size variation, why do not birds that lay three eggs increase in frequency in the population? Lack observed that females laying two eggs fledged more young than those that laid three, on average, because most parents were unable to adequately provision three young. Lack suggested that for birds with extensive post-hatch parental care, clutch size was no larger than the maximum number of young which the parents could likely raise per year.

Lack (1967) reached a different conclusion when considering the regulation of clutch size in waterfowl. Waterfowl produce relatively large eggs which hatch as well-developed precocial self-feeding young. Lack argued that post-hatch provisioning was less critical in such species, and that the availability of nutrients to females prior to or during egg formation was the most important limit to total offspring production. While specific proximate and ultimate mechanisms of clutch size limitation in waterfowl continue to be hotly debated (Arnold and Rohwer 1991; Ankney *et al.* 1991), consensus remains that clutch size itself, rather than limitations on post-hatch parental care, limits annual fledgling production.

Limitation of annual reproductive output by clutch size versus post-hatch provisioning produces different predictions about the evolutionary trajectory of a population's clutch size, assuming a non-zero heritability of the trait, as appears to be the case for Snow Geese (Chapter 8). Limitation at the parental care stage imposes normalizing selection on clutch size itself, since smaller and larger clutches are less productive. We might expect heritable variance to decrease, but the mean should not change over time. In contrast, limitation at the egg-laying stage imposes directional selection favouring larger clutch sizes, and mean clutch size might increase over time. This logic holds for a population even if individuals adjust their clutch size to mesh with their relative ability to produce eggs and/or provide for young (Nur 1988). While individual optimization may occur, those individuals with higher optima will contribute more genes to subsequent generations, and thus larger clutch sizes should evolve in a population.

Clutch sizes obviously don't continually increase in size over time. Life history theory, dating back to Fisher (Fisher 1930) and G. C. Williams (1957, 1966), suggests that total annual reproductive effort should evolve to maximize lifetime, not annual reproductive success. Life history strategies evolve such that the level of current reproductive effort is determined by its effect on future survival and reproductive effort. Given variation in components of fitness within an equilibrium population, microevolutionary trade-offs (sensu Stearns 1992) between levels of current and future survival and reproduction would select for genomes which optimize levels of annual reproductive effort with respect to lifetime success. We do not expect to observe such trade-offs at the phenotypic level in the wild, however, primarily because variation among individuals in individual optima means that correlations between current and future success, or between other components of fitness, may often be zero or positive (Reznick 1985; van Noordwijk and de Jong 1986), and because of complications introduced by genetic correlations and genotype-environment interactions (Stearns 1992). Negative correlations, however, would provide some evidence of a cost of reproduction. Thus we cannot necessarily test the applicability of life history theory directly with our phenotypic data. What we have done is to characterize the selection regime and potential phenotypic trade-offs on clutch size, using the components of fitness model described in Chapter 4, as a step towards understanding the evolutionary potential of the trait in this population.

Our components of fitness model is well suited for characterizing the selection regime operating on clutch size. If other components of fitness vary independently of clutch size, directional selection would favour geese that lay as many eggs as possible, which would then produce relatively more recruits. Alternatively, birds laying overly large clutches might have lower mean values of downstream fitness components, such as gosling or parental survival, and thus an intermediate 'optimal' clutch size might be favoured overall, as in the parental care limitation case. This approach allows us to learn something of the mechanism of selection and the pressures that the geese are under at different stages of their life. At a more practical level, if annual clutch size is a good estimator of a downstream fitness component which is more difficult to measure, such as recruitment of breeding young, then we can legitimately use clutch size to assess the relative reproductive success of a goose with respect to other characters. Clutch size has a significantly positive repeatability and heritability (Chapter 8; Findlay and Cooke 1983, 1987), thus to some degree annual clutch size reflects a property of an individual. If survival and breeding propensity are independent of clutch size, then those birds laying a higher clutch in one year should have a higher lifetime fecundity.

10.2 *The selection regime on clutch size*

We calculate the selection regime on clutch size using our components of fitness model.

10.2.1 Fecundity components

Clutch size and covariates

We tabulated breeding attempts by total clutch size, and assessed the values of downstream components of fitness among classes (Rockwell *et al.* 1987; Chapter 4). The raw data were completed clutches initially found at the one-egg stage, 1973–1984, except for 1979 when sample sizes were small. The annual distributions are all fairly normal, with a modal clutch size of four eggs, in 8 years, and five in 3 years (Fig 10.1). The sample includes continuation clutches (renesting in a second location after loss of the earlier eggs at the original nest), which are not readily detectable as such, and parasitized clutches, where total clutch size is usually larger than the physiological clutch laid by the nest attendant. Direct analysis of post-ovulatory follicles on 47 females collected immediately after the beginning of incubation, in two years, showed that physiological clutches varied from two to six (Hamann *et al.* 1986), as found earlier by Ankney and MacInnes (1978) at the McConnell River Colony. From hatching gosling colour ratios, approximately 38 per cent of the six egg clutches and 96 per cent of the seven egg clutches contain eggs laid by two or more females (Rockwell *et al.* 1987). Thus nearly all of the few one egg clutches are likely continuation clutches, and clutches larger than six are virtually all laid by more than one female.

Clutch sizes change with age, nest parasitism, year, and seasonal

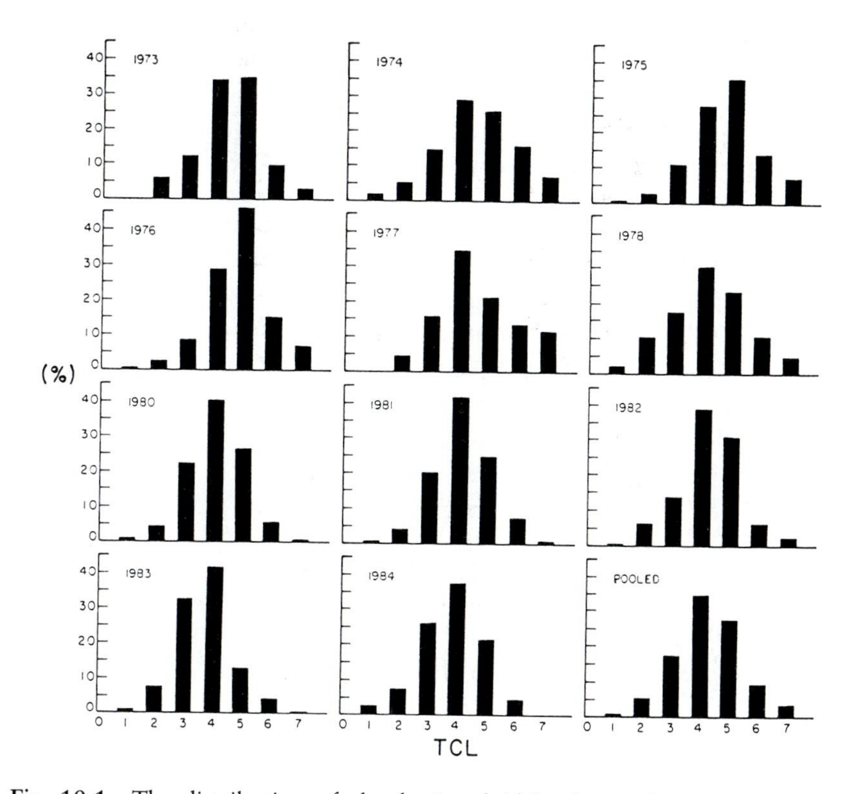

Fig. 10.1 The distribution of clutch sizes laid by Lesser Snow Geese at La Pérouse Bay, 1973–1984. (Adapted with permission from Rockwell *et al.* 1987).

timing of laying, none of which are controlled for directly in the final analysis presented below. Mean clutch sizes increase between ages 2 and 5 years, and other fitness components also change with age (Chapter 6). We cannot control for age for the initial fitness component of total clutch size itself because few old birds lay small clutches and few young birds very large ones. Thus age differences may be invoked to account for any patterns observed. The nests in this analysis have not been screened for nest parasitism, which is increasingly likely to have occurred in larger clutch sizes. For clutch sizes 2–7, approximately 6, 9, 10, 16, 40, and 96 per cent of clutches contain parasitic eggs (Cooch *et al.* 1989; Lank *et al.* 1989*b*), which have lower probabilities of hatching. Where possible, we tested for year effects and pooled to maximize power when factors of interest did not depend on them. While clutch size declined *c.* 3–4 per cent over the years analysed (1973–1984), the total loss and transition probabilities changed little if at all during this period (Fig. 6.2), thus year does not substantially bias the analysis. Finally, there is a strong correlation between clutch size and laying date. Birds nesting early in the season lay on average more eggs than those nesting later, and in early nesting seasons, the average clutch size is higher than in late seasons. We will tease this relationship apart in the next chapter.

Egg and brood survival

Total clutch loss changes with total clutch size (Table 10.1). Total nest failure is much more frequent among one- and two-egg clutches, with no significant differences among larger clutch sizes. Most total nest failure during incubation is due to predation, although it is often difficult to distinguish nest predation from nest abandonment followed by scavenging. Three factors can account for the higher nest failures among one and two egg clutches. First, loss of one- or two-eggs to predators will wipe out a small clutch, whereas such a predation would be classified as partial clutch loss among the larger clutches. Second, some one- and two-egg clutches may be continuation re-nesting attempts in poor sites. Third, and best documented, small clutches are more prevalent among young birds which also have higher nest failure rates in general (Chapter 7; Rockwell *et al.* 1993). In our sub-sample of nests of known-aged birds, multidimensional contingency analysis showed no dependence of nest failure on clutch size *per se* (G^2 = 1.54; d.f. = 4, P = 0.82), but a significant dependence on female age, with lower nest survival in 2-year-olds contributing most to the heterogeneity (birds classed as 2, 3, 4 years, or older; G^2 = 9.14; d.f. = 3, P = 0.03). Almost 20 per cent of the two-year-old birds failed, whereas about 5 per cent of the older birds were unsuccessful. Thus, young parents are a major reason for higher nest failure rates of small clutches. This may be due to inexperience of how to deal with predators or to a lower ability to cope with the entire incubation period due to an inadequate accumulation of nutrient reserves for incubation.

For nests surviving through incubation, egg survival and hatching success were significantly higher for smaller clutch sizes, however (Table 10.1). Nest parasitism is a large contributor towards this pattern, as parasitic eggs laid and remaining outside of nests are more vulnerable to predation, and about 25 per cent of parasitic eggs are laid after incubation has begun, causing parasitic eggs to have substantially lower hatching success in general (Lank *et al.* 1990). Age effects are of less importance for these components (Chapter 7). Although hatchability declines in birds older than 9 years, such birds contribute negligibly to this data set. Finally, the lower survival and

Table 10.1 Components of fitness with respect to total clutch size. Values are mean values derived from polynomial regression and total clutch size for all components except pre-reproductive survival, adult survival, and adult probability of mate change. The significance values and patterns of variation are taken from more detailed analyses presented in Rockwell *et al.* (1987)

Fitness component	Clutch size							Significant variation	Pattern of variation	Significant covariates
	1	2	3	4	5	6	7			
Fecundity										
Nest failure (%)	44	24	13	8	9	9	9	yes	1,2 > 3–7	age
Egg survival	1.0	0.99	0.95	0.94	0.94	0.93	0.91	yes	2–5 > 6,7	ISNP[2]
Hatching success	1.0	0.94	0.93	0.93	0.92	0.88	0.79	yes	2–5 > 6,7	ISNP
Fledging success	1.0	0.95	0.84	0.78	0.75	0.72	0.66	yes	2 > 3–7	age
Total brood loss (%)	16	12	10	8	7	6	6	yes	1–3 > 4–7	age
Offspring viability										
First-year survival (% band recoveries)	0	7	6	10	9	5	19	no	1–7 =	
Recruitment (% breeding females)	14	14	14	15	17	18	18	no	1–7 =	
Adult viability										
Adult survivorship (% detected returning)	53	48	51	51	50	45	46	no	1–7 =	
Mate change[1] (% with new mates)	100	14	32	35	29	26	17	no	1–7 =	

[1] Nests classed by clutch size at hatch rather than total clutch size because of small sample sizes.
[2] ISNP = intra-specific nest parasitism.

hatching success of large clutches may reflect a genuine decreased parental ability to incubate or defend clutches of six or seven eggs.

Post-hatch gosling survival presents a mixed picture with respect to clutch size. Total brood loss is higher among small broods (Table 10.1), as might be expected since a loss of one or two goslings puts the brood into this category. Total brood loss is higher for 2- and 3-year-olds, which may partially account for this pattern. Among successful broods, goslings from broods of two and three have a higher than average probability of survival, while those from broods of six have a lower survival probability. More than 70 per cent of large broods lose at least one gosling. This is also consistent with the age-specific pattern of partial brood loss (Chapter 7).

To summarize the nesting stages, the smallest clutches fail completely most often, but among incubated nests, egg survival and hatching success appear to be independent of clutch size in the range of two to five eggs. For clutches (broods) greater than five, egg and gosling survival declines.

10.2.2 Viability components

Is post-fledging survival and recruitment as breeders influenced by family size? We indexed pre-reproductive survival from hunter band recoveries during the first 4 years of life, and, for females only, directly indexed recruitment from re-encounters with returning breeders at the colony (Table 10.1). While these methods do not provide absolute survivorship or recruitment estimates, they may be used for relative comparisons, assuming that the recovery or re-encounter detection probabilities are not influenced by the size of the family from which the bird came. The fates of fledged goslings were tabulated by clutch size, but we evaluated the relative probability of re-encounter per fledged gosling, regardless of how many goslings fledged from particular clutches (Chapter 4; Rockwell *et al.* 1985*a*).

For the band recovery data of goslings banded 1973–1982, we partitioned the recoveries into two reporting intervals, covering year 1 and years 2–4. For each period the number of recovered offspring was contrasted with the number not recovered using multidimensional contingency analysis. Preliminary analyses showed no significant differences between sexes, measures from goslings banded in different years, or first year and second through fourth year survival, nor were there any significant interactions. Consequently, we pooled the data over sex and natal year. Pre-reproductive survival, as measured by recovery data, is independent of the brood size from which the gosling came. This means that families that fledge more goslings are likely to have more goslings surviving to breeding age.

As a direct measure of survival and recruitment, we tallied females resighted or recaptured on the colony as a breeding adult at least 2 years after her original banding. We assume that the rare cases of female emigration are randomly distributed among the various clutch sizes. Resighted or recaptured birds were contrasted with the number of fledged females not re-encountered. Data were pooled over years. Recruitment was not significantly different among brood sizes, despite a trend towards higher recruitment from larger brood sizes, which may result from an age effect and/or the long term changes in recruitment probability (Chapter 7). Thus, recovery and recruitment analyses provide similar answers: post-fledging survival and recruitment probabilities are independent of initial clutch size.

10.2.3 Expected annual reproductive success

The effects of clutch size variation on various fitness components can be combined to produce an empirical estimate of the relative fitness of each clutch size phenotype, accounting for both total and partial losses. We calculate the composite number of recruited offspring expected per breeding attempt (*ERS*) for each clutch size (i) using the formula:

$$ERS_i = i \times (1 - TNF_i) \times P1_i \times P2_i \times P3_i \times (1 - TBF_i) \times PR_i$$

where PR = probability of detected recruitment of a fledgling and the other acronyms are defined as in Chapter 4.

The composite expectation of reproductive success increases through clutches of six, but no further (Fig. 10.2). Since few, if any, of the seven egg clutches were laid by one female, females laying more eggs per clutch are likely to produce more recruits in the breeding population. Thus, within the observed range of individual clutch size variation, there is directional rather than normalizing selection, at least for the fecundity component of selection.

10.2.4 Adult survivorship and lifetime reproductive success

Before accepting that our conclusions based on annual clutch size have implications for lifetime reproductive strategies, we must test for survival effects on breeding adults. Do birds that produce larger clutches have lower fitness values for viability components? As outlined earlier, we are not intending to test life history theory with these analyses, but rather to more fully assess the selection gradient on clutch size.

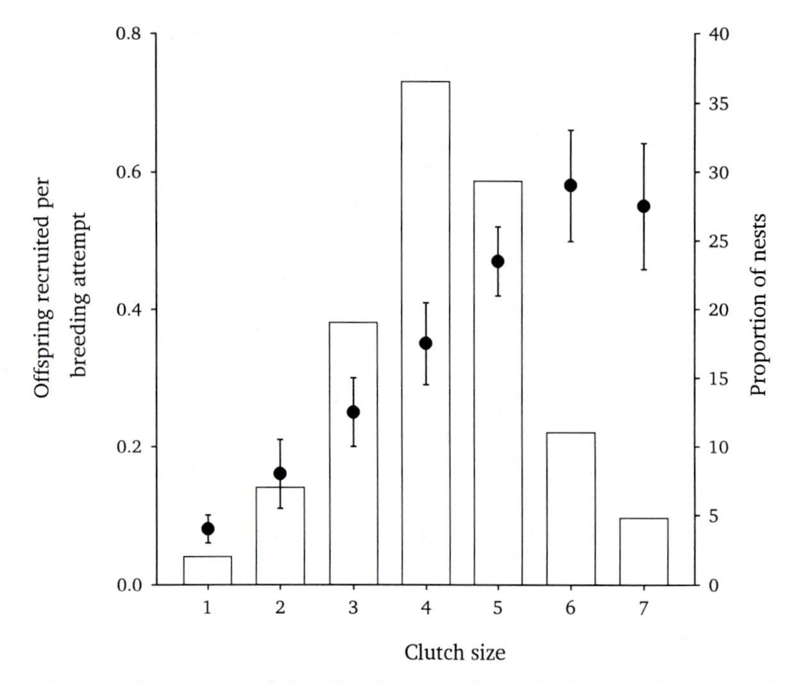

Fig. 10.2. A histogram of the distribution of clutch sizes, and estimated recruitment in relation to clutch size (curve). (Adapted with permission from Rockwell *et al.* 1987).

To examine the relationship between clutch size and adult survivorship, we investigated a bird's clutch size in year t and the probability of its return in year $t + 1$. We pooled data over sexes and years, and contrasted the number of returning adults (either resighted or recaptured) to the number that could have returned but did not or were not detected (Table 10.1). We assumed that detection probabilities were independent of clutch size in the previous year. Adult return is statistically independent of clutch size in the previous year; the apparently lower rate for clutch size 1 reflects a sample of three birds. There is no evidence that larger clutches are associated with any lower viability of the breeding adults.

A second way to to test for relationships between adult viability and clutch size involves tabulating mate changes. Nearly all pair bond breakage results from the death of one of the members of the pair (Cooke and Sulzbach 1978). Thus, the proportion of females returning to the colony in two successive years with new mates, per clutch size class, indexes their previous mates' mortality rates. Because our sample of successive total clutch sizes was small, we have presented results as a function of clutch size at hatch (Table 10.1). As with return rates, there are no significant differences in mate change as a function of clutch size.

Since there appear to be no survivorship trade-offs for individuals laying larger clutches, the expected number of recruits plotted as a function of annual clutch size (Fig. 10.2) represents the selection gradient on clutch size. Within the clutch size range laid by individuals, selection favours larger clutch sizes. Birds with the highest annual reproductive success will also have, on average, the highest lifetime reproductive success. The selection gradient shown in Fig. 10.2 is real, but the covariation of clutch size with age and intraspecific nest parasitism affects its shape. The age effect will depress the recruitment rates for smaller clutches, while nest parasitism depresses that of the larger clutches. These biases thus cancel each other out to some degree. While the shape and slope of a pure clutch size effect might differ from that shown, the directional selection for larger clutch sizes will certainly persist.

10.3 *What limits clutch size?*

Annual reproductive effort by Lesser Snow Geese at La Pérouse Bay appears to be limited by the number of eggs laid. Those females laying larger clutches of eggs fledge more young, on average. The data suggest that clutch sizes larger than six, which are rarely if ever laid by individual females, may be less productive, but a careful manipulative study would be needed to confirm this, since the hatching success of most of these larger clutches are affected by nest parasitism. Thus, selection may not favour females which lay clutches larger than six, and we can characterize the selection regime as truncated or normalizing in this range. Few females approach this apparently limiting clutch size, however. The most common clutch size falls well short of the most productive (*sensu* Charnov and Krebs 1974). Nearly all females fledge fewer young than those laying six eggs, on average, and thus we characterize the selective regime as directional favouring larger clutches within the range of two to six eggs. We found no evidence of negative phenotypic trade-offs between fecundity and viability, in common with most other observational studies (Nur 1990), as expected if individual females adjust or optimize their clutch sizes in relationship to current physiological state and/or future survival and reproductive prospects. Nevertheless, since different birds within a population apparently optimize at different clutch sizes, those producing larger clutches will have higher fitness. Birds producing smaller clutches are, in a sense, making the best of a bad job. Since there is a genetic component to the clutch size variation (heritability is not zero, Chapter 8), the directional selection documented in this chapter should lead to an increased clutch size over time. Since we don't know the genetic covariance of clutch size and other characteristics, predicting an evolutionary response by the population must be qualified (see discussion in Chapter 1). Nonetheless, the fitness payoff for laying larger clutches is so strong that we are confident in asserting that the clutch size mode should shift from four towards five or six (Figs 10.1 and 10.2).

Two issues need to be addressed as a result of our analyses. If selection favours birds laying larger clutches, (1) why is the average clutch size a third less than the most productive clutch size, and (2) why is there variation in clutch size? These questions are clearly interrelated and are based, in part, on the assumption that there exists or has existed in the Snow Goose population genetic variation that may have been reduced or even eliminated given the strengths of the directional selection that is occurring.

Explanations can be listed under four general headings that are not necessarily exclusive.

1. Evolutionary lag: directional selection is occurring at present from a formerly different local equilibrium.
2. Gene flow: directional selection at La Pérouse Bay is balanced by gene flow from populations under different selection regimes.
3. Environmental constraint: the variation in clutch size is largely due to the environment.
4. Selection on correlated traits or 'genomic environment degradation': the predicted response to selection has been negated by selection of correlated traits and/or intrinsic 'genomic environmental degradation'.

Evolutionary lag: Both the distribution and composition of the Hudson Bay Snow Goose populations have recently changed (Chapter 3). The blue and white populations merged during the present century, and there has been a considerable expansion of breeding colonies in the southern part of the breeding range, where summers are longer and weather conditions probably more benign. Thus selection may have formerly favoured a four egg clutch, and the population at La Pérouse Bay may be evolving towards a higher mode. Given a selection differential of 0.33, calculated from Fig. 10.2 (Rockwell *et al.* 1987), and a clutch size heritability of approximately 0.1, we should expect an increase in clutch size of 0.033 eggs per generation (Rockwell *et al.* 1987). Although we are dealing with overlapping generations, if we assume a stable age distribution and a mean generation time of 5 years, clutch size should have increased by approximately 0.1 egg over the course of our study. Instead, we observed the opposite: clutch size declined *c.* 0.4 eggs over that time period (Fig. 6.1). Thus, to the best of our ability to measure it, there is no response to selection in the current population.

Balancing selection and gene flow: The modal clutch size at La Pérouse Bay could be out of local evolutionary equilibrium due to gene flow from other populations, assuming that males contribute genetically towards clutch size variation (Chapter 3). There is massive gene flow between birds of the various breeding colonies, mainly because of male dispersal. If selection at other colonies favoured birds that produced mid-sized or smaller clutches, or a smaller maximum clutch size, a balance between gene flow and local selection might maintain an average clutch size lower than the most productive at La Pérouse Bay. This would also generate local genetic variance for the trait, as suggested by the heritability calculation, in the face of directional selection. Without a detailed study similar to our own from another Snow Goose colony, it is difficult to evaluate this hypothesis thoroughly. We commend such a study to future researchers! Nonetheless, we find this explanation unlikely. The available data suggest that modal clutch sizes at more northern colonies may be smaller, perhaps half an egg smaller, than those at La Pérouse Bay (for example Cooch 1961; Syroechkovsky and Kretchmar 1981; S. Johnson, personal communication). Assuming that these populations are near their genetic equilibrium, gene flow from such populations would not be sufficient to account for a modal clutch size of four rather than 6 at La Pérouse Bay.

Environmental constraint: Lack (1967) argued that clutch size in nidifugous species depends on nutrient acquisition during laying. Ryder (1970) suggested that in arctic geese, such acquisition occurred south of the breeding grounds, in both in wintering areas and staging sites on spring migration. Ankney and MacInnes (1978) showed that clutch size and the nutrient reserves of Snow Goose females arriving at the breeding grounds were positively related. For La Pérouse Bay Snow Geese, hyperphagia associated with nutrient accumulation for clutch production occurs mainly on the staging areas of southern Manitoba and the Dakotas (Chapter 4). Thus, at the proximate level, pre-laying foraging success controls clutch size. Identification of such a proximate mechanism increases our understanding of the situation, but does not in itself resolve the problems of clutch size mean and variance in relation to the selection gradient. Individuals with particularly effective foraging strategies, for example, may acquire the most nutrient, transport it to the north, and subsequently lay, hatch, fledge, and recruit the most young. If variation in any aspect of foraging strategy had a heritable component, such strategies should increase in frequency, resulting in larger clutch sizes, unless countered by gene flow or correlated counter-selective regimes.

Recognition of nutrient limitation as a clutch size controlling mechanism suggests that mean clutch sizes of Snow Geese usually are not maximal simply because most individuals most of the time are incapable of generating the level of reserves needed to lay and care for six eggs without undue risk to personal survival. As Lack suggested, for such species, reproductive effort is limited at the egg-laying stage. As outlined above, six may be a maximal clutch size either because laying more eggs is not profitable, either annually or on a life-time basis, or because storing and transporting sufficient nutrient to do so is beyond a Snow Goose's capabilities. When brought into captivity and given food *ad libitum*, however, first or second generation Snow Geese do not increase their maximal clutch sizes (M. A. Bousfield, personal communication). This suggests that normalizing selection may have operated to limit Snow Geese to maximal clutch sizes of six.

Nutrient acquisition depends on both food gathering ability and on food availability. The latter changes from year to year and varies spatially within years. If individuals can produce their maximum clutch size only when they have reached some minimum nutritional level (for example Ryder 1970; Ankney and MacInnes 1978; Findlay and Cooke 1983; Hamann *et al.* 1986), then yearly differences in food availability will prevent individuals from consistently achieving their maximum output. Similarly, local variation will produce heterogeneity within years. Over a life span, then, we expect an individual's average annual clutch size to be below the maximum it is capable of and generally less than that associated with maximum fitness. Averaging over individuals, the population's mean clutch size is thus substantially lower than the most productive one. Thus, most of the variation in clutch size should be environmental, as is the case, since at least 90 per cent of the phenotypic variance is not heritable (Chapter 8).

Despite our demonstration that larger clutch sizes are more productive, mean clutch size declined over the course of our work (Cooch *et al.* 1989). In predicting a response to selection as we did above, one assumes that the environment remains constant. Foraging habitat degradation related to increased population size and overgrazing both at La Pérouse Bay and/or at wintering or migratory staging areas probably accounts for the decline (Cooch *et al.* 1991b; Cooch and Cooke 1991; Chapter 6), and thus our failure to observe a phenotypic response given our selection gradient and heritability.

Selection on correlated traits and 'genomic environmental degradation': One might fail to observe an evolutionary response because selection on correlated traits, either genetic or environmental, cancels out the expected population response (Lande and Arnold 1983; Price and Liou 1989). It is not possible for us to analyse potential genetic covariates, but the correlated phenotypic trait often considered with respect to selection on clutch size is a trade-off against the timing of breeding. This will be examined in the the next chapter.

A final, more subtle point involves a fundamental evolutionary dynamic. Evolutionary biologists measure fitness relative to other members of a population. In one sense, this explains why mean fitness itself does not increase, by definition, despite being under directional selection, although adaptation to the environment is constantly refined. When members of a population compete for shares of a resource, phenotypes expressing particular genotypes may be more successful, and the next generation would contain an increased proportion of individuals with traits which make them more competitive in the current environment. Since the next generation's 'environment' contains a higher proportion of 'more competitive' individuals, however, the performance of those individuals may be no better than that of the current generation. In this sense, the 'environment' of succeeding generations is always automatically becoming more difficult, or, 'deteriorating' as populations evolve (Fisher 1930; Frank and Slatkin 1992). If clutch size in Snow Geese is regulated by female-female competition for nutrients, it too might be expected not to change over evolutionary time, despite continued directional selection and evolutionary change in phenotypic competitive ability (Cooke *et al.* 1990*b*).

10.4 *Conclusion*

Few female Lesser Snow Geese, in few years, lay clutches of six, which is both the maximal and the most productive clutch size produced by individuals. Maximal performance apparently is possible only under extremely good environmental conditions and/or for the best competitors for pre-laying nutrients. Individuals may optimize their lifetime performance by adjusting their annual reproductive investment, but this still provides an evolutionary advantage to those individuals with larger optimal clutch sizes. Although selection favours larger clutch sizes, we may fail to observe an evolutionary response because of changes in external environmental conditions and/or within-population competition for limiting resources.

10.5 *Summary*

1. There is considerable variation in clutch size, with natural clutches varying from two to six eggs. One egg clutches are probably partial clutches and clutches of seven or more reflect laying by more than one female.
2. Total nest failure is more frequent among one- and two-egg clutches, but most of these are probably incomplete.
3. Egg loss and hatching failure are more frequent among clutches with more than four eggs.
4. Partial brood loss increases with clutch and brood size, whereas total brood loss is higher among small broods. Overall, goslings from broods of two and three have higher than average, and broods of six lower than average survival.
5. Recruitment of young and pre-reproductive survival is independent of clutch size.
6. Overall reproductive success increases monotonically with clutch size up to a clutch size of six.
7. There is no correlation between the clutch size of a pair and their survival and return to the colony in the following year.
8. Variability in clutch size is mainly due to variability in environmental factors such as nutrient availability.
9. There has been a decline in clutch size despite directional selection favouring birds that lay larger clutches, strongly suggesting a decline in per capita nutrient availability over time.

11 *The timing of reproduction*

Ornithologists have long accepted that the seasonality of breeding by birds is driven in large measure by the timing of availability of food for young. The extraordinary breeding biology of Emperor Penguins *Aptenodytes forsteri*, which involves a 3–4 month incubation fast by males through the Antarctic winter, testifies to the extreme adaptations which have evolved to ensure that eggs hatch at a time when chicks may be provisioned (Le Maho 1977). Lack (1968) suggested that laying dates, clutch sizes, and chick growth rates were coadapted packages that ensure maximal fitness. Snow Geese start to nest as soon as snow melts on the breeding grounds, and by the end of incubation, suitable food is usually available to support gosling growth. In this chapter, we examine variation in laying date within the normal nest initiation period.

Snow Goose females laying larger clutch sizes have higher fitness, and we argued that clutch size variation reflected primarily differences in females' abilities to transform nutrients into eggs (Chapter 10). This assumes that eggs must be laid within a limited time window to have a reasonable probability of producing surviving offspring. Otherwise females would do well to delay nesting until they had accrued sufficient stores on migration to lay maximally-sized clutches on the breeding grounds. All but perhaps the best-provisioned females thus face a real trade-off between laying fewer eggs now versus more eggs later.

The timing of laying is constrained by three interacting phenomena: the seasonal phenology of the availability of potential nest sites, gosling forage plant growth, and the behaviour of conspecifics. The primary constraint is the short time available for chick growth in arctic summer. Snow Goose chicks hatching earlier in the season have a higher probability of survival and recruitment than those hatching later, apparently due to enhanced post-fledgling survival, favouring early breeding (Cooke *et al.* 1984). Competition with conspecifics may play a role in this process, but the external environment clearly limits the total time available and selects for earlier breeding. However, the earliest nesting geese are not in all ways the most productive, due to social competition and its associated predation regime. Females initiating clutches earlier or later in the season experience higher losses at the nesting stage than those initiating clutches during peak initiation (Findlay and Cooke 1982*b*; Collins 1993). This population dynamic produces a degree of normalizing selection favouring synchronous breeding. Thus, a combination of directional selection, favouring early breeding, and normalizing selection, favouring synchronous breeding, operate at different stages of reproduction in this population.

We re-examined timing of breeding with respect to clutch size to closely describe and distinguish the effects of these two modes of selection. If directional selection predominates, we predict that early nesting birds have the highest fitness for any given clutch size. If synchrony is more important, we predict different maximally productive clutch initiation dates for birds laying clutches of different sizes. Birds laying smaller clutches would have later optimal initiation dates than those laying larger clutches because by delaying laying, they avoid higher rates of early nest loss. Later initiation dates for geese with smaller clutch sizes also increases the synchrony of hatching, which may be advantageous due to predator swamping, that is, avoiding a hypothetical window of increased gosling vulnerability to predators among the earliest hatched broods. By examining the fitness consequences of variation in laying date for birds of different clutch sizes, we can assess the relative importance of directional and normalizing selection. In this chapter, we re-examine the selection gradients of various fitness components with respect to seasonal timing of laying, and document the net fitness consequences of female breeding strategies with respect to the timing and number of eggs.

11.1 *Seasonal timing of nest initiation*

Several general aspects of Arctic goose biology suggest that Snow Geese nest as early as practicable. First, geese bring the bulk of their nutrient reserves with them from more southerly locations. Thus eggs may be produced as soon as sites are available, although whatever food can be obtained at the time of laying may also contribute to the production of eggs (Chapters 2 and 6). Secondly, at La Pérouse Bay, in every year, at least some females have begun to lay eggs as soon as snow melts and secure nesting sites become available. Barry (1962) pointed out that early laying was critical to successful reproduction for geese nesting in the short arctic summers. In late seasons, the production of young was lower than in early seasons. In a delayed breeding season, young birds have actually been found frozen in the ice with incompletely developed flight feathers. Harvey (1971) reported adult Snow Geese in emaciated condition towards the end of incubation in severe seasons, and on rare occasions we have noted this at La Pérouse Bay. Finally, Arctic geese in general have shorter incubation periods than closely related temperate species, although this effect is less dramatic when incubation periods are adjusted for body size effects, a possible adaptation to the short season.

11.2 *Annual variation in laying date*

First and mean laying dates have varied between 12 May and 8 June, and 18 May and 11 June, respectively, during our study (Table 2.1). This variation reflects primarily differences in the onset of snow disappearance and flooding, which is a function of the depth of accumulated snow and the temperature regime in spring. Mean clutch sizes are lower in years with later mean initiation dates, except for the few years when birds bred immediately upon arrival, when clutch sizes were also small (Fig. 6.3; Davies and Cooke 1983*a*). It may be that the usual pattern of feeding locally while waiting for nest sites to emerge enables birds to enhance their clutch size.

Laying of first eggs takes place over a *c*. 10 day period. Standard deviations average only 2.2 days (Table 2.1), thus about two thirds of the nests are begun within a 5 day period, and more than 90 per cent of nests are usually initiated within 4 days of the mean. The initiation date distribution usually shows a slight positive skew (Fig. 11.1, Findlay and Cooke 1982*b*). Differences in the shape of the distribution are largely determined by the pattern of snow disappearance. When snow disappears slowly, as in 1973 and 1978, potential nest sites emerge slowly and a negative skewness is found. If there is no snow at all, as in 1977 and 1985, there is a slight positive skew characteristic of many nest establishment regimes (Gochfeld 1980), and expected when there is directional selection for early nesting. While the distribution is usually unimodal, we have had some seasons in which severe weather during initiation lead to a prolonged and bimodal onset of laying (for example Fig. 11.1: 1989).

11.3 *Intra-seasonal decline in clutch size*

Seasonal declines in clutch size have been widely, though not universally, documented in birds (Klomp 1970). Despite our short laying season, this is one of the strongest and most robust patterns in our data set. Birds laying larger clutches initiate nests earlier than those

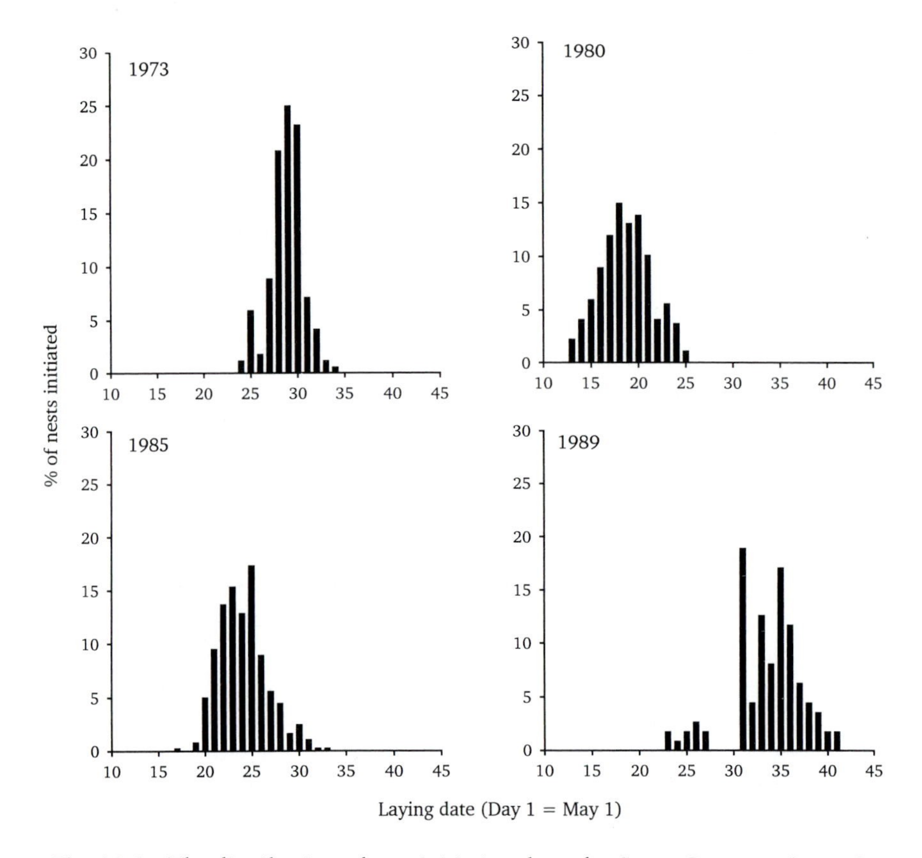

Fig. 11.1 The distribution of nest initiation dates for Snow Geese nesting at La Pérouse Bay in selected seasons, showing a range of different patterns.

laying smaller clutches in every year of our study (Fig. 11.2). Although slopes vary annually, this phenomenon occurs in early and late seasons over a range of environmental conditions (data for 1973–1990 except 1979; n = 3594 nests, within-year relative initiation date by year interaction: f = 4.31, d.f. = 16, $P < 0.0001$).

The change in average clutch size over the season is primarily caused by differences in the timing of laying of birds producing different clutch sizes, rather than being driven by other processes which might change the number of eggs in a nest during or after laying. Two such processes, differential partial nest predation and nest parasitism, contribute in a minor way to the pattern. Jean Hamann, over two seasons, collected 98 female geese, whose nests had been monitored daily, immediately after their clutch completion (Hamann *et al.* 1986). The number of post-ovulatory follicles, which show the actual number of eggs laid by a particular female, declined

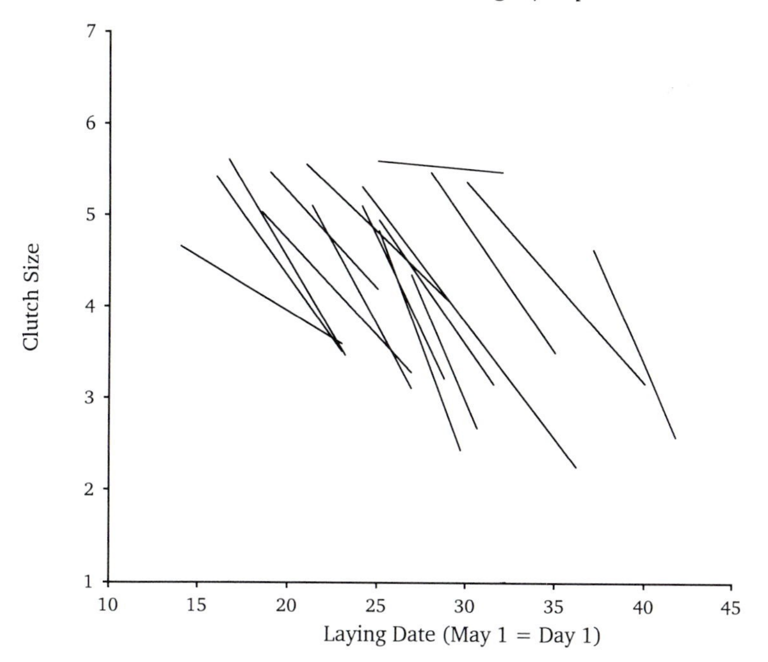

Fig. 11.2 Slopes of the seasonal decline in clutch size of Snow Geese nesting at La Pérouse Bay with respect to initiation date, 1973–1992.

as the season progressed with a slope similar to that of the observed clutch size. Thus birds laying smaller clutch sizes really did lay later. On the other hand, within one year's sample, females laying prior to the mean initiation date had significantly more eggs in their nests than post-ovulatory follicles, while those laying later than the mean had fewer eggs in their nests than they had actually laid (a mean of 0.3 more eggs than follicles in 20 pre-peak nests versus 0.4 more follicles than eggs in 27 post-peak nests, J. Hamann, unpublished data). The former result is consistent with the pattern of higher nest parasitism rates in nests initiated earlier in the season (Lank *et al.* 1989*a*), while the latter may reflect continuation clutches initiated in second nests following nest loss or earlier parasitism (Ganter and Cooke 1993). These processes will somewhat increase the slope of the decline, but are not fundamentally responsible for it.

Lemieux (1959) proposed that the seasonal reduction in clutch size of Greater Snow Geese might be due to the fact that young females, laying smaller clutches, nest later than adults. If this explanation alone accounts for the seasonal reduction in clutch size, there should be no seasonal decline in clutch size within individual age classes. In fact, the decline occurs within each age class, with similar slopes as the population as a whole (Hamann and Cooke 1989).

Ryder (1970) proposed that geese which are not able to find a nest site immediately used up some of the reserves which would have been allocated to eggs, for their own maintenance, resulting in smaller clutches later in time. This hypothesis addressed both the intra-seasonal decline and the pattern of clutch size variation among seasons. It assumes that feeding opportunities are highly restricted at the nesting colonies during the laying period. If females satisfied their maintenance requirements by drawing nutrients from reserves already allocated to egg production, they should increase the number of atretic, rather than post-ovulatory, egg follicles. If the time between arrival on the colony and nesting is on average greater for later nesting females, early laying females should have fewer atretic follicles than those which lay later. Hamann's data (1986) show just the opposite: early nesting birds in 1981 had more, not fewer, atretic follicles than those which laid later (0.30 ± 0.13 versus 0.07 ± 0.05). Apparently birds breeding early in the season had over-committed themselves to a greater extent than those breeding later.

All these findings suggest that both clutch size and laying date relate to some quality of the bird, probably related to nutrient acquisition, storage, or mobilization. As with clutch size, individual Snow Geese both increase their clutch sizes and advance their laying dates as they age at least up to the age of 5 years (Hamann and Cooke 1987), suggesting that changes in the two traits are manifestations of a single underlying physiological state.

Drent and Daan (1980) pointed out that 'In ultimate terms, the adaptive value of the seasonal decline in clutch size depends on the decline in returns, i.e. survival probability of the nestlings as the season progresses.' Why should changes in survival probability lead to a seasonal decline in clutch size? If it is advantageous to lay early should it not be equally advantageous for those laying small clutches as for those laying large? Most researchers (for example Drent and Daan 1980; Perrins and Birkhead 1983) have resolved this problem by assuming that there is some underlying characteristic of individual birds which allows them both to (1) lay a clutch of a particular size and (2) to start nesting at a particular time. If the 'quality' of a bird influences both these attributes, a bird of good quality can lay a large clutch early, a bird of poor quality will lay a small clutch late. Clutch size and laying date, which from the population geneticists' point of view can be analysed as separate characters, may in fact be different manifestations of a common underlying character.

In the Weaverbird (*Quelea quelea*), for example, protein stores of individual birds provide a proximate mechanism for the determination of both clutch size and laying date (Jones and Ward 1976).

Drent and Daan (1980) argue that birds vary in their ability accrue nutrients; those which are efficient feeders can acquire nutrients rapidly which allows them to lay a large clutch and lay early. Gauthier *et al.* (1992) showed that Greater Snow Geese left the staging areas of the St Lawrence River when they had reached a threshold body weight. Some birds reached this weight rapidly and left early, others took considerably longer before reaching the weight and departing.

A model by Reynolds (1972) to explain timing of nesting in Mute Swans (*Cygnus olor*) provides a simple proximate mechanism by which timing and clutch size may be determined (Fig. 11.3). This model postulates an internal condition threshold for laying which declines as the season progresses. As individual birds reach the threshold, breeding is initiated and a clutch of a particular size laid down. Birds which reach the threshold quickly lay a large clutch early, those which reach the threshold slowly lay a small clutch later. Perhaps such a threshold is mediated through hormone levels, triggered by some environmental '*Zeitgeber*'.

However useful a model such as Reynolds' is, it does not come to grips with the ultimate mechanisms. If there is a seasonal decline in

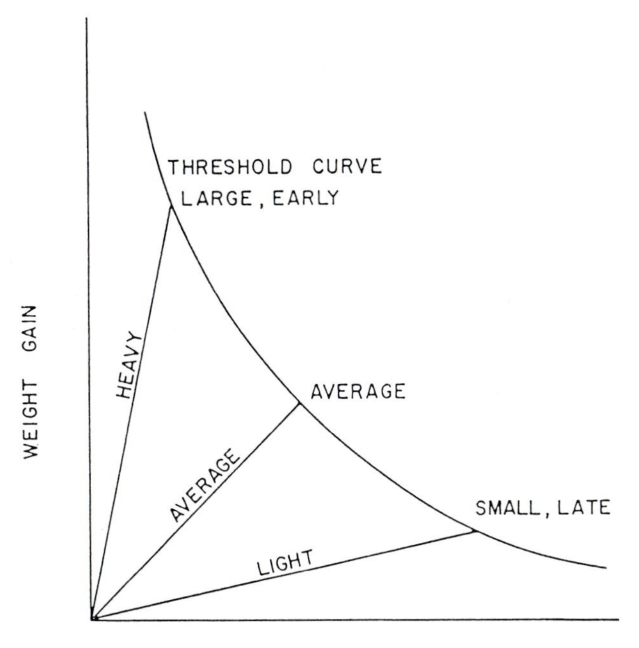

Fig 11.3 'Reynold's model' of time-specific clutch size and timing thresholds (from Hamann 1983).

clutch size, what selective pressures have led to this pattern? Are there indeed advantages to early nesting, beyond those due to the covariance between clutch size and laying date? Is there some trade-off between clutch size and timing of nesting? Is early nesting advantageous to birds of all clutch sizes or is late nesting the most advantageous to those birds laying small clutches? In this chapter we document patterns of laying and the effect of timing on different components of fitness. We look at timing of laying for each clutch size separately, in order to assess the relative importance of clutch size and laying date to overall fitness, and then compare among clutch sizes to assess the potential for trade-offs between clutch size and time.

11.4 *Fitness consequences of timing of laying*

The goal of this analysis is to calculate and compare the recruitment success of nests as a function of initiation date and clutch size. We restricted the analysis to clutch sizes 2–6, since larger clutches are virtually all laid by more than one female (Chapter 5). For the most critical analyses, data from 4326 nests with known laying date were available for the period 1973–1987.

Snow Geese lay only one clutch of eggs, but they may establish second nests and lay continuation clutches if a nest site is lost early in the laying period (Ganter and Cooke 1993; Chapter 2). Thus, our sample of small, late clutches, especially, may not in every case reflect the original laying date for that female. However, in general, the laying of the first egg in a nest is a good measure of the timing of reproduction which is not confounded by the problem of true second clutches.

Within clutch size classes, age has little or no effect on nest initiation date. In the sub-sample of nests of known age birds, initiation dates for younger birds (ages 2–3 years: $n = 81$) were only 0.1 days later than those of older bird (ages 4+ years, $n = 283$), controlling for clutch size, a small and not significant difference (2-way ANOVA, F for age term predicting laying date = 0.01, d.f. = 1, $P = 0.94$; age × clutch size interaction term: $f = 0.66$, d.f. = 4, $P = 0.62$). This lack of an age bias allowed us to analyze the larger sample of nests from birds of unknown age without considering potential age biases in the sample.

Nests with known initiation dates were assigned laying dates relative to the annual mean for their clutch size. Nests laid more than 4 days on either side of the annual clutch size mean were merged into the extreme categories; these nests (n = 146 early, 172 late) averaged −4.6 days (±0.88 SD) from the clutch size and year specific means for early nests and +4.5 days (±0.57 SD) for late nests. In the sample pooled over years, the mean relative laying dates of nests of clutch size 2 through 6, with day 0 being the global mean, were: 1.87, 1.01, 0.05, −0.79, −1.52, respectively (least-squared means from an ANCOVA model of laying date as a function of year and clutch size). Thus on average, a bird laying a two-egg clutch started to lay only 3.4 days later than one laying a six-egg clutch.

For each fecundity component of fitness in each year, we calculated clutch size-specific daily mean values. These daily values were combined to produce composite estimates of hatching success by clutch size class of nests initiated on each day of each year. These annual values were themselves averaged to produce an overall picture of the long term selection regime. In the figures presented below, each clutch size and date specific point is a mean based on 8–15 years, except where otherwise noted. We refer to these values as global estimates. Data for all components of fitness were not necessarily available for all combinations of clutch sizes and dates in all years.

We calculated fecundity fitness components for each class of nests from incubation through hatching in our standard manner (Chapter 4), but we developed novel ways of handling pre-incubation failure and post-hatch survival and recruitment. Since the potential clutch size of a nest which fails prior to incubation is unknown, as are the fates of eggs remaining to be laid by females that fail at this stage, we bracketed the potential effects of pre-incubation failures under a set of assumptions detailed below. For the post-hatch components, we also adopted a new methodology to avoid bias. Our measure of total brood loss may be biased with respect to time because families hatching earlier in the season seem a priori more likely to disperse away from the area where we capture and band birds. To minimize bias, we calculated a recruitment index which transits directly from goslings marked at nests to detected recruitment. By doing so, we sacrificed the potential for dissecting the stage at which differences might occur, but in this case we had no other practical methodology for estimating the relative post-hatch fates of goslings.

11.4.1 Pre-incubation nest failure

A surprisingly high proportion of the nests we find are abandoned or predated before incubation begins. Of 4326 nests with fewer than seven eggs found at the one egg stage, 26.8 per cent fail during laying.

Table 11.1 The percentage of nests failing at the pre-incubation stage containing one to six eggs

Clutch size	Percentage of sample
1	73.8
2	16.6
3	5.9
4	2.6
5	0.7
6	0.3

n = 1160 nests.

Seventy-three per cent of these failures occur when only one egg is in the nest, and over 90 per cent with two or fewer (Table 11.1). The mean annual pre-incubation nest failure rate throughout the laying season is shown in Fig. 11.4. Birds nesting early have a significantly higher failure rate than those nesting in the middle of the season, and rates may rise slightly towards the end of the season, although the data are noisy as sample sizes per year decrease. This seasonal pattern occurs in all but two years, despite large differences in the absolute annual frequencies of pre-incubation failure. In one of these exceptional years, a higher failure rate at mid-season occurred due to depredation caused by the arrival of a herd of Caribou at mid-laying.

We face two problems when trying to integrate pre-incubation failure into our calculation of fitness as a function of timing of breeding and clutch size. First, we do not know the potential clutch size of the failed nests. We have dealt with this by assuming that within each day, pre-incubation failure is independent of total potential clutch size. Under this assumption, we can apportion each day's failed nests among clutch size classes in proportion to the clutch size distribution of successful nests started that day. We restrict this apportionment to clutch size classes as large or larger than the clutch size in the failed nests. This procedure estimates the number of females of each potential clutch size which failed during laying each day.

Our second problem is that we do not know the fates of eggs which a failed female has yet to lay. Birds which lose their nests early in the breeding cycle are seldom seen by the field workers, and thus resightings of the females involved is seldom possible. Such females may relay or lay parasitically (Chapter 2, 5), as suggested by the relationships between eggs and post-ovulatory follicles in Hamman's data, outlined above. We have windowed the potential effect of residual eggs by running our analyses under best and worst case

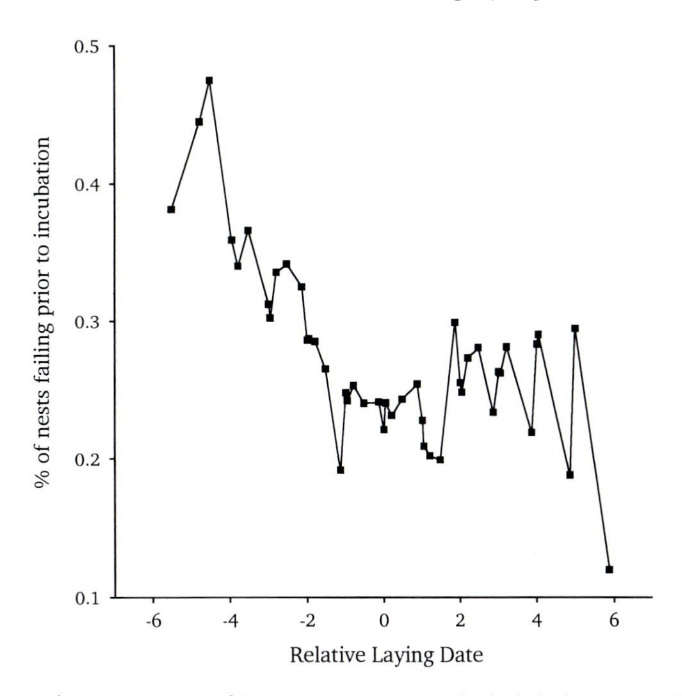

Fig 11.4 The proportion of Snow Goose nests which failed prior to the onset of incubation as a function of nest initiation date. Each point is a sample of nests initiated on a particular date in one year, plotted relative to its annual mean.

scenarios. For the best case, we assume that all of a females' eggs are incubated, whether in her own continuation nest or as parasitic eggs. Under the total incubation assumption, a female's clutch size is unaffected by nest loss, pre-incubation failure does not occur again, and that her remaining eggs have the same fates as other eggs of a particular clutch size initiated on the original nest initiation date. For the worst case, we assume that residual eggs are worth nothing. Reality must lie in between, but we believe that it is strongly biased towards total incubation.

Under these assumptions, we can estimate the effects of pre-incubation failure on birds which originally would have laid clutches of two through six on any particular date by distributing failed eggs among clutch sizes in proportion to the clutch size distribution of relevant successful nests established that date. Suppose that 25 of 100 nests initiated on a particular day fail with one egg, and that 20 of the 75 competed successful nests begun that day had a clutch size of two. Under the total loss case, the adjusted clutch sizes are simply: 40 eggs / 26.7 clutches = 1.5 eggs per clutch, a 25 per cent reduction from two eggs, directly reflecting the 25 per cent rate of pre-incubation failure that day. For the total incubation case, the expected mean clutch size for females initiating 2-egg clutches that day would be calculated as: $(20 \times 2 + (20/75) \times 25 \times (2-1)$ eggs / $(20 + (20/75) \times 25)$ clutches 46 eggs/26.7 clutches, or 1.7 eggs per clutch. For larger clutch sizes, additional terms are added for nests failing with two or more eggs.

Global estimates of pre-incubation failure rates and adjusted clutch sizes under both pre-incubation failure assumptions are shown for each clutch size in Fig. 11.5. Within each clutch size, failure rates decrease as the season progresses, driven by the pattern shown in Fig. 11.4. The effect is strongest for the five and six-egg clutches because they are disproportionately laid earlier in the season, while the failure rates for two- and three-egg clutches do not have significantly negative slopes. If re-laying occurs, the negative effect of pre-incubation failure in early nesters is significantly mitigated.

In summary, relatively high rates of pre-incubation failure penalize birds which nest early. Although large numbers of birds lose one or two eggs at the onset of laying, the cost of such failure may be mitigated if they lay the rest of the clutch elsewhere. The intensity of the effect depends on the subsequent fate of birds which lose their nests prior to incubation.

11.4.2 Nest failure during incubation

A sample of 3165 incubated nests were available for determination of components of fitness through incubation. Nest loss during the incubation period is far less frequent than pre-incubation loss, but its consequences are probably more severe insofar as no further nesting can take place if it occurs. Nest loss may occur due to nest abandonment by a female, perhaps due to emaciated condition, or as a result of nest predation by Caribou, wolves, foxes, or large groups of Herring Gulls (Chapter 2). Nest failure is significantly higher among late laying females (Fig. 11.6). This general pattern may well occur because birds which lay late are in poorer condition and may be unable to replenish their reserves before completing the incubation process. In a few years (for example 1972) one or two females were discovered dead on the nest when a late snowstorm hit the colony late in the incubation period.

11.4.3 Egg and hatchling survival

A sample of 2779 surviving nests were available to determine the fitness components of egg and hatching survival. Egg survival rates in successful nests are high throughout the season, but birds nesting

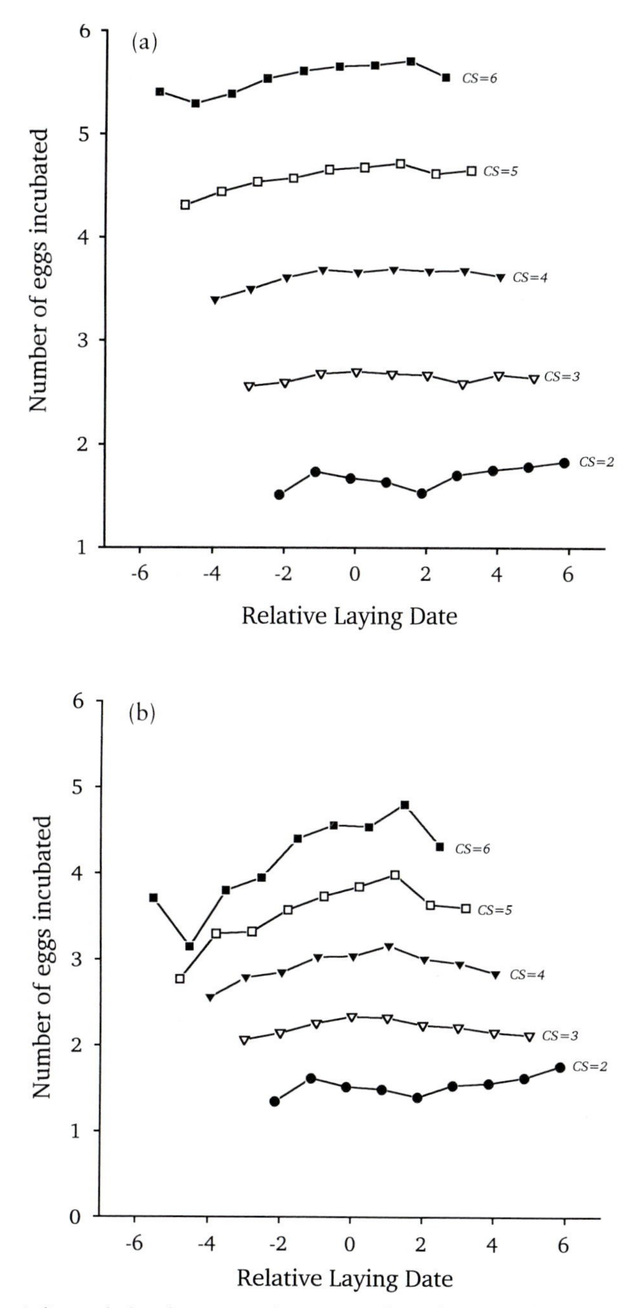

Fig. 11.5 Adjusted clutch sizes at the onset of incubation relative to laying date for Snow Goose females laying two to six eggs, allowing for losses at the pre-incubation stage. Graph (a) assumes that all residual eggs of a female's clutch following partial nest loss are incubated in a different nest. Graph (b) assumes that all residual eggs are lost.

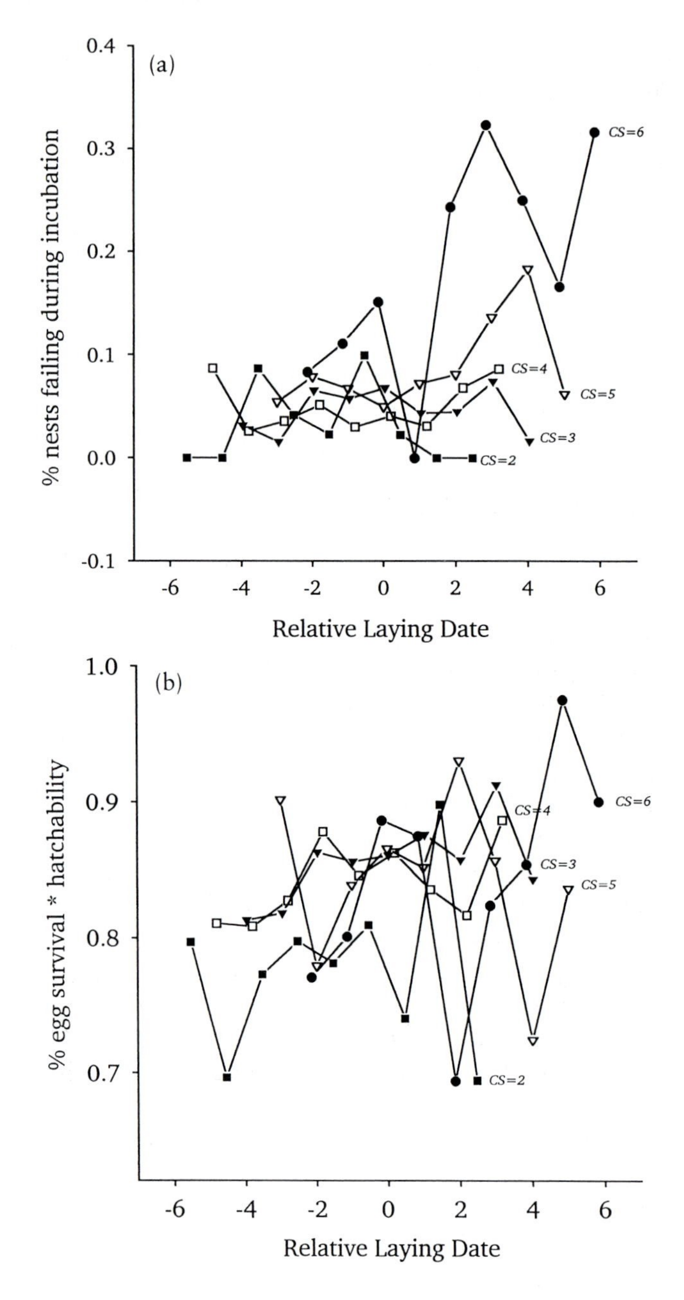

Fig. 11.6 Total nest failure rates (a) and composite egg survivorship and hatching success (b) as a function of relative nest initiation date for Snow Goose nests at La Pérouse Bay. Each point is a mean of annual means (see text).

early were somewhat more likely to lose eggs than those nesting later. There were no yearly differences detectable in the pattern.

Hatching success was also relatively high throughout the season and for all clutch sizes, but losses were again somewhat higher amongst early laying birds. Figure 11.6 presents the global estimates of the product of egg and hatching survival for different clutch sizes throughout the season.

11.4.4 Composite success through hatching

We generate two composite measures of the expected number of goslings leaving nests from clutches of a given size initiated on each day, one estimate for each assumption about the fate of residual eggs following pre-incubation failure (Fig. 11.7). The estimates are calculated within initiation dates and years as:

total loss: $GLN_1 = TCL \times (1 - PIF) \times (1 - TNF) \times P1 \times P2$
total incubation: $GLN_i = TCL_{pif} \times (1 - PIF) \times (1 - TNF) \times P1 \times P2$

where TCL_{pif} is the adjusted clutch size under the assumption that residual eggs are incubated, and the other acronyms have their standard definitions (Chapter 4).

Under the incubation assumption, the expected brood size leaving the nest increases slightly among the larger clutch sizes, and is uniform across the season for the smaller clutch sizes laid later in the season. Under the total loss assumption, larger clutch sizes are disadvantaged early in the season, and perhaps also suffer towards the end of the season. In either case, thus far through our analysis, there is no advantage to early nesting for any clutch size. The advantages of nesting at mid-season found by Findlay and Cooke (1982*b*) are weaker, in this re-analysis, primarily because our treatment of pre-incubation failure differed markedly from theirs. Nonetheless, a degree of normalizing selection on laying date, with both early and late nesters somewhat disadvantaged, may operate at this stage of the reproductive cycle. Before we conclude that it is advantageous for birds to lay in synchrony, however, we must follow the goslings through to the recruitment stage. Does the fact that more goslings leave the nests of mid-season layers mean that more are recruited into the breeding population?

11.4.5 Recruitment

For most of the traits considered in this book, we measured post-hatch survival and recruitment using fledgling success, total brood loss, and recruitment components outlined in Chapter 4. For timing

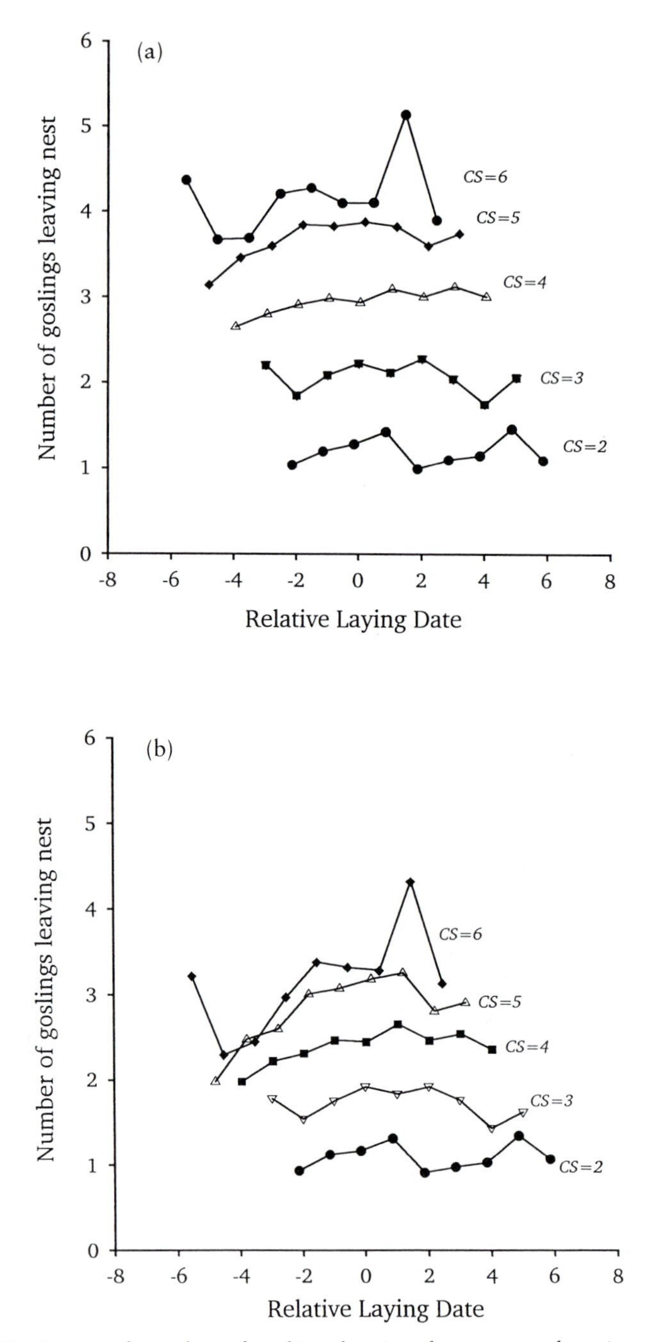

Fig. 11.7 Expected number of goslings leaving the nest as a function of relative nest initiation date and clutch size for Snow Goose nests at La Pérouse Bay. (a) assumes that residual eggs from females whose nests fail prior to clutch completion are laid in other nests. (b) assumes that residual eggs are lost.

of breeding, previous work suggested that broods hatching earlier in the season had a lower probability of being re-encountered in post-fledgling banding drives, presumably because they had more time to travel away from the breeding colony (Cooke *et al.* 1984). A bias in total brood loss estimates would seriously distort our results. In place of these three components, we calculated a single recruitment index (*RI*) based on the proportion of tagged goslings leaving the nests each day of hatch which were eventually re-encountered at the colony at ages 2 years or older. By using hatch dates to calculate *RI*, rather than laying dates, we greatly increased the number of tagged goslings contributing towards the index, since 'non-intensive' nests with unknown laying date could be included in the sample (Chapter 2). The hatch date *RI*s may then be appended to the sample of nests with known laying dates on which the rest of the analysis had been based to calculate laying-date specific relative recruitment rates, as detailed below.

There are three problems with calculating a recruitment index based on webtags. First, the annual *RI*s calculated were pooled over brood sizes. Since goslings in larger broods grow more quickly, controlling for hatch date (Cooch *et al.* 1991*a*), they may have a higher survivorship rate, which may complicate our subsequent interpretation of the results, but we have insufficient data to subdivide the calculations by brood size and year. Second, goslings banded in their natal year have higher resighting probabilities because of their leg bands. Computing an index based on raw re-encounter rates of web tagged goslings will thus be biased towards birds banded in their natal year, thereby potentially over-representing later hatching broods. We eliminated this bias by restricting re-encounters to those tallied in banding drives, when feet are examined for web tags. The third problem is that goslings banded in their natal year have a decreased chance that their identity will be lost. The probability of losing both metal and plastic bands prior to re-encounter as a recruit is negligible (Seguin and Cooke 1983), but 49 per cent of geese lose legible web tags in their first year, and about a 76 per cent in a random sample of older birds (Seguin and Cooke 1985). To minimize this detection bias, we computed an *RI* which weighted re-encounters by the estimated detection probability at the gosling's age when first banded, thereby compensating for the lower detection rate of goslings not banded in their natal year. The *RI* is obviously far lower than the real recruitment rate, since our detection of recruits is incomplete.

Recruitment rates decline with hatch date, falling about 20–30 per cent from the start to the end of the season (Fig. 11.8). This

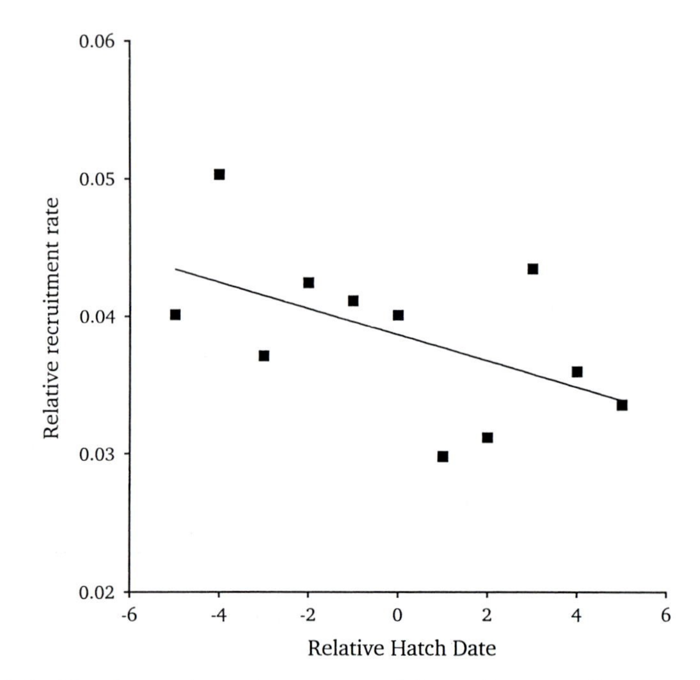

Fig. 11.8 The detected recruitment rate of Snow Goose goslings hatching on different relative hatch dates at La Pérouse Bay. Each point is a mean of annual mean values (see text).

component of fitness provides a strong advantage for females laying and goslings hatching earlier in the season.

11.4.6 Composite estimate of relative recruitment

We complete our analysis by applying hatch date specific recruitment indices to each known initiation date hatching nest (n = 2736), and computing two composite relative expected recruitment rates (ER) for each clutch size and initiation date sample per year, one under each assumption about the fate of residual clutches following pre-incubation failure. Assuming total loss,

$$ER_1 = GLN_1 \times \text{mean } RI \text{ per day.}$$

Assuming incubation,

$$ER_i = GLN_i \times \text{mean } RI \text{ per day.}$$

Global estimates of both relative recruitment rates are shown in Fig. 11.9. The higher rates of recruitment for nests hatching earlier in the season more than offset the disadvantages of early nest loss, producing higher net relative recruitment rates for earlier nesting

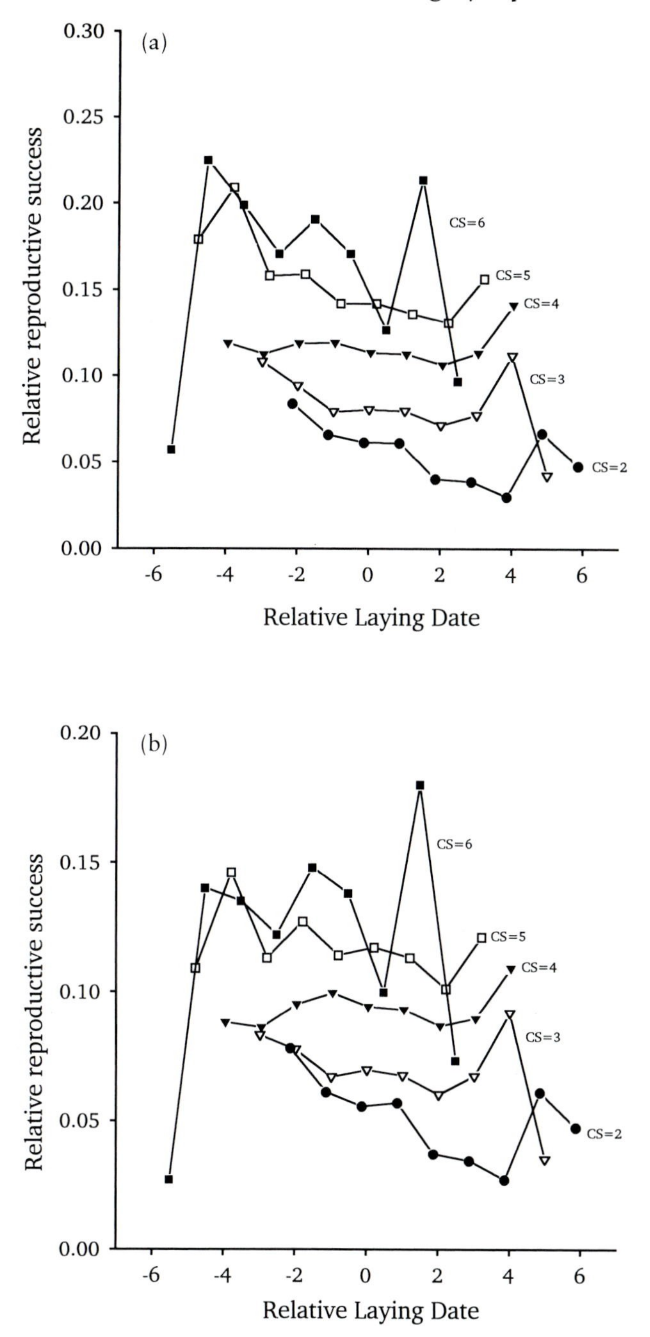

Fig. 11.9 Expected relative number of recruits as a function of relative nest initiation date and female clutch size for Snow Goose nests at La Pérouse Bay. (a) assumes that residual eggs from females whose clutch fails prior to completion are laid in other nests. (b) assumes that residual eggs are lost.

birds, assuming incubation of residual eggs in failed partial clutches. Birds laying four-egg clutches shows no seasonal trend. The data for six-egg clutches are the noisiest, as expected from their small sample sizes, but the leading, exceptionally low point deserves further consideration. Birds laid four or more days prior to the mean of all six-egg clutches in only three years, while all other points in the graphs are means of 8–15 years of data. It may be that the very earliest nests are in fact less productive, but it is also possible that this point is unfairly compared with others on the graph. If we assume that all residual eggs are lost, there are no strong trends except within clutch sizes 2 and 3, where earlier clutches have higher recruitment rates. We believe that birds generally do re-lay if their first clutch is lost prior to incubation, based on limited direct observation of re-nesting (Ganter and Cooke 1993), the prevalence of nest parasitism (albeit with somewhat lower probabilities of hatching success (Lank *et al.* 1990)), and analyses of egg sizes and sequences (Williams *et al.* 1993*a*; Collins 1993; Chapter 12). Assuming that 75 per cent of residual eggs are incubated, we have summarized our best guess of the relationships between laying date, clutch size, and expected fitness in Fig. 11.10. For four out of five clutch sizes, earlier breeders are more productive, and no seasonal pattern is found for the fifth.

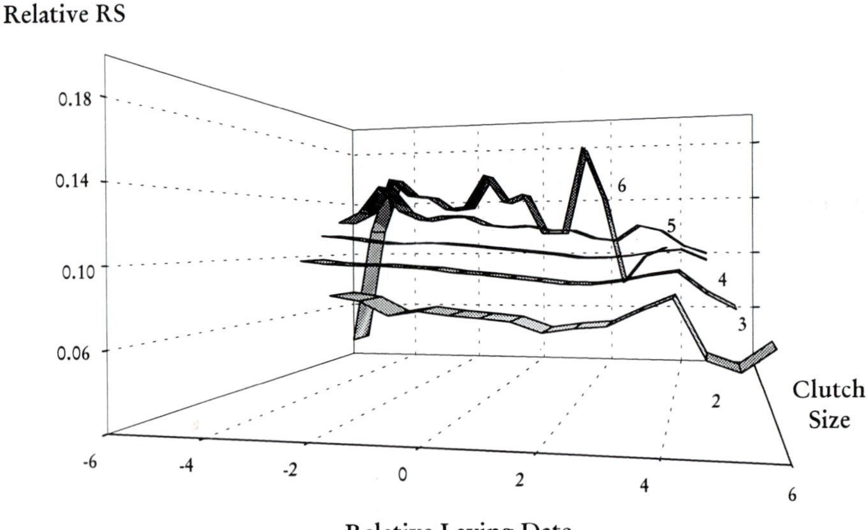

Figure 11.10 Expected relative number of recruits as a function of relative nest initiation date and female clutch size for Snow Goose nests at La Pérouse Bay. Values assume that 0.75 of the residual eggs from females whose clutches fail prior to completion are laid in other nests.

11.5 *Do Snow Geese trade-off clutch size and timing of breeding?*

Within clutch size classes, there is strong directional selection favouring early laying, except for clutches of four, which show little seasonal trend. In general, geese benefit by laying a clutch of a given size as early as possible. Cooch *et al.* (1991*a*) provided a functional explanation of why early laying might be advantageous. Goslings hatching earlier in the season grew more rapidly than later hatching goslings, which may enhance their pre- or post-fledgling survival probabilities (Chapter 13). Despite indications that synchrony may be advantageous during the laying and incubation period (Findlay and Cooke 1982*b*; Collins 1993), the net advantage usually accrues to those birds which nest early and with a larger clutch size. The directional selection imposed by the selection gradient on recruitment as a function of laying date more than offsets the higher risk of egg losses experienced by early nesters, under our best guess at the probable fates of the remainder of the clutch.

There may be no way for the earliest laying geese to avoid relatively high rates of nest loss. Birds avoid laying in snow, thus the earliest nesting time is limited by habitat. Like penguins standing at the edge of seal infested waters, however, someone has to go first. Collins (1993) carefully examined the causes of partial and total nest loss among Snow Geese at La Pérouse Bay, and argued that the synchrony of nesting was consistent with models of predator swamping, and that the first nests were intrinsically more vulnerable to predators. The mechanism of early nest loss is thus partially intrinsic, determined by intra-specific social behaviour, as opposed to being extrinsically driven by weather or predator pressure alone. Thus, there may be no pay-off for the earliest birds to wait until later in the season.

Both early laying and larger clutch size enhance overall recruitment, but for most females, one can be gained only at the expense of the other. This was nicely illustrated by Drent and Daan (1980, Fig. 6(a)). What should the behavioural strategy be for females which appear to be constrained from laying maximal clutch sizes as early as possible, probably due to insufficient nutrient reserves, as discussed in the previous chapter? The estimates in Fig. 11.10 suggest that birds might achieve equivalent fitnesses either by initiating their clutch early, with a clutch size of n, or waiting and breeding later with clutch size $n + 1$. For all clutch sizes, the recruitment estimates for birds laying late in the season approach those of birds laying $n - 1$ eggs at the start of the season. Such adjustment could be an essential part of females' breeding strategies if they could change their clutch size over the 10 days which bound the average initiation date season. Drent and Daan (1980) posed the problem this way:

> Let us reduce the female's problem of determining clutch size to her decision to make one extra egg or not. When she has reached a condition sufficient to lay x eggs, the curves for the numbers of offspring surviving from clutches with size x and size $x + 1$ will ultimately determine whether she should delay egg laying until she has attained the condition needed to produce an clutch one egg larger. A finite time is needed to produce an extra egg and there must be a trade-off of time against clutch size.

Breeding females commonly prepare to develop more eggs than they lay (Hamann *et al.* 1986). Using data on female body condition and clutch size published by Ankney and MacInnes (1978), we can estimate the apparent cost of an additional egg in terms changes in body mass, a protein index, and a fat index (Lank *et al.* 1990; Table 11.2). To assess the potential for physiological or behavioural tactical flexibility, the remaining question is: how long might it take for a Snow Goose to obtain the nutrients necessary to develop a follicle rather than having to abort the process? A female's timing and clutch size decisions are determined not only at the breeding grounds, but also in the last stages of migration. When a flock leaves, should an individual stay and fatten or leave and breed? Data on the time course of changes in body mass, protein and fat reserves of Snow Geese during the last stages of spring migration, collected and analyzed by Alisauskas (1988), are presented in Table 11.2 and converted into potential egg equivalents. During the first 10 days of May, an average female Snow Goose increased her body mass, protein, and fat at rates sufficient to produce 0.5, 0.8, or 2.0 additional eggs, respectively, assuming that each measure is limiting. Thus the 10 days which span the laying season in an average year translate into the time a female needs to make from half to two eggs. Since eggs come in integral

Table 11.2 Estimation of the time needed to obtain nutrients to make eggs. Cost per egg estimated by Lank *et al.* (1990) from data in Ankney and MacInnes (1978). Acquisition rates taken from Alisauskas (1988, Fig. 6.7)

Variable	Cost per egg (g)	Acquisition rate in May (g)	Potential eggs per 10 day laying period
Body mass	200	10/day	0.5 eggs
Protein reserve	12.8	1/day	0.8 eggs
Fat reserve	56.8	11.5/day	2.0 eggs

units, the time needed to make even half an egg may be sufficient to push a female into the next clutch size category. Obtaining equivalent fitness by trading off pre-laying nutrient acquisition time against clutch size, as hypothesized by many, appears to be a real ecological option for female Snow Geese.

Females with fewer reserves may delay nest initiation while continuing to feed. The size of a bird's clutch may affect its probable laying strategy. Birds laying the largest clutches accrue the greatest success, despite the costs of early nesting, by early laying. Those with fewer reserves may well do better to delay initiation until after the most vulnerable period at the start of the season. Since continuation nesting prospects decline, females with fewer eggs may adopt a less risky initiation date strategy.

Our results emphasize that natural selection will operate jointly on laying date and clutch size. While favouring birds capable of laying early and producing large clutches, a range of equally productive timing and clutch size options are available to birds in less than maximal condition. Selection should act on those genetic changes which enhanced both characters, and we would expect relatively low, and similar genetic contributions to the variance of the two characters, as was found (Table 8.1). As with clutch size, it seems that variation in some environmental factor accounts for much of the variation in laying date, and a bird's condition probably plays a major role determining both.

11.6 *Summary*

1. There is much annual variation in the onset of laying, which coincides closely with snow melt.
2. Nest initiation in Snow Geese is highly synchronized and shows a slight positive skew.
3. Clutch size declines with increasing laying date in all years. This decline occurs not only in observed clutches but also for real clutches as detected through examination of post-ovulatory follicles.
4. Early nesting birds suffer significantly higher pre-incubation failure than late nesting birds. Birds which fail during laying probably continue to lay the rest of the clutch in a new nesting location or become nest parasites.
5. Total nest failure and partial clutch loss are somewhat higher among nests initiated later in the season, but trends in these components contribute little to overall seasonal differences.

6. Through the hatching period, the relative productivity of early nests is similar to or lower than those initiated later, within clutch sizes, if most eggs from females who lose early incomplete nests are laid elsewhere. If most of such eggs are lost, a pattern of normalizing selection, with highest values for birds initiating during peak laying is found.
7. Per capita recruitment of offspring into the breeding population is highest among early nesting birds and lowest among late nesters.
8. Overall, selection favours those birds which lay and hatch earliest, despite some disadvantages of early laying.
10. There is low repeatability and heritability of laying date suggesting that most differences in laying date among females are due to non-genetic differences.
11. In general, an early nesting bird with clutch size n will recruit as many offspring into the breeding population as a late nesting bird of clutch size $n + 1$. This suggests a phenotypic trade-off between laying date and clutch size.

12 *Egg size*

Egg production by females appears to be the rate-limiting step in Snow Goose reproduction. This chapter focuses on the phenotypic and genetic patterns of egg size variation in the Snow Geese of La Pérouse Bay and develops a model to account for them. Snow Geese are under strong directional selection for larger clutch size, and nutrient reserves, mainly accumulated during the spring migration, determine in large part the number of eggs laid (Ankney and MacInnes 1978). If birds are allocating limited nutrient reserves to egg production, one might imagine that egg size as well as clutch size would be affected. By adjusting egg size, a fixed amount of reserves could be apportioned either to x eggs of a certain size, $x - 1$ larger eggs, or $x + 1$ smaller eggs. In theory, then, birds could increase their clutch size, as favoured by selection, without changing the proportion of their reserves allocated to reproduction, by laying smaller eggs. On the other hand, smaller eggs might produce smaller goslings with a lower chance of survival, at least in some years (Ankney and Bisset 1976). What do our data from Snow Geese show?

At the population level, Snow Goose eggs vary greatly in size. The largest eggs measured by Ankney and Bisset (1976) at the Snow Goose colony at McConnell River NWT were 67 per cent heavier than the lightest. We have found an even larger range at La Pérouse Bay: the smallest egg which hatched had a fresh weight of 86 g. compared with 166 g. for the largest, which was 95 per cent heavier. The mean egg weight of 22 562 eggs was 124.5 g, with SD of 9.4 g. The median egg weighed 124 g, but the mode was only 120 g. The distribution was thus not normal ($D = 0.03$, $P < 0.01$), but instead showed more birds laying heavier-than-average eggs. We conclude that a strategy of making more, smaller eggs, could at least produce viable eggs in many cases.

Egg size variation occurs at three levels: the variation in egg size within a clutch, the variation in mean egg size within a bird in clutches laid in different years, and the variation in mean egg size among birds. The first two levels show the extent to which individual birds change the size of their eggs, variation due to ontogenetic physiological and/or environmental effects. The among-bird variation comprises both genetic and non-genetic components (for example Larsson and Forslund 1992). Most of the variation in egg size is in fact attributable to among-female differences (repeatability within females = 0.80; Lessells *et al.* 1989), and the heritability of egg size is relatively high, *c.* 0.51 (Chapter 8). These data suggest that heritable differences account for about half of the variation among females, a higher proportion of phenotypic variance attributable to genetic variation, on which selection might operate, than was present for the other quantitative traits considered thus far. Conventional logic in population genetics predicts that we should find relatively weak selection on characters with high heritabilities (Fisher 1930; Falconer 1989; Mousseau and Roff 1987), however several more complex models suggest different ways in which heritable variation may persist in the face of selection (Chapter 8; Price *et al.* 1988; Cooke *et al.* 1990*b*; Price and Schluter 1991; Schluter and Gustafsson 1993).

Average egg size does not differ among clutch size classes (Table 12.1), and the range of egg sizes is large. Thus, some clutches of size n eggs actually have a greater total clutch weight than some clutches with $n + 1$ eggs. Ankney and Bisset (1976) found that approximately 30 per cent of five-egg clutches had a smaller total clutch weight than the largest four-egg clutch. Ankney (1980) later showed that yolk, albumen and shell weight was proportional to total egg weight. Thus, for given clutch size, females allocate more nutrient to reproduction when they produce larger eggs. The overlap in total clutch mass, combined with a similar nutrient composition over the range of egg sizes, suggests, theoretically at least, that nutrient could be packaged into clutch weight in a number of possible different ways in terms of egg number and size. Evolutionarily, females do face the trade-off between laying fewer larger eggs or more smaller eggs, and they may also do so at the proximate level.

Lessells *et al.* (1989) searched directly for phenotypic and genetic correlations between clutch size and egg size in Snow Geese, and

Table 12.1 Mean egg size with respect to clutch size of Snow Goose females nesting at La Pérouse Bay. Egg weight is expressed as deviations from annual means and the data have been pooled over the years 1976–1986

Clutch size	Mean egg size (g)	SD	*n* nests
2	−0.4	10.1	121
3	0.6	9.5	814
4	0.3	9.0	2342
5	0.2	8.9	2180
6	0.3	9.1	654

were unable to detect them. This result suggests that the hypothesized trade-off between clutch size and egg size does not occur, and that the logic leading to the prediction is an inadequate description of the situation. The most obvious possibility is that goslings from larger eggs have higher probabilities of survival or fecundities as adults, thereby offsetting the advantage of producing more, smaller eggs. In support of this, Ankney (1980) found by experiment that starved goslings from heavier eggs survived on average 17 h longer than goslings from lighter eggs. In a preliminary investigation of the La Pérouse Bay data, Cole (1979) found that light goslings survived to fledging better than heavy ones in 1976, but that the situation was reversed in 1977. Cole failed to control for the effects of egg sequence and parental age, however, and as we will show later, this makes his findings difficult to interpret. Cargill (1979) showed that egg size varies with sequence, and that last eggs in a sequence had lower hatching success. Despite these complications, Cole's result suggested that the selective regime on gosling size might vary from year to year (Ankney and Bisset 1976), which could provide a mechanism for the maintenance of genetic variation for egg size within the population and account for the lack of a trade-off between egg and clutch size.

This chapter details patterns of variation in egg size within clutches, within individuals across years, and among individuals, and assesses the relative fitness of eggs of different sizes. In our discussion, we integrate our results into an evolutionary genetic model which is consistent with all available information. As with other characters elsewhere in this book, we analyse fitness components with respect to egg size. Differences in mean egg size between successful and unsuccessful eggs would suggest directional selection may be operating. If the variance in egg size is larger among unsuccessful eggs, normalizing selection may be operating. Larger variance among the successful eggs might suggest diversifying selection. Failure to find differences would suggest that selection on egg size is weak or nonexistent over the wide range of sizes found in the population.

12.1 *Egg measurements*

The measurement analyzed here is the weight of freshly-laid eggs. Weight correlates well with measures of volume (Newell 1988), and probably most accurately reflects the allocation of nutrients to the egg. Eggs were weighed on daily nest visits, thus within 24 h of being laid. One complexity of the data set is that at least 6.6 per cent of the eggs are laid parasitically (Chapter 5). Because these are sometimes laid after incubation has begun, parasitic eggs have a lower hatchability, on average, than parental eggs. Since birds differ considerably in the size of eggs which they lay, nest parasitism increases measured variance in egg size within a clutch and lowers our heritability estimates and their reliability (Chapter 8).

12.2 *Patterns of egg size variation*

12.2.1 Variation within clutches

Several authors have documented variation in egg size within Snow Goose clutches. Syroechkovsky (1975) found that fresh egg size declined with laying sequence among Snow Geese nesting on Wrangel Island. Cargill (1979) found a significant decrease in egg weight of the last egg in clutches of three, four, and five for Snow Geese nesting at La Pérouse Bay in 1978. The last egg had a significantly lower probability of hatching, but within a laying sequence position, there was no effect of egg weight on the probability of hatching. Both studies noted a strong positive correlation between order of laying and order of hatching. In Canada Geese, Lessells (1982) found the second egg to be largest, on average, with egg weight decreases of about 4 g per egg from the second to the last eggs in clutches of six eggs. In a different population of Canada Geese, first and last eggs are significantly smaller than the rest (Leblanc 1987).

All of these studies show that last eggs are smaller, but they differ somewhat with respect to the size of first and later eggs. The apparent differences may simply reflect the resolving power of the data. Williams *et al.* (1993*a*) performed a comprehensive analysis of 12 years' data from adult Snow Geese at La Pérouse Bay on egg size with respect to laying sequence, clutch size, hatching success, and gosling survival. Figure 12.1 plots the mean weights of eggs, expressed as residuals from annual mean egg sizes, laid by adult females with respect to clutch size and sequence for clutch sizes 3–6. A general pattern is clear. The first egg is relatively small, the second egg is the largest, and subsequent eggs decline in weight. In two-egg clutches, the first egg is the larger (egg 1, relative mean size = 1.1; egg 2, −2.6). Egg 1 of two-egg clutches had the highest variance of eggs in any clutch size-sequence position. Some of the two-egg clutches are the last two eggs of larger clutches, thus size patterns with respect to sequence are difficult to interpret. For all clutch sizes, the average last egg is the smallest. For both first and last eggs, there is a trend towards smaller egg size in larger clutches, but also a trend towards larger eggs in the middle of the sequence.

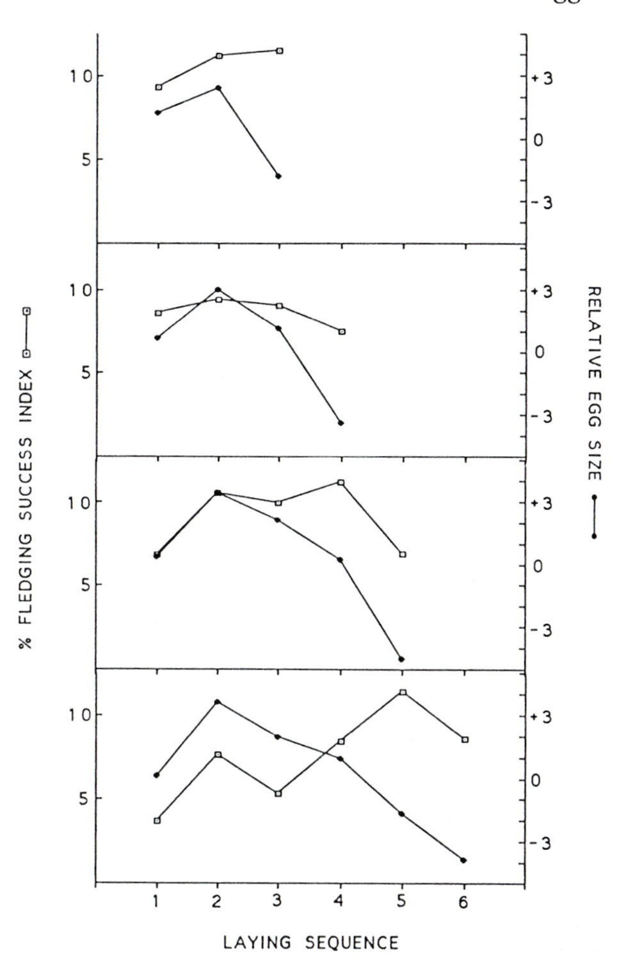

Fig. 12.1 Egg size, scaled relative to annual mean, and the probability of each egg producing a gosling which survives to fledging in Snow Geese nesting at La Pérouse Bay with respect to clutch size and egg laying sequence. (Reproduced with permission from Williams *et al.* 1993*a*).

While this pattern is clear, the magnitude of the variation is small relative to egg size variation in general. The largest difference between sequence positions occurs in six-egg clutches, where egg 2 is on average 7.6 g heavier than egg 6. This represents about a 5 per cent difference in egg size, in contrast to the nearly 2-fold difference in egg sizes found among females in the population.

In general then, egg-weight, and thus nutrient allocation varies with sequence. Leblanc (1987) interpreted this pattern at the proximate level as a reflection of hormonal changes during egg formation.

Flint and Sedinger (1992) documented a similar egg size and laying sequence pattern in Black Brant *Branta bernicla nigricans*, and suggested that a failure to find increased variation in the size of last laid eggs argued against direct nutrient stress as a physiological cause of the decline. Williams *et al.* (1993*a*) considered the adaptive hypothesis that nutrient allocation within a clutch varies according to the average expectation of fitness value eggs have by virtue of their sequence *per se*. If so, one would predict a positive correlation between the mean egg size of a particular sequence position within a clutch and its mean fitness relative to other sequence positions. We can evaluate whether fitness differences might be due to egg size, rather than sequence, by testing for egg size effects on fitness within sequence positions. Finding a positive relationship between mean egg size and fitness among positions, and that egg size has no effect within positions would be consistent with the adaptive hypothesis.

The fitness value of eggs was partitioned into two components: (1) what is the probability of hatching and leaving the nest (Table 12.2), and (2) what is the probability that a gosling leaving the nest survives

Table 12.2 Fates of eggs with respect to clutch size (clu) and egg sequence (seq). Percentages of each category are given with total sample size. *PIF* = pre-incubation failure. From (Williams *et al.* 1993*a*)

		Egg Fate						
Clu	Seq	*PIF*	Total failure	Incubation fail	Bad egg	Abandon egg/chi	Hatch and leave	*n*
2	1	31.3	12.0	1.2	1.2	7.2	47.0	83
	2	4.5	22.7	2.3	2.3	11.4	56.8	44
3	1	27.1	3.8	1.4	3.8	2.9	61.0	346
	2	8.1	4.6	2.8	3.5	4.6	76.5	285
	3	1.6	4.8	3.2	4.0	7.1	79.4	252
4	1	25.2	3.2	2.8	3.2	1.8	63.9	684
	2	7.9	3.9	2.9	2.7	2.9	80.7	560
	3	4.7	3.7	2.8	3.3	2.3	83.2	614
	4	0.9	3.9	3.5	3.3	7.4	81.1	570
5	1	26.8	2.3	1.6	3.1	3.3	62.9	485
	2	8.9	3.0	3.5	3.7	2.0	78.7	400
	3	4.1	2.5	2.8	3.2	3.0	84.4	435
	4	2.6	2.8	3.1	3.1	3.7	84.7	457
	5	0.2	2.8	4.3	3.6	9.2	79.9	422
6	1	40.9	6.6	2.9	3.7	2.2	53.7	136
	2	12.7	6.6	2.8	2.8	1.9	73.6	106
	3	6.2	8.0	3.5	4.4	2.7	75.2	113
	4	2.8	8.4	1.9	2.8	4.7	79.4	107
	5	1.9	8.6	2.9	3.8	4.8	78.1	105
	6	0.0	11.0	6.1	3.7	13.4	65.9	82

Table 12.3 Proportion of goslings leaving the nest which were subsequently encountered in pre-fledging banding drives, with respect to clutch size and laying sequence (from Williams *et al.* 1993*a*)

	Clutch size									
	2		3		4		5		6	
Sequence	%	*n*	%	*n*	%	*n*	%	*n*	%	*n*
1	10.3	39	15.2	211	13.5	437	10.8	305	6.9	73
2	0.0	25	14.7	218	11.7	452	13.3	316	10.3	78
3	—	—	14.5	200	10.8	511	11.7	367	7.1	85
4	—	—	—	—	9.3	462	13.2	387	10.6	85
5	—	—	—	—	—	—	8.5	336	14.6	82
6	—	—	—	—	—	—	—	—	13.0	54

to fledging (Table 12.3)? Five types of failure at nests are recognized and tabulated. *PIF* refers to eggs which fail during the laying period, either because the egg was removed by a predator or because the entire nest was abandoned prior to the start of incubation. Since clutch size is unknown in the latter case, eggs in failed nests were assigned among clutch sizes under the assumption that total failure was independent of potential clutch size, as done in the previous chapter. *PIF* occurs for about 30 per cent of first-laid eggs and declines dramatically thereafter. Eggs may also fail during incubation, either because the whole nest fails or because a predator removes one or more eggs from an otherwise successful nest. Both these types of failure affect all eggs in a sequence equally. Infertile or rotten eggs account for 3–4 per cent of all eggs, and are also independent of sequence. Eggs or goslings may be abandoned in the nest after the rest of the brood has left, and last eggs are two to four times more likely to be abandoned than other eggs in the sequence.

Up to the time when goslings leave the nest then, first laid eggs fail most often, middle eggs are the most successful and similar to one another, and last eggs fail at relatively high rates. The reasons for failure differ for first and last eggs: the lower success of first eggs is mainly due to PIF, while the lower success of last eggs is due to higher abandonment of later-hatching eggs and goslings.

Sequence effects persist after goslings leave the nest. Eggs which hatch last have a marginally lower probability of being detected at fledging ($P = 0.05$), but other eggs in the sequence do not differ in their success rates. At Wrangel Island, Syroechkovsky (1975) also found higher gosling loss occurred among last hatching goslings, which were presumably predominantly last laid egg because of the close correlation between laying and hatching sequence (Cargill and Cooke 1981).

The mean fledging success of an egg for each clutch size and sequence are given in Fig. 12.1, where these curves may be compared with those for mean egg weight. First and last eggs are lighter and less successful than middle sequence eggs, except for the last egg of six-egg clutches, which is relatively successful, but there is no close relationship between success and mean weight for the middle eggs.

One might argue that the first and last eggs are less successful because they are smaller, but there appears to be no effect of size itself. This was tested by asking whether egg weight covaries with hatching success *within* clutch and sequence categories (Table 12.4). Only two of the 20 clutch-size and sequence categories approached significant difference; in one case successful eggs were heavier, while in the other unsuccessful eggs were heavier! These strong, negative results show no evidence of higher success for larger eggs, once clutch size and sequence are taken into account in samples from adult birds.

Table 12.4 A comparison of the fresh egg weights of hatching and non-hatching eggs laid by adult females with respect to clutch size (clu) and laying sequence (seq). From Williams *et al.* (1993*a*)

		Unhatched			Hatched				
Clu	Seq	Mean	SD	*n*	Mean	SD	*n*	*t*	*P*
2	1	127.19	10.77	21	125.72	10.37	39	0.52	0.60
	2	121.65	11.05	20	122.32	7.02	25	0.24	0.81
3	1	125.07	9.12	56	125.69	9.66	211	0.43	0.67
	2	127.80	10.88	49	126.44	8.93	218	0.91	0.36
	3	121.44	8.87	52	122.58	8.90	200	0.82	0.41
4	1	125.32	9.60	107	125.21	9.07	437	0.11	0.91
	2	125.97	8.79	78	127.83	8.64	452	1.75	0.08
	3	125.29	8.84	86	125.68	8.54	511	0.39	0.70
	4	120.78	7.34	109	120.93	8.65	462	0.18	0.86
5	1	126.01	9.34	79	124.90	9.63	305	0.93	0.36
	2	128.18	7.82	62	128.19	8.90	315	0.01	0.99
	3	125.59	8.89	54	127.04	8.11	367	1.21	0.23
	4	124.64	8.42	61	125.14	7.90	387	0.45	0.65
	5	119.60	8.92	85	120.10	8.19	336	0.49	0.62
6	1	124.44	10.10	36	125.44	10.05	73	0.48	0.63
	2	126.77	8.60	22	128.92	9.31	78	0.97	0.33
	3	127.25	7.59	24	126.60	8.05	85	0.35	0.72
	4	126.21	10.62	19	125.60	8.43	85	0.27	0.79
	5	124.05	9.17	21	123.06	8.59	82	0.46	0.64
	6	124.68	8.43	28	119.04	7.92	54	2.99	<0.01

To summarize this section, fresh egg weights differ systematically with clutch sequence in a pattern which is generally consistent with the egg's probability of survival. In general, first eggs are small, second eggs largest, and successive eggs decline in weight, with the last egg being the smallest of all. The small first and last eggs of the clutch have the lowest probability of producing hatching goslings which survive through fledging, although for different reasons. The differences in survival probabilities appear to be due to sequence *per se*, rather than egg weight. Within middle eggs, there is no correlation between egg size and gosling survival. Within each sequence category, there is no difference between the fresh weights of successful and unsuccessful eggs. The adaptive argument that birds may be allocating more resources within a clutch to those eggs which have a higher probability of success is supported by the data, but not strongly. On the other hand, we have seen no evidence thus far that by laying a larger egg, females enhance the probability that an egg hatches successfully or increases the gosling's chance of survival, at least to the fledgling stage. Our analysis thus far has failed to find an advantage to laying larger eggs.

12.2.2 Variation within individuals

Snow Goose females have different mean egg sizes. There is high repeatability of egg size, with about 80 per cent of the variance in mean egg weight per clutch attributable to differences among individuals (Lessells *et al.* 1989). High values for repeatability of egg size are typical of birds (Boag and van Noordwijk 1987), and contrast markedly with the low values for clutch size variation. Variation within birds is much less than the variation among birds. Despite this, individual birds do vary their egg weights in several ways. We have already considered the patterns of variation within a clutch. Mean egg weight increases from ages 2 through 4 years (Fig. 12.2), within individuals as well as in the population analysis (Robertson *et al.* 1994). Eggs laid by younger females are longer and thinner than those of older birds. As with the sequence effects, the age differences are small, changing about 2–3 per cent of mean weight for the first two years. The pattern of change with age is strikingly similar to that documented for clutch size (cf. Fig. 7.1), and as with clutch size, there is no evidence of a decline in mean egg weight with old age. This result suggests that younger birds are less capable of laying larger eggs, as previously argued for age-related increases in clutch size.

Egg weights of adult females also vary from year to year. There are small but significant annual population differences (Figure 12.3;

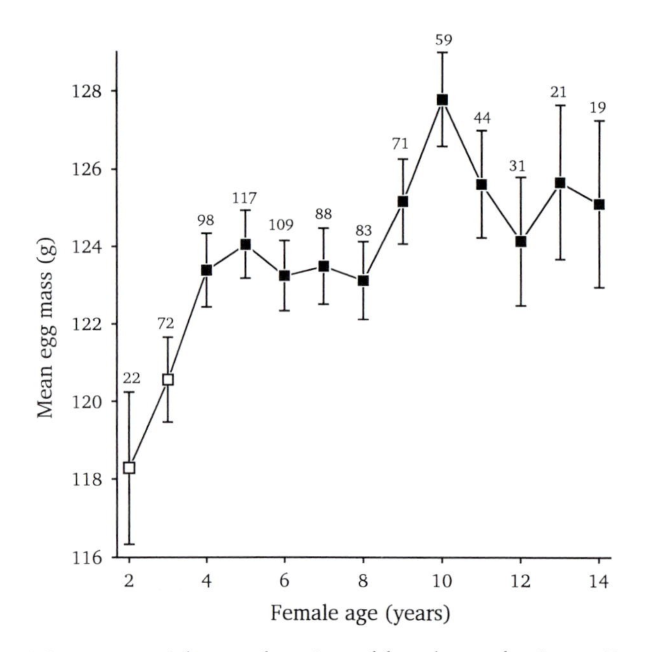

Fig. 12.2 Mean egg weight as a function of female age for Snow Geese nesting at La Pérouse Bay. The values presented are least-squared means of clutch egg size means, with standard errors, adjusted for annual variation. Open boxes are significantly different from closed boxes. n = number of clutches.

ANOVA: $F = 5.66$, d.f. = 13,1966, $r^2 = 0.026$, $P < 0.0001$), and mean egg size has declined slightly, but systematically, over the years (linear regression on annual means: $r^2 = 0.35$, $F = 6.59$, d.f. = 1, $P = 0.02$). As with age and within-clutch differences, the magnitude of the variation is relatively small: annual means range between 120.2 and 127.5 g, and the decline represents about 2 per cent of the total egg weight. Seasonal variability cannot be explained by differences in the age structure or genetic constitution of the populations (Newell 1988), and thus may reflect annual differences in the breeding condition of individual birds. Differences in clutch mean egg weights of individual adults between particular pairs of years correlate positively with the general annual differences of the rest of the adult population ($r = 0.19$, $n = 157$, $P = 0.02$), suggesting that to some degree, these annual changes do result from annual variation in individuals' mean egg sizes.

We assume that the long term decline in egg size is in some way related to the concurrent decrease in fecundity, which we have attributed to a deterioration in breeding conditions at the colony and/or at spring staging sites (Chapter 6). The sharp dips in egg size in 1977

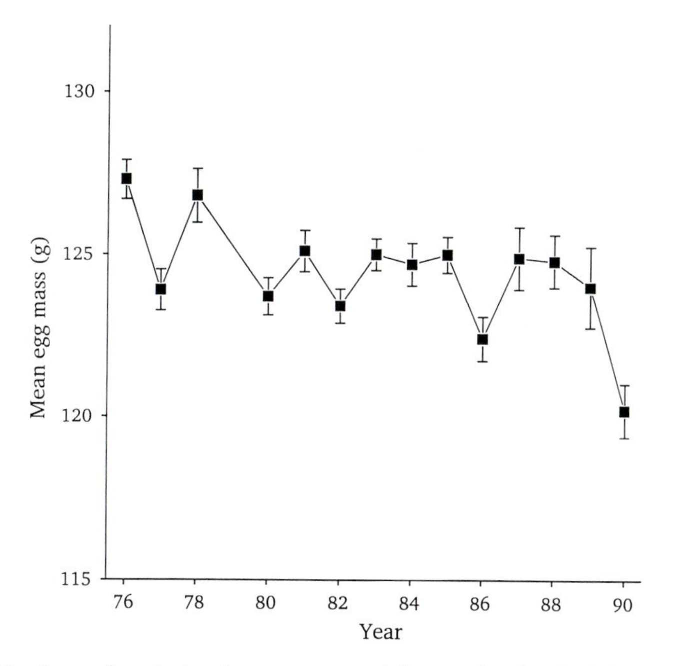

Fig. 12.3 Annual variation in mean egg weight per clutch of adult Snow Geese nesting at La Pérouse Bay. The number of clutches for years 1976–1990 (except 1989) are: 138, 125, 68, 195, 173, 193, 214, 129, 192, 114, 60, 96, 51, 67.

and 1980 correspond to years of vernal nutritional stress, concurrent with decreased clutch sizes and combined with unusually early breeding by the geese (Fig. 10.1; Davies and Cooke 1983*a*). Spring 1990 was also dry at migratory staging areas. However, we cannot ascribe the dip in egg-size in 1986 to this same source of nutritional stress, as migratory staging areas were exceptionally wet that year. If nutrient condition influences egg size, we might expect a correlation between egg size and both clutch size and laying date. We argued in the previous two chapters that birds in the best condition laid earliest in the season and laid the largest clutches. We might predict therefore a negative correlation between laying date and egg size, and a positive correlation between clutch size and egg size. However, no such correlations were found when the effects of age were removed, even in models controlling for clutch size.

Lessells *et al.* (1989) also found no correlation between clutch size and egg size, but found a weak negative relationship in the subset of the data with multiple observations per female. This tested for whether individual females altered mean egg size when they laid a clutch of a different size. We re-examined this relationship among individuals using a more comprehensive sample of adult females aged 5 years or older with clutches in two or more years, where both variables were computed relative to the female's own mean values and adjusted for year effects. There was no hint of a significant relationship ($n = 260$, $r = 0.05$, $P = 0.43$). When adult females lay larger clutches they do not change their mean egg weight in any consistent way.

In summary, eggs laid by individual birds vary in size, but to a rather limited degree. Female age, year, and the egg's position within a clutch all affect egg size within individuals. The lack of a within-season trend in egg size contrasts with the strong one for clutch size. Despite their parallel increases during maturation, this non-concordance suggests that at some level, different mechanisms control clutch size and egg size.

12.2.3 Variation among individuals

Female Snow Geese differ substantially in the mean size of eggs which they lay. The high repeatability and heritability values in mean egg weight suggest that about half of the differences in mean egg weight among birds can be ascribed to genetic or developmental differences, rather than more proximal environmental causes. Do differences in egg size among females result from differences in body size? At the inter-specific level, larger bodied species lay larger eggs (Heinroth 1922; Huxley 1927; O'Connor 1980). Intra-specifically, there is a weak positive correlation between egg size and body size in some species (r^2 *c.* 0.05; Ojanen *et al.* 1979; Otto 1979) however ages of breeding females have been unknown in most field studies, and age or cohort effects may confound these results.

Does a Snow Goose female's egg size relate to measures of her body size? Small but significant simple correlations between a set of measures of body size and egg size were found by Alexander (1983), based on 36 females collected at nests (for tarsus, culmen, and body mass: r_s^2 of *c.* 0.03–0.09). Using a larger data set of known-aged birds for whom we had egg size information from resightings at nests, and body size information from recapture and measurement in banding drives, Cooch *et al.* (1992) failed to find any significant egg size–body size relationship. The most important difference between the studies is not the method of data collection, but the correction for age and cohort (year of birth) effects on body size possible in the latter study. Since both egg size and body size have declined over time (Fig. 12.3, Chapter 13), simply tallying data from birds of unknown age and cohort produces a significant, but potentially spurious correlation. If egg size is in part determined by body size, correcting for cohort, and thus in part body size, eliminates much of the covariance we are testing for. An unbiased test of body size effects can be done within cohorts. While our sample sizes become small, we have failed to find significant covariation of body size and egg size within cohorts. Variation in egg sizes of female Snow Geese is unrelated to variation in body size.

12.3 *Fitness consequences of egg size variation*

In our analysis of egg sequence effects, we found egg size unrelated to hatching success, within sequence position. This section summarizes a comprehensive fitness components analysis of egg size. Is there any fitness advantage to laying large eggs? Since fluctuating selection pressures, with fewer larger eggs being favoured in some years and more smaller eggs being favoured in others is a real possibility, and sample sizes were relatively large, we analysed the data on an annual basis (Table 12.5). Sample sizes and further details of these analyses are given in Williams *et al.* (1993*c*).

12.3.1 Pre-incubation nest failure

We assessed whether egg weight was related to the probability of pre-incubation failure by comparing first egg weights among nests reaching the incubation stage and those which did not. The mean egg weight of nests which failed prior to incubation was lower than that of nests which survived through clutch completion in 9 of 12 years, and in 3 years the differences were significant. Egg size variance was greater among failed nests in 7 of 12 years, three of which significantly so. We do not know the ages of the parents who lose their nests at this stage, but it could be a result of young birds, which on average lay smaller eggs, being more susceptible to failure. These results are probably an effect of age or breeding experience rather than rather egg size *per se*.

12.3.2 Survival to hatch

We used two components to test whether egg size was related to the probability of eggs hatching in successful nests. First, we determined whether total nest failure was correlated with mean egg size per clutch. Total nest failure was unrelated to mean egg size except in one year, 1977, when eggs were smaller, based on a sample of only eight failed nests. As with pre-incubation failure, we lack information

Table 12.5 A summary of results from analyses of annual fitness components as a function of egg weight for a 12 year period (summarized from Williams *et al.* 1993*c*). Within years, data were tested for both directional effects on means (top) and for normalizing or diversifying selection on variances (bottom). The summary over years at the end shows how many individual years had higher mean values for successful eggs (top), or lower variances for successful nests (bottom), even if the differences were not significant within years

	Fitness component differences between successful and unsuccessful eggs						
Year	*PIF*	*TNF*	*HSuc*	*TBL*	*FSuc*	*Recr*	*MSur*
			Tests on means				
76	**+	ns	ns	ns	ns	ns	ns
77	ns	***+	ns	ns	ns	ns	ns
78	ns	ns	**+	ns	ns	ns	ns
80	ns	ns	ns	ns	ns	ns	ns
81	ns	ns	***+	ns	ns	ns	*+
82	ns	ns	ns	**−	ns	ns	ns
83	ns	ns	ns	ns	ns	ns	ns
84	ns	ns	**+	ns	ns	ns	ns
85	ns	ns	ns	ns	ns	ns	ns
86	ns	ns	**+	ns	ns		ns
87	*+	ns	ns	ns	ns		ns
88	*+	ns	***+	ns			
76–88	9/12+	7/12+	8/12+	7/12+	4/12+	3/9+	6/11+
			Tests on variances				
76	ns	ns		ns		ns	ns
77	ns	***+		ns		ns	ns
78	ns	ns		ns		ns	ns
80	ns	ns		ns		ns	ns
81	ns	ns		ns		ns	ns
82	**−	ns		ns		ns	ns
83	ns	ns		ns		ns	ns
84	*x*−	ns		ns		ns	ns
85	**−	ns		ns		ns	ns
86	ns	ns		ns			ns
87	ns	ns		ns			ns
88	ns	ns		ns			
76–88	7/12−	6/12−		7/12−		5/9−	5/11−

PIF = preincubation failure, *TNF* = total nest failure, *ESur* = egg survival through incubation, *HSuc* = hatching success, *TBL* = total brood loss, *FSuc* = fledging success, *Recr* = fledgling recruitment success, *MSur* = relative female survival probability.

ns, $P > 0.10$; *x*, $0.10 > P > 0.05$; * $0.05 > P > 0.01$; ** $0.01 > P > 0.001$; *** $P < 0.001$.

+, Higher survival from larger eggs (top) or more variable nests (bottom). −, Higher survival from smaller eggs (top) or less variable nests (bottom).

on the age of females at this stage, but younger females have higher rates of nest failure (Rockwell *et al.* 1993; Chapter 7). This weak result may simply reflect a higher proportion of young females among the failed nest sample that year. By pooling over years, we had a sufficient sample to test for a relationship among older females only, and found none ($P > 0.20$).

We have already seen that within sequences, egg size was not related to hatching probability (Table 12.4). Nonetheless, we examined whether the difference in mean weights of hatching and non-hatching eggs within successful nests differed from zero. This within-nest comparison intrinsically controls for any potential biases due to differences among sampled females. The weights of hatching eggs within nests were greater than those failing to hatch in 8 of 12 years, and significantly so in five. Since there are no systematic trends for eggs within sequences, however, we attribute this result to sequence effects and conclude that hatching probability is independent of egg size.

12.3.3 Fledgling survival

The most likely stage of the life cycle when egg size might affect survival is during the fledging period. Ankney's (1980) starvation experiments pointed to the possibility that eggs with less nutrient might provide the newly hatched goslings with less of a 'buffer' if severe weather inhibited feeding. On the other hand, Thomas and Peach Brown (1988) argued that larger goslings may be more vulnerable to starvation because they have smaller reserves proportional to their body size. We examined the potential effect of egg weight on gosling survival using a measure of total brood loss and one of fledging success within successful broods. The mean egg weights of those clutches where at least one gosling was found in the banding drives were compared with with those clutches where no goslings were found. For successful broods, we assessed fledging success by comparing the mean weight per brood of eggs which hatched goslings encountered in the banding drives versus those which were not encountered, a similar analysis to that done for egg survival within clutches.

Families caught in banding drives and those not caught did not differ in either mean or variance of mean egg weight, in all years except 1982. In 1982, families caught in the banding drives had a significantly smaller mean egg size than those which were not. Within families captured at fledging, in no year did we find a significant difference in the mean fresh egg weight of those goslings which were still alive at banding compared to those which were not present (and presumably dead or adopted). The mean egg sizes producing encountered goslings were actually smaller in eight of the twelve years. Although the samples each year are relatively small (mean $n = 27$ families, range = 4–61), there is no significance in a pooled analysis of 342 broods (paired $t = 0.77$, $P = 0.48$). In neither of our tests is there any indication of fitness differences associated with egg size, except in 1982.

12.3.4 Gosling recruitment

The fresh weights of eggs from which females fledged and subsequently recruited into the breeding colony were similar to those which fledged but were not subsequently encountered. With weights expressed as deviations from annual means and pooled over years, the 838 non-recruited birds had a fresh egg weight of -0.2 ± 9.0 g (SD), compared with -0.560 ± 9.76 g for 197 recruited birds.

12.3.5 Laying female survival

Our final analysis examined whether we could detect a difference in the survivorship and/or return rate of females as a function of their mean egg size. We found no overall differences in female return rates, relative to mean or variance in mean egg size, but returning females had a lower mean egg size in one year, and a larger variance in another.

12.3.6 Composite selection regime

We have summarized the selective regime on egg size for adult females in Fig. 12.4 (Cooke *et al.* 1991). The figure shows the frequency distributions of eggs laid by adult females, those which hatched, those whose goslings survived and were captured at fledging, and those whose goslings eventually recruited into the population. Egg weights were adjusted for annual variation in means and have been pooled over years. The lines are third-order polynomials. There were no significant differences in means or variances of the distributions at any stage, nor patterns in the transitions between stages (Cooke *et al.* 1991). The similarity of frequency distributions across the range of egg sizes dramatically demonstrates the lack of directional, diversifying or normalizing selection operating on this character.

We have found no consistent pattern of directional, normalizing or diversifying selection on egg size itself. We conclude that a large

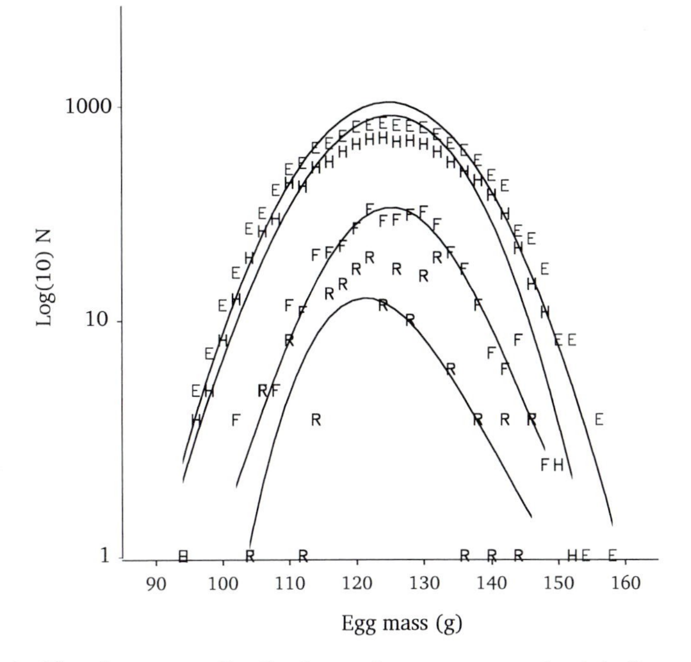

Fig 12.4 The frequency distributions of egg masses of adult Snow Geese nesting at La Pérouse Bay when laid (E), which produced hatchlings (H), and were known to have produced fledglings (F) and recruits (R). The lines are third order polynomials. (Adapted with permission from Cooke *et al.* 1991).

number of different mean egg size phenotypes exist in our population. Given relatively high heritabilities, this variation reflects, in part, a wide range of equally fit genotypes. The lack of an obvious selection gradient on egg size means that these various genotypes have roughly equivalent fitnesses.

12.4 *Why does egg size vary among birds?*

Characters closely related to fitness are usually thought to have low heritabilities (Mousseau and Roff 1987; Falconer 1989), yet the heritability of egg size is high. There are a number of potential explanations for the maintenance of such genetic variability, including: (1) little or no correlation between egg size and fitness, (2) temporally variable selection pressures (3) spatially varying selection pressures coupled with gene flow, or (4) antagonistic pleiotropy between genetic variance for egg size and other correlated traits of the bird (Rose 1982; Lande and Arnold 1983). The detailed examination of the relationship between egg size and fitness components described above narrows the possibilities.

A wide range of egg weight phenotypes and genotypes occurs within our population, and we have found very little evidence of fitness differences among size categories. At the beginning of the chapter we outlined reasons for expecting egg size to have substantial fitness consequences, but these seem not to have occurred during our study. The consistent patterns of variation as a function of egg size for fitness components in Table 12.5 are attributable to covarying age or sequence effects rather than egg size *per se*.

The Snow Goose is similar to other birds in having a high repeatability and heritability of egg size (Boag and van Noordwijk 1987), and one might infer that egg size variation is not under strong selection in this or other species. Evidence for fitness differences among birds of different egg sizes in other species is ambivalent. Most studies addressing this question failed to control for environmental covariates such as habitat, laying sequence, or maternal age (Williams 1994*b*). Our results are consistent with the first hypothesis of equivalent fitnesses, but as outlined in the introduction of this chapter, such a result raises the question of why more birds don't lay a larger number of smaller eggs. If egg size is irrelevant, why differentiate in this way? We have seen that individuals do not make such a trade-off, but the evolutionary potential for such a trade-off remains. This result is also somewhat inconsistent with our suggestion that the within-clutch pattern of egg size variation may be an adaptive strategy for investing less in individual eggs with a lower probability of success. We will return to this question after evaluating additional hypotheses of temporal or spatial heterogeneity in fitness gradients.

We found only weak evidence, at best, that annually fluctuating selection pressures might maintain egg size variation. Ankney and Bisset (1976) argued that selection could favour larger egg size, even at the expense of clutch size, in years when conditions during the fledging period were harsh. In more benign years, there might be a selective advantage for individuals laying large clutches at the expense of egg size. In analyses of 12 years' data and seven fitness components, we did find a few year-specific fitness differences; we would expect a few due to chance alone. Moreover, detailed examination of the specific effects occurring in particular years does not support the hypothesis that larger eggs produce goslings better able to cope with difficult conditions at hatch (Williams *et al.* 1993*a*), the most plausible mechanism of fluctuating selection. One may interpret the scattered significant points, or the relatively even distribution of positive and negative effects, as evidence of year-specific variation in some of the components of fitness, but we believe that our results more closely resemble random variation. We do not feel that our data support a strong role for temporal variation in directional selection as a mechanism for the maintenance of genetic variability in egg size.

If there were spatial variation such that selective regimes were different in shape or mean values at other colonies, genetic variation could be maintained through the large amount of gene flow among the various breeding colonies. We cannot seriously evaluate this without additional studies, but the selection regimes would have to be drastically different, in different directions, from that operating at La Pérouse Bay to sustain such variation where we have measured it. It is difficult to believe our results are completely unrepresentative of other Snow Goose colonies.

Population geneticists have only recently developed quantitative techniques to consider how selection on one trait may affect the phenotypic response of genetically correlated characters (Lande and Arnold 1983). In the next paragraphs, we recapitulate the salient features of our analyses, outline questions raised by the results, and synthesize our finding in a conceptual model which suggests that the selective regime and heritability of egg size may be best explained through processes involving selection on correlated characters.

There is relatively little variation in egg size within individuals, but that which exists suggests that within limits, larger eggs are favoured. Two of the patterns of egg size variation within individuals indirectly support this conclusion. First, in adults, first and last eggs within a clutch are smaller and have lower fitness than other eggs. Through partial correlation, we showed that the fitness effects were accounted for by sequence itself, rather than egg size per se. We interpret this pattern as an adaptive mechanism in which more resources are allocated to those eggs within a clutch which have the highest probability of surviving. This suggests that larger egg size is important within individuals. The second pattern of egg size variation within individuals is age related. Older birds lay larger eggs, in addition to laying larger clutch sizes, and are more successful nesters than younger birds (Rockwell *et al.* 1993). The increase with maturity again suggests that within individuals, larger eggs are favoured, if they can be afforded. These results point towards there being some value for individuals to produce larger eggs within their individual range of variation, but we were unable to detect such effects from field observational data. Of course, more proximal physiological or anatomical explanations may be invoked to account for both the sequence and age effects.

Among birds, there are relatively large differences in mean egg sizes, about half of which may be attributable to genetic differences. We found no obvious directional or normalizing selection over the observed range of mean egg sizes. This selection regime is in marked contrast to the strong directional selection for larger clutches (cf. Fig. 10.2). Parallel to the selection regimes, the heritabilities of egg size and clutch size are at different ends of the spectrum of heritability values. Finally, there are no detectable phenotypic or genotypic correlations between egg size and clutch size (Lessells *et al.* 1989). Mean egg sizes do not vary with clutch size, and there are no indications of antagonistic pleiotropy between genes for the two traits.

These results leave us with a number of questions. In a species whose reproductive output may be limited by the amount of resources it can acquire, there would appear to be a substantial advantage to laying smaller eggs, since these are just as successful as large eggs. Why do birds laying larger eggs persist? Larger eggs produce larger goslings which could survive starvation longer; in theory this could provide a compensating selective advantage to producing more, smaller young. Why, then, do we fail to find such a selective advantage for larger eggs at the population level? Why are the patterns of selection and heritability values found for egg size so different from the patterns documented earlier for clutch size, since both are thought to reflect a commitment to reproduction from limited resources? In short, why are individual birds apparently constrained to a narrow range of egg sizes? Why is there apparently little adaptive flexibility?

We postulate that the between-bird variation in egg size is constrained by selection operating on some correlated trait (Williams *et al.* 1993*c*). Variation in egg size among individuals may be an epiphenomenon resulting from selection on more fundamental correlated physiological traits. Although selection may favor individual birds which produce larger eggs, genes for this character may, through antagonistic pleiotropy, have a negative effect on some other attribute of the bird. For example, if alleles coding for large egg size also resulted in slightly higher metabolic rates which were selectively disadvantageous, then large egg size would not be favoured because of the countervailing selection operating on metabolic rate. Antagonistic pleiotropy could lead to the maintenance of a variety of equally fit egg size/metabolic rate genotypes. We postulate that there is relatively little egg size variation within individuals because it is advantageous for a bird to lay as large an egg as possible within the constraints of the correlated trait.

What correlated trait might be in antagonistic pleiotropy with the genes for egg size? We postulate that the trait is a physiological one but can only speculate on what it might be. Basic metabolic rate (BMR) may constrain egg surface area (and therefore size) through gas exchange processes, which may affect the successful development of the embryo, so individuals with different BMRs might produce different sized eggs for maximal fitness. Alternatively, variation in individual BMR may constrain the rate of egg formation, and thus eventual egg size, within a given egg-laying interval. This would explain why there is relatively little variation in egg size within individuals. The question of why there is so much population variation in egg size cannot be answered until we know what the correlated trait might be, but we suggest that more attention be given to possible physiological variables associated with egg size variation.

This hypothesis is admittedly speculative, but several testable expectations follow from it.

1. We expect relatively little phenotypic variation within individuals. The high repeatability of egg size confirms this.
2. If individuals cannot maximize the sizes of all of their eggs, because selection favours larger clutch size and resources are limited, they should allocate fewer resources to eggs with an intrinsically lower probability of survival, and lay larger eggs when they are capable of doing so. We have argued above that both of these processes may occur.
3. If maximal egg size were favoured by individual females within the constraints of a correlated character, one would predict that the provision of ad lib food prior to nesting would not enable geese to lay larger eggs, but would lead to the production of larger mean clutch sizes. This appears to be the case among captive geese derived from eggs collected at the La Pérouse Bay colony (M. A. Bousfield, personal communication).
4. We attribute the apparent lack of directional selection for larger egg size to countervailing selection on some correlated trait.

Our hypothesis suggests why geese living in a nutrient limited environment do not appear to trade-off egg size with clutch size in order to increase their fecundity. While clutch size for an individual is relatively plastic, egg size, for selective reasons given above, is not. We suggest that future research should focus not on the selective regime on egg size per se, but on an examination of other characters, particularly physiological ones, which are correlated with egg size. Basal metabolic rate and gas exchange rates of eggs are two characters which come to mind.

12.5 *Summary*

1. Within a clutch, egg weight is low for the first egg, increases for the second egg and then declines to the last egg. Egg success is lowest among first and last eggs of a clutch. First eggs are frequently lost due to pre-incubation failure and last eggs are abandoned more frequently at hatch or lost as goslings. Thus hatching success is lowest among those sequences in the clutch which have smallest egg sizes, but analyses of egg size within sequence positions show that this lower fitness is due to sequence per se rather than egg size.
2. Egg weight increases with age at least up to 4 years of age, at the individual and population levels.
3. The variation in egg weight within birds is small.
4. Mean egg weight varies significantly among seasons, apparently due to environmental effects on the population as a whole. There has been a slight but highly significant long term decline.
6. The heritability of egg size is *c.* 0.53, thus differences in mean egg masses among individuals result in part from genetic differences.
7. There is no correlation between egg weight and clutch size, nor between egg weight and laying date.
8. There is no relationship between female body size and egg size.
9. There are no fitness differences among birds of different mean egg mass, suggesting that selection is not operating or operating only weakly on egg size.
10. There is little evidence for temporally varying selection pressures despite 12 years of suitable data.
11. We suggest that the high heritability and lack of selection on egg weight is due to selection acting on a correlated physiological trait.

13 Body size

Variation in body size has been widely studied by evolutionary biologists, in part because it is relatively easy to measure. In diverse animal taxa, especially those with indeterminate growth, an individual's size is an overwhelming predictor of its social status, relative fecundity, or probability of survival. This is not necessarily so for adult birds. At certain ontogenetic stages, such as when nest-mates compete for food brought by parents, relative body size may be a life or death matter. As adults, however, phenotypic variation in body size is apparently usually unrelated to components of fitness. In the exceptionally well studied Galapagos Large Cactus Finch, where information was available on both genetic and phenotypic variation, body size was generally unrelated to fitness, although there was selection on beak shape (Grant and Grant 1989). Even in highly polygamous species, in which sexual dimorphism is often attributed to directional selection for larger members of the limited sex, there is little evidence from field studies for directional selection for larger adult body size. In one case, the aerial acrobatic displaying Long-tailed Manikin *Chiroxiphia linearis* (McDonald 1989), the most successful males may actually be lighter or smaller, but most such studies find no correlations between body size and male success. Alatalo and Lundberg (1986) found normalizing selection on tarsus length in Pied Flycatchers (*Ficedula hypoleuca*).

Body size is more complex than the traits we have already considered. Size changes dramatically over the life of a goose and there are many ways to measure and quantify it, both in theory and practice. We are interested in structural size rather than daily or seasonal fluctuations in body weight and composition (Campbell 1979; Alisauskas 1988). In species studied thus far, avian adult body size variation has both a large heritable component (Boag and van Noordwijk 1987; Grant and Grant 1989) and substantial environmental variation, primarily due to variation in food availability during juvenile growth (James 1983; Boag 1987; Larsson 1993). Lesser Snow Geese have exceptionally fast growth rates for geese and for precocial birds in general, growing from *c.* 95 to 1400 g in a month, giving them size-specific growth constants similar to those of altricial birds (Aubin *et al.* 1986; Whitehead *et al.* 1990). One might imagine two sets of genes, those for growth rate and those for asymptotic size *per se*, potentially independent, but normally interacting with each other and with the environment input to determine adult body size (Lynch and Arnold 1988). The interplay of genetic and environmental components of variance makes body size an especially complex, but potentially fruitful, area of study.

We have organized our inquiry into body size variation as follows. We start with methodology. Second, we describe the development and degree of adult sexual size dimorphism. Third, we examine the ontogeny of body size variables measured at different stages of the life cycle, by correlation of an individual's sizes as an egg, gosling, fledgling, and adult. Fourth, we consider within- and between-season environmental effects on gosling growth rates and subsequent adult body size. We focus on this because variation in gosling growth rate has been invoked as the mechanism responsible for fitness differences with respect to timing within seasons (Chapter 11) and for long term changes in juvenile survival rate (Chapter 6; Cooch and Cooke 1991; Francis *et al.* 1992*b*; Cooch *et al.* 1993; Cooke and Francis 1993). Fifth, we consider the genetic side in a more detailed examination of the heritable component of variation in body size than that presented in Chapter 8. The final data sections examine the relationships between body size and fitness at different stages of the life cycle. We start with survivorship analyses as functions of body size at hatch, gosling growth rates, and pre-fledging weight. We use a genetic marker for variation in growth rate, namely sex differences, to assess the potential interplay of genetic and environmental effects on growth rate at these stages. We test for differences in age of first breeding and adult survivorship with respect to body size. Finally, we consider adult fecundity, using our standard set of fecundity fitness components, with a special emphasis on clutch size.

13.1 *Measures of body size*

In this chapter, we consider only three body size variables: body weight, length of the tarsometatarsus (tarsus henceforth) and length of the exposed culmen (culmen). All three variables were measured on goslings just prior to fledging and adults during our annual banding operation, typically conducted 4–5 weeks after hatch. We measured goslings which were web tagged and thus of known age. We adjust gosling metrics for age when captured in our analyses with linear regression, which is valid over the short time period of the banding drives (Cooch *et al.* 1991*a*). In our data set, body weight is as repeatable a measurement at this time as are the skeletal metrics, and despite its obvious potential for seasonal variability, weight is in some respects the best summary of an individual's size. Only weight was measured on hatchlings, due to the imperative of processing and web tagging as many goslings as possible at this time of the season. weights of hatchlings have been adjusted for their hatching stage when measured (for example pipped egg, wet chick, or dry fluffy gosling), and fledgling weights were adjusted for the age of the gosling when measured. Further details on measurements and procedures are summarized in Davies *et al.* (1988), Cooch *et al.* (1991*a*), and Dzubin and Cooch (1992).

These variables can be simply and repeatably measured on live birds and are commonly enough used that comparisons to other studies are straightforward. They are not, however, interchangeable measures of 'size'. The three measures correlate significantly, but not strongly, with one another (tarsus–culmen, 0.30; tarsus–weight, 0.38; culmen–tarsus, 0.36; partial correlation coefficients between measurements, adjusted for differences among cohorts, $n = 3114$, all $P < 0.0001$). In some cases, we have combined the three measures using Principle Components Analysis (PCA) and extracted a single composite variable, *PC*1, which is often considered to be a general indicator of structural size. In our work, however, we have generally examined the relation of fitness to each of the measured variables as well as to the derived composite. Although the conclusions are generally independent of the variable used, there are some subtle differences that help illuminate the evolutionary role and dynamics of different aspects of body size. In essence, we feel that while each variable measures some attribute of the abstraction called body size, no two measure the same thing nor does one measure the totality.

These subtleties, and the importance of considering these variables separately, have been further resolved in a recent, more detailed analysis of structural size of Snow Geese (Ratner *et al.*, in preparation). Our initial assessments of structural size combined the three variables measured on live birds with PCA to generate *PC*1, a composite measure of structural size (Davies *et al.* 1988; Cooke *et al.* 1990*a*). In those analyses, variance in culmen, tarsus and weight (or its cube root) contributed approximately equally to the *PC*1 variable. Ratner *et al.* used measures of more than 25 bones from skeletonized specimens to generate a more generalized measure of structural size for this species. While tarsus contributed strongly to their *PC1*, culmen did not. Most revealing, perhaps, was the relation of this general, skeletal indicator of structural size to the lengths of tarsus and culmen measured on the specimens before they were sacrificed. The correlation of *PC*1 with 'live tarsus' was $r \cong 0.73$ ($P \leq 0.01$)

while that with 'live culmen' was only $r \cong 0.03$ ($P > 0.80$). Clearly, tarsus and culmen are not equivalent indicators of structural size. Both are, however, commonly and historically used morphometrics and since they may indicate different things about body size, we have included them both here.

13.2 *Sexual size dimorphism*

Male and female Snow Geese begin life with identical body weights. The means and variances of egg size and hatched gosling size do not differ significantly by sex (Table 13.1). By 5 weeks of age, males are 1.06 times heavier than females. The rate of change in gosling size does not differ between the sexes over the period in which we measure pre-fledging weight (28–45 days), so the sexual dimorphism principally reflects differences in growth rate during the first four weeks of life. The values for adults in Table 13.1 are not strictly comparable with those for earlier stages, since the sample of male adults consists almost entirely of immigrants. We know little of the selective mechanism which maintains sexual dimorphism in Snow Geese (Chapter 5). The low philopatry rate of males severely limits our ability to determine the consequences of variation in male growth rate or body size from fledging through breeding age. However, differential changes during the study in growth and survival

Table 13.1 Sexual dimorphism in egg, hatching, fledgling, and adult masses, and in adult tarsus and culmen. Values for egg, hatching and fledging mass are least-squared means from models correcting for year, hatching condition, and age at banding (n = number of birds, 1976–1992). Values for adult characters are means of annual means based on annually balanced samples of raw data for both sexes (n number of years, 1976–1986); number of birds = 2200 males and females for mass and tarsus, 1891 for culmen)

Variable	Male $x \pm$ SE	Female $x \pm$ SE	Dimorphism (M/F)	n
Egg mass (g)	125.3 ± 0.8	125.6 ± 0.8	1.00	797
Hatching mass (g)	95.1 ± 1.5	94.9 ± 1.5	1.00	2310
Fledging mass (g)	1342.8 ± 7.8	1264.9 ± 8.5	1.06	2638
Fledgling tarsus (mm)	95.2 ± 0.2	91.4 ± 0.3	1.04	2031
Fledging culmen (mm)	41.4 ± 0.1	40.5 ± 0.2	1.02	2636
Adult mass (g)	2180.5 ± 39.7	1952.5 ± 48.5	1.11	11
Adult tarsus (mm)	99.6 ± 0.6	96.4 ± 0.7	1.03	11
Adult culmen (mm)	56.6 ± 1.1	54.8 ± 1.3	1.04	19

rates of goslings by sex provide an intriguing tool for assessing selection on growth rate (Cooch *et al.* submitted, see below).

13.3 *Correlation of body size measures at different stages*

What are the relationships among measurements taken at different stages in the life of a goose? Once the birds become breeding adults, they show little change during their lives. To what extent does egg size, size at hatch, or fledgling size correlate with adult size? Table 13.2 shows the correlations between the weights of individuals at these four stages. There is a strong correlation between fresh egg weight and hatch weight. The correlations between egg or hatch weight and fledgling weight are far weaker, but still significant. Most of the variation in fledgling and adult weight must be explained elsewhere.

Table 13.2 Correlation between the mass of individuals as eggs, hatchlings, fledglings, and adults, and between fledgling and adult tarsus and culmen. Correlation coefficients are shown above the diagonal, with sample sizes given below. All correlations are significant at $P < 0.001$

Stage	Egg	Hatching	Fledging	Adult
Egg	—	0.56	0.27	
Hatching	9793	—	0.23	
Fledging	402	2434	—	0.56
Adult			345	—

Pre-fledging body size measurements are close to asymptotic adult sizes for tarsus (95 per cent), but only 74 per cent and 63 per cent for culmen and weight (Table 13.1). The correlation between fledgling and adult weight is reasonably strong (0.56), and those for tarsus and culmen sizes are stronger (tarsus, 0.83; culmen, 0.64; $n = 174$, $P < 0.0001$; Cooch *et al.* 1991*a*). Thus, whatever determines the size of a goose during its first 35 days of life has a lifetime relationship to adult size. The extent to which the variance among geese is due to the differences in genes which are expressed at this time, or due to variation in the environment during the early growing period, is considered in the next sections.

13.4 *Environmental effects on gosling growth rate*

We have analysed variation in gosling growth rates within seasons and among years in relation to environmental variables.

13.4.1 Changes within seasons

There is no seasonal trend in the size of goslings at hatch (Cooch *et al.* 1991*a*), as expected from the similar lack of a trend in egg size. However, goslings which hatch later in the season grow more slowly than those which hatch earlier, as evidenced by their smaller age-adjusted pre-fledging weight (Fig. 13.1). Some of the residual variation in the relationship between fledgling and adult size may be due to the timing of hatch, which was not corrected for in the simple correlations given above. Late hatching goslings could grow more slowly, but eventually reach the same mean asymptotic size as early hatching birds. Although there is some compensatory growth, it is insufficient to allow late hatching goslings to completely catch up with the early hatching goslings (Cooch *et al.* 1991*a*).

There are a number of possible explanations for the seasonal decline in growth rate. (1) Phenotypically larger adults may nest and hatch earlier than smaller ones, perhaps passing on genes coding for faster growth rates. (2) Birds may lay larger eggs which produce larger goslings earlier in the season. (3) Environmental conditions for favourable growth may decline during the season, for example food availability may decline as the season progresses. Cooch *et al.* (1991*a*) argued against the first two hypotheses by showing that neither female body size nor mean egg size correlated with hatch date. They concluded that changes in fledgling size as the season progressed were due to a slower growth rate of later hatching goslings.

The salt marshes around La Pérouse Bay provide much of the food supply for the growing goslings during the first few weeks of life. The earliest hatching goslings have earliest access to this food, whereas those which hatch later have to compete with other and larger families. Bazely (1988) showed that grazing pressure over the course of the season reduced the total amount of above-ground forage at La Pérouse Bay, and Hik and Jefferies (1990) showed a reduction in the capacity of vegetation to show compensatory growth following severe grazing. Thus later hatching goslings are at a competitive disadvantage.

13.4.2 Changes among years

Mean pre-fledging body sizes differ significantly from year to year (Fig. 13.2, 1976–1988). The three characters vary in parallel from

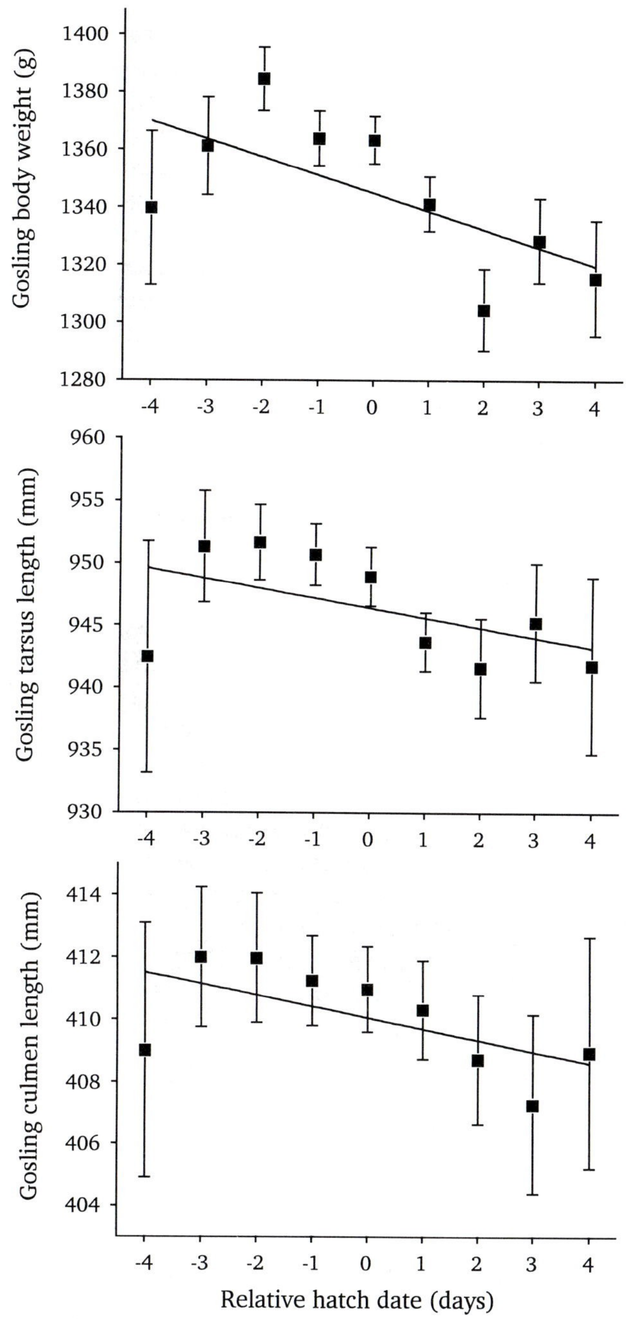

Fig. 13.1 Partial regression of pre-fledging measurements adjusted for brood size and gosling age with respect to hatch date scaled relative to annual means. Means are plotted with standard errors. (Adapted with permission from Cooch *et al.* 1991*a*).

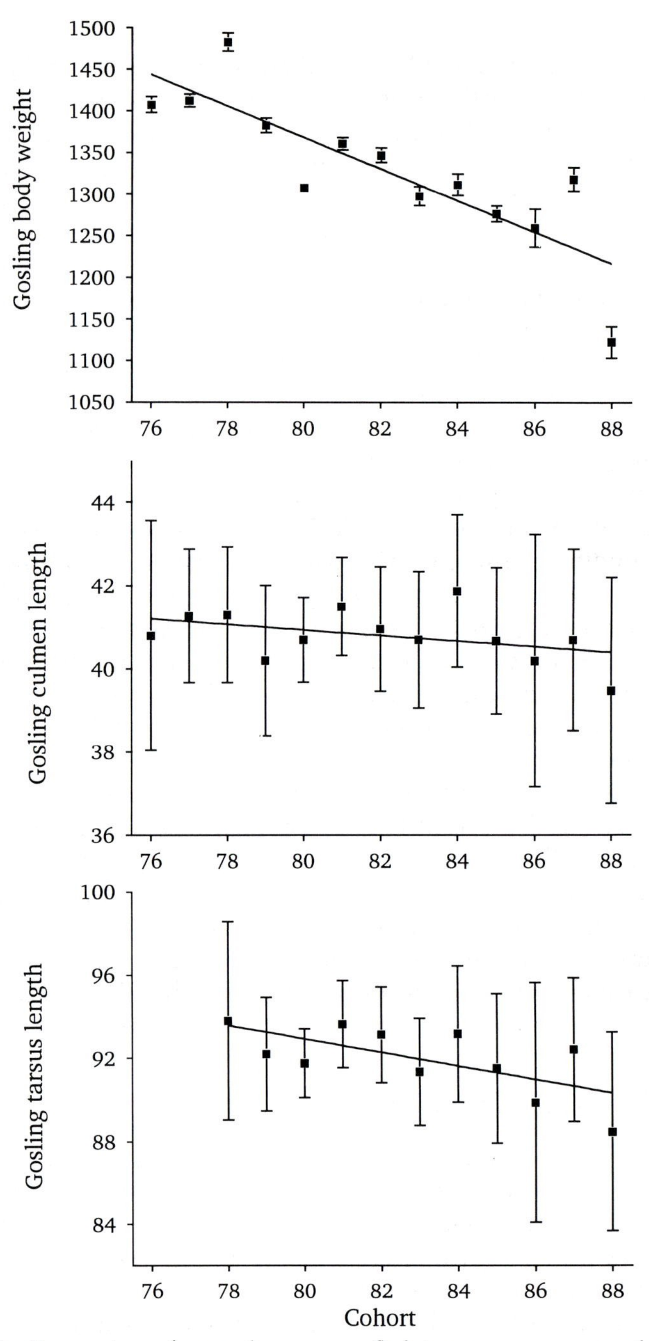

Fig. 13.2 Regression of annual mean pre-fledging measurements of goslings. Values are least-square estimates controlling for hatch date, brood size, and weather in the natal year. Means and standard errors are shown. (Adapted with permission from Cooch *et al.* 1991*b*).

year to year, and all declined during the course of the study, with the largest proportional effect in weight. This annual variation in fledgling size is uncorrelated with fluctuations in hatch weights (Cooch *et al.* 1991*b*). Such annual variation could reflect birds of different genetic constitutions breeding in different years, variations in environmental conditions during the period of gosling growth, or some combination of the two.

Several observations favour attributing the bulk of this variation to environmental causes (Cooch *et al.* 1991*b*, 1993). The long term decline in gosling growth rate parallels the long term decline in the availability of local salt marsh vegetation during the course of our study (Chapters 2 and 6). Families which dispersed from La Pérouse Bay into an area with lower grazing pressure produced faster growing offspring than did families which stayed in the bay. When the effect of the long term decline is corrected for, post-hatch weather conditions also correlate with fledgling size, with lower growth rates in cold wet summers. Finally, measurements of broods of goslings of individual females from several years are smaller for goslings produced in later years, and the degree of size difference was greater with more years between broods. This powerful result proves that the decline is attributable to environmental, rather than genetic, change.

To summarize, gosling growth rates differ within and among years, and much of this variation can be attributed to differences in the feeding environment during the period when the goslings are growing rapidly. Growth rate declined significantly during the course of the study, concurrent with the decreased per capita availability of favoured salt marsh forage. Weight shows the largest proportional variation, but measures of structural size are also affected.

13.5 *Variation in adult size*

Annual cohorts of adult females differ in size (Fig. 13.3; Cooch *et al.* 1991*b*). Since adult size reflects in part the gosling growth rate in the natal year, a cohort's mean adult body size can be used as a *post hoc* index of the growth rate in the year in which the goslings hatched. Although we only started to weigh and measure geese in 1978, we have measured adult females from cohorts as early as 1970. Interpreting these data as reflections of goslings growth rates assumes that adult mortality is unbiased with respect to size, which is the case for years for which we do have the data (see below). Cohort-specific tarsus, culmen and weight of adults have declined significantly from

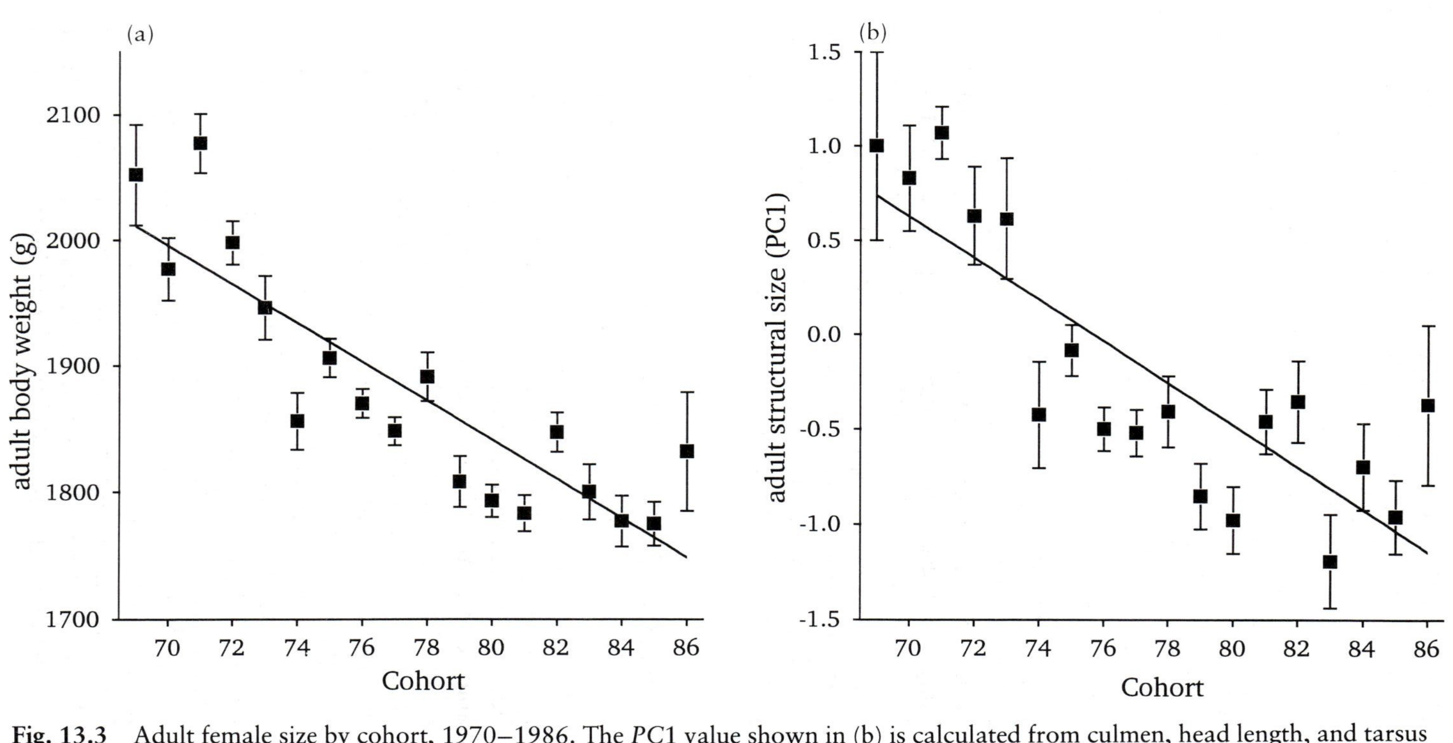

Fig. 13.3 Adult female size by cohort, 1970–1986. The *PC*1 value shown in (b) is calculated from culmen, head length, and tarsus (see text). (Reproduced with permission from Cooch *et al.* 1991*b*).

1970 to 1986, with the largest proportional change occurring in body weight. We cannot rule out the possibility that gene flow is also bringing genes for small size into the colony, but a good working hypothesis is that as nutrient availability has declined, the birds have grown more slowly and reached a smaller adult size.

We can make a rough partitioning of the proportion of environmental variation in adult body size by comparing the ranges of seasonal and annual variation with the total variation in the population. For weight, which appears to be the most plastic character, annual means differed by about 200 g, and seasonal variation averaged about 100 g from earliest to latest hatching goslings (Cooch *et al.* 1991*a*,*b*). The standard deviation in the population, pooled over years, is 138.0 g, thus 95 per cent of the population falls within about a 275 g range. These numbers suggest that annual and seasonal variation could easily account for more than half of the variation in body size present in the population; genetic variation could account for the balance. The heritabilities of body size characters presented earlier (Table 8.1) give a similar picture: genes and environment contribute roughly equally to the body size variation present in the population. A closer examination of assumptions behind this interpretation of the heritability values is given in the next section.

13.6 *Heritable effects on body size*

Adult body size in Snow Geese is clearly affected by environmental conditions during growth. Nonetheless, a significant proportion of the phenotypic variance in all three body size variables can be attributed to heritable variation (Table 8.1). As explained there, 'heritable' could include sources of mother/daughter (M/D) similarity other than the additive effects of nuclear alleles. These included common environments (owing to female natal philopatry) and various factors generally lumped under the term maternal effects (Falconer 1989; Chapter 8).

A standard approach to gauge the relative contributions of these sources to M/D estimates of heritability is to compare them to estimates based on father/daughter (F/D) regressions. 'Maternal effects' related to factors such as mitochondrial DNA, egg quality and maternal behaviour obviously would not contaminate F/D estimates. Moreover, since fathers at La Pérouse Bay were generally hatched and fledged at other colonies, the effects of common environment should be minimal (Davies *et al.* 1988). Fortunately, we can estimate the heritability of our body size variables with this

Table 13.3 Body size heritability estimates for female Lesser Snow Geese at La Pérouse Bay

Method	Trait	Lower confidence limit	Heritability	Upper confidence limit	Sample size
Mother/daughter	culmen	0.06	0.38	0.67	144
	tarsus	0.33	0.70	1.05	144
	mass	0.31	0.58	0.86	144
Father/daughter	culmen	−0.00	0.36	0.73	62
	tarsus	0.06	0.50	1.00	62
	mass	0.46	0.91	1.31	62
Midparent/daughter	culmen	0.03	0.30	0.56	54
	tarsus	0.24	0.50	0.79	54
	mass	0.26	0.55	0.88	54

In all cases, the weighted regressions of mean daughter score on mother, father or midparent were significant indicating heritability differed significantly from 0. The point estimates and confidence limits are based on 10 000 bootstrapped, family-size-weighted regressions of mean daughter scores on the scores of mothers, fathers, and midparent. All scores were adjusted for annual variation with covariance procedures. Within a method, the same random sample of repeated individual measures was used for each variable.

approach since, unlike our reproductive success components, the variables can be measured for both sexes.

For weight, tarsus, and culmen, the estimated heritability based on F/D regression differs significantly from 0 (Table 13.3), a result consistent with that found using M/D regression. Although there are numerical differences for each variable in the point estimates from the two methods, the respective confidence limits overlap substantially and we do not take this as evidence for significant differences. Since the M/D and F/D estimates do not differ and since F/D estimates are not contaminated by maternal or common environmental effects, one might argue that the major contributor to the heritability of body size is additive genetic variance.

Aside from the statistical problems with such arguments (β error), there is a biological complexity of Snow Geese that must be considered. Males contribute substantially to brood rearing in this species, especially during the early period where females are replenishing incubation weight losses and goslings are growing rapidly. This raises the potential for 'paternal' effects contributing to the similarity of fathers and daughters. Suppose, for example, that larger males were able to sequester better feeding areas or provide more protection from the disruptive activities of predators. Their daughters would be able to feed more on, perhaps, better forage and should grow both faster and larger (above). The result would be that males that are larger, for whatever reason, would have larger daughters and the heritability of body size estimated from F/D regression would be contaminated by paternal effects.

The similarities of body size heritabilities estimated from M/D and F/D regressions do not necessarily mean that additive genetic variance is the primary contributor to the heritable portion of body size variation in this species. It is possible, at least in Snow Geese, that variance due to maternal and paternal effects could, in essence, 'substitute' for each other. It is our view, however, that perfect substitution is unlikely and that some portion of the heritable variance is due to variation in genes with additive effects. Unfortunately, determining the precise contribution of each factor requires cross-fostering manipulations that are exceedingly difficult in studies of this type.

Larsson and Forslund (1992) calculated heritabilities of body size in a population of Barnacle Geese. In contrast to our study, M/D heritability of tarsus was significantly higher than F/D heritability, although no such differences were found for heritabilities of weight or head length. Larsson and Forslund argued that maternal effects through cultural transmission of foraging sites accounted for the higher M/D values. The mother–daughter correlation in quality of foraging sites may be larger in the island population of Barnacle Geese studied than has been the case as the environment changed for Snow Geese at La Pérouse Bay.

Finally, we must comment on the relation of heritability and assortative mating. Davies *et al.* (1988) correctly explained that heritability estimates must be adjusted if there is assortative mating based on the phenotypic value of the trait being considered. By this they meant, for example, that large females selected large males because they were large. While there is an overall positive correlation in body size among mated pairs of Snow Geese at La Pérouse Bay, the correlation reflects the facts that mates tend to be chosen from the same cohort and that the mean sizes of cohorts differ (Chapter 6). Such cases are termed 'primary environmental resemblance' by Falconer (1989) who explains that in such cases adjustments to heritability are not warranted. The similarity of heritabilities estimated from the regression of daughters on mid-parent values to those from M/D and F/D regressions (Table 13.3) supports our decision not to adjust for intra-pair body size correlation since the former are uninfluenced by such correlations.

13.7 *Body size and fitness*

How do the patterns of body size variation outlined above relate to variation in survival and reproduction? We first examine survival probabilities as a function of body size and growth rate, from hatching through adult life. Second, we examine fecundity components of fitness of adults with respect to body size, with an emphasis on clutch size.

13.7.1 Survival components

Hatchling size

Does selection during gosling growth to fledging differ with respect to initial hatchling size? Since egg weight was unrelated to the probability of producing a fledgling (Chapter 12), and hatch weight covaried strongly with egg weight, it will be surprising if we find effects in this analysis. We have much larger samples of weighed hatchlings than fresh eggs, however, and since the egg size-hatch size correlation was far from perfect, a direct analysis is warranted.

We cannot perform a within-brood analysis of size effects, as was done for egg size (Williams *et al.* 1993*c*) because we lack sufficient data to correct for hatching sequence effects, which would likely confound our results. Thus, our analysis compared fledging success among broods of different sizes. A set of multivariate models examined the relationship between the mean gosling weight per brood and the proportion of goslings known to have survived at least to the fledging stage. To minimize biases, only families where all tagged goslings were weighed were used in the analysis, which also controlled for year, brood size at the nest, and hatch date. Data for 15 951 nests (1976–1988) were available. Hatch weight was not significant as a linear (directional) or quadratic (non-linear) main effect, and no model contained a term with a significant interaction involving weight. Thus, in the data set as a whole, we detect no consistent differences in survival probability with respect to mean gosling size at hatch. Nonetheless, analyses of individual years suggest that hatch weight effects may occur (Table 13.4). In 1983, the year when breeding was latest, due to slow snow melt, broods known to produce one or more fledgling(s) had slightly higher mean hatching weights. Although not quite statistically significant in Table 13.4, a hatch weight effect was significant in an ANCOVA model with brood size and hatching date as classification variables ($P = 0.02$). In 1987, when broods which fledged goslings had significantly lighter hatching weights than those not known to, the ANCOVA model

Table 13.4 Mean weight of goslings at hatch, among broods which were known or not known to have produced at least one fledgling. Mass is expressed as a residual from ANOVA on hatching condition. Differences in means tested with *t*-tests, *F*-tests for variances

	No known fledgling			Known fledgling(s)			*P*	
Year	Mean mass	SD	*n*	Mean mass	SD	*n*	Mean	Variance
76	0.10	6.56	173	−0.42	6.50	80	0.56	0.94
77	−0.78	6.49	469	−1.07	6.61	312	0.54	0.71
78	−3.74	8.58	722	−2.97	7.47	213	0.23	**0.02**
79	−0.78	6.39	128	−1.58	6.21	55	0.44	0.83
80	1.12	7.41	774	0.97	7.06	348	0.73	0.31
81	1.34	7.28	719	2.15	7.36	228	0.15	0.82
82	0.41	8.13	1050	0.12	6.72	337	0.56	**0.00**
83	−0.52	7.47	1601	0.38	7.22	285	0.06	0.46
84	1.11	7.50	1719	0.65	7.29	297	0.32	0.54
85	1.15	7.13	1837	0.50	7.24	247	0.18	0.73
86	−0.96	8.10	1419	−1.09	8.20	120	0.86	0.83
87	−1.07	6.79	855	−2.88	7.26	74	**0.03**	0.40
88	−0.16	8.36	1485	0.21	7.58	68	0.72	0.30

again confirmed an effect of hatchling weight. Among recent years, goslings both grew and survived well at La Pérouse Bay that year (Figs. 6.2, 13.2), apparently due to weather conditions which were favourable for plant growth. Two years in Table 13.4 have significantly higher variances among nests with no recruits, providing weak evidence of normalizing selection.

One might interpret our results as evidence that annual fluctuations in selection differentials do contribute to the maintenance of genetic variation in egg size (Ankney and Bisset 1976; Ankney 1980), which would result in variation in hatchling size. In a particularly stressful year, we appear to have detected a higher survival rate in goslings with larger hatching weight. It is more difficult to see why smaller goslings might survive better than larger ones in a benign year, and the fluctuating selection pressure argument is usually couched in terms of a trade-off against increased egg number, which seems not to occur (Chapter 12). Finally, normalizing selection suggested in 2 years is at variance with fluctuating selection pressures. We do not take our results as evidence that fluctuating selection pressures are strong contributors towards the maintenance of the genetic variation in egg and thus hatchling size. We argue simply that the selection gradient of survivorship is flat over the range of hatchling sizes in the population.

Growth rate

The period between hatching and fledging has relatively high mortality rates (Table 4.1), thus selection could be acting strongly at this stage. Four correlational analyses suggest that pre-fledging size, determined largely by gosling growth rate, strongly affects survival probabilities both prior to and after fledging. Over years, fledging success has decreased significantly in parallel with the decreased mean fledgling size (Figs 13.2 and 6.1; Cooch *et al.* 1993; Williams *et al.* 1993*b*). Smaller fledglings in more recent years appear more vulnerable to the stress imposed by our banding drives, implying lower viability in nature (Williams *et al.* 1993*d*). Within years, we could relate the slower growth rate of broods hatching later in the season to the seasonal decline in recruitment (Figs 13.2 and 11.8). Finally, a within-year spatial analysis shows that mean brood sizes are larger in foraging areas where fledgling sizes are larger, suggesting differential mortality by size (Cooch *et al.* 1993). While the inference that slower growth rates cause lower survival rates is reasonable, we cannot directly test survivorship probabilities as a function of growth rate because we lack body size data from dead goslings.

The analysis in the previous section suggests that mortality at this stage is random with respect to initial hatchling size, a character with a reasonably high heritable component, but it may not be random with respect to genetic variation for growth rate itself (Lynch and Arnold 1988). Suppose that each bird has an optimal potential growth trajectory which is partly determined by genotype. If environmental conditions are suitable, this trajectory can be followed, but in an adverse environment, the bird grows less well, is in poorer condition relative to its optimal developmental rate, and has a lower probability of survival. Birds with different trajectories may be differentially advantaged or stressed in different environments, causing their relative survivorship probabilities to change in different environments. Plots of changes in phenotype of a single genotype in differing environments are called 'reaction norms' of phenotypic plasticity (Stearns 1992). In the situation imagined above, the reaction norms of two growth rate genotypes might cross, such that one genotype had a higher survivorship probability in one environment while a second genotype was favoured in a different environment. In our case, birds predisposed to more rapid growth might be more stressed under poorer feeding conditions, while birds with slower optimal trajectories would be less affected.

To assess the potential applicability of the process outlined above, we would need a marker of a bird's genetic growth rate trajectory. Given the heritabilities for body size (Table 13.3), one potential source of genotypic information would be parental body size. In theory we might assess a gosling's condition relative to an optimal genotypic trajectory by comparing its phenotypic deviation from parental size (van Noordwijk *et al.* 1988; Alatalo *et al.* 1990). However, this is not possible in our system, primarily because of the significant environmental component of adult size itself. However, we are fortunate in having large samples of fledglings which carry a genetic marker for different ranges of growth rate genotypes, namely all males, genotype ZZ, and females, genotype WZ. Since the distributions of developmental trajectories of the sexes differ for genetic reasons, with males growing more rapidly than females, we can look for evidence of selection on growth rate genotypes by comparing the fates of the two sexes under different environmental conditions (Cooch *et al.*, submitted).

Sex ratios at hatch appear to be equal (Cooke and Harmsen 1983). When our study began, the sex ratios of fledglings caught in our banding drives were slightly biased towards males (50–52 per cent male), whereas in latter years, they were slightly female biased (48–50 per cent male), producing a significant long term trend (Fig. 13.4). Male goslings, with faster growth rates, apparently had higher survival rates under prime feeding conditions, but as food became scarcer, males became less likely than females to survive. Although the mean body sizes of both sexes have declined, males have declined more quickly than females (*c.* −100 g versus −50 g, 1976–1991), resulting in less dimorphic goslings.

We know little of the selective mechanisms which maintain sexual dimorphism in Snow Geese (Chapter 5), and the low philopatry rate of males precludes our direct assessment of potential size-specific benefits for males. Our data imply that the degree of sexual dimorphism maintained in Snow Geese is in part determined by an interaction of sex and growth rate or size differential gosling survival. Sex differences unrelated to size might be responsible for the changes, however growth rate is the most obvious sexually dimorphic characteristic of goslings at this stage. This situation creates a selection gradient on variation in 'degree of dimorphism' genes which parents pass on to their offspring. The most productive set of genetic instructions which parents pass to their offspring with respect to sex-differential growth strategies will change with environmental conditions. Under highly productive conditions, genes for relatively larger males will be favoured, assuming that larger surviving males have some advantage in mating competition, predator defense, or whatever process

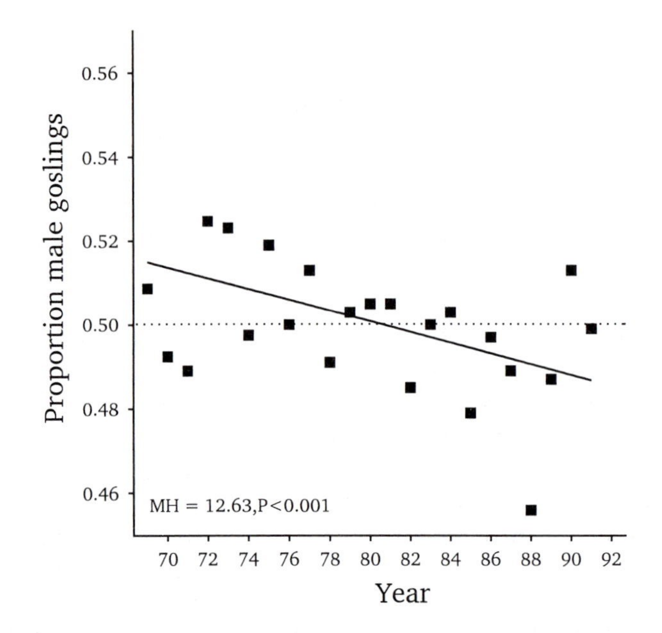

Fig. 13.4 Sex ratios of Snow Goose goslings captured in pre-fledging banding drives at La Pérouse Bay, 1969–1991.

favours dimorphism. Under less productive conditions, these favoured genotypes will have lower survival probabilities as goslings, and a smaller degree of dimorphism might be favoured. Studies of size and sex specific variation in mating competition, reproductive performance, or adult survival may demonstrate differential benefits for the two sexes with respect to body size, but the differential costs imposed at the growth stage will have to be considered to provide a complete picture of the selective regime on male and female body sizes.

For the more general argument, the importance of the sex ratio analysis is that birds in poorer condition relative to their potential growth trajectory in a richer environment appear to be disadvantaged at this developmental stage. We have documented directional selection and a genotypic response, namely more W and fewer Z chromosomes, resulting in slower growth rates among surviving fledglings, which are now biased towards females. We expect that a similar selection gradient on growth rates exists within sexes as well as between them, although we lack the genetic markers to prove it.

Survival from fledgling to breeding adult

Does a fledgling's size influence its likelihood of surviving to breeding age? As with survival to fledging, correlational studies document such a relationship. First-year survival has dramatically decreased (Figs 6.6 and 6.7). Among years, there is a strong positive relationship between mean annual pre-fledging weight and annual survival rates, based both on recovery data (Francis *et al.* 1992*b*), and on local survival rates, based on recapture data (Fig. 6.9; Cooke and Francis 1993). The parallel declines in body size and survival need not be causally related, of course. An analysis examining body size effects within cohorts would provide a stronger test of direct effects of fledgling weight on survivorship.

We performed such tests using our most sensitive methods, estimates derived from recapture models (Lebreton *et al.* 1992). One uses these techniques by contrasting the the goodness-of-fit of the data to various models constructed under different assumptions about the pattern of variation in survivorship and recapture among different classes of individuals (for example, age, cohort, size). Following the principle of parsimony (*sensu* Lebreton *et al.* 1992), we searched for the model which maximized the amount of variation explained while minimizing the number of model parameters. The variables involved were: yearly cohort, age class, time class in the season hatched, and body size class as predictors of survival and recapture rates. Data were available for 2543 female goslings which were measured in banding drives between 1976 and 1987.

We are interested in the survival rates of goslings, but we have few recaptures of yearling birds, who are rarely captured with the post-breeding flocks we band. Thus our first age class spans ages 0–2 years. The estimated survival rate of the first age class was a compound of the survival rate of gosling to yearling and yearling to 2 year-old. Since juvenile mortality rates are *c.* 2–4 times those of older birds (Chapters 4 and 7), most of the differences in cohort survival rates may be attributable to effects in the first year.

We categorized body size into large and small categories (66 and 33 per cent of the data) relative to the distribution of absolute pre-fledging gosling weight pooled over all years. The cut-offs for each category were determined by preliminary analysis, and reflect the approximate point of significant departure in the results between categories. We used weight rather than other characters since it was measured over the greatest span of years and probably reflects variation in gosling condition better than any other character.

We divided the data into two hatch date classes since hatch date significantly influences both gosling growth rate and the probability of recruitment, and we wanted to assess body size effects independent of hatch date. We categorized the hatch period into early and late periods (60 and 40 per cent of the data, respectively) relative to the annual mean and variance, with the cut-off chosen in the same manner as for body sizes.

For both early and late, the most parsimonious model fitting the data was one which used a two age-class model of survival rate (ages 0–2 years versus older). The most parsimonious model in all cases had a constant survival rate across years within the first age-class, annually varying survival rates of the second age-class. Adult survivorship (age $\geq$ 2 years) was held constant, consistent with the survivorship analyses of adults (see below and Chapter 8). To test for significant differences in first-year survival among size classes, we compared the goodness-of-fit of the most parsimonious model which included varying parameters for size class with the nearest neighbouring model which assumed equal parameters for large and small goslings, using a likelihood ratio test (Lebreton *et al.* 1992). Significant deviation between the models was interpreted as indicating significant differences among size classes.

What did we find? Within the early hatch period, we detected no significant differences among large or small goslings in first-year survival ($\phi_{lrg} = 0.256 \pm 0.030$; $\phi_{sml} = 0.249 \pm 0.075$; $\chi^2_1 = 0.09$, $P > 0.8$). However, among late hatching goslings, bigger goslings had significantly higher first-year survival than did smaller goslings ($\phi_{lrg} = 0.341 \pm 0.057$; $\phi_{sml} = 0.130 \pm 0.041$; $\chi^2_2 = 0.6.89$, $P < 0.01$).

Selection appears to be acting against small goslings hatching late in the season, presumably because these individuals cannot compensate sufficiently given the time and/or nutrient base available later in the season. However, small size per se appears to be no disadvantage, since there are no size differences earlier on. It is doubtful that the difference in size-specific survivorship pattern between early and late hatching groups has genetic consequences with respect to body size. There are no seasonal trends in the body sizes of breeders, nor in gosling sizes at hatch, as might be expected if early and late breeders differed with respect to body size genetics. Thus, the size-specific difference in local survival rates operates on an environmental component of body size variance (van Noordwijk *et al.* 1988).

Can this pattern of selection account for the drastic long term reduction in juvenile survivorship (Figs 6.6 and 6.7)? Figure 13.5 plots the proportion of 'large' birds in our samples as a function of year and time period. Keep in mind that our size classification in this analysis is absolute, not relative to goslings hatched each year. On average, goslings have been getting smaller each year during both early and late time periods, but the change during the late hatching period is drastic. Whereas such chicks were 90 per cent classed as large in 1976, only <5 per cent were so in 1987! This increase in the

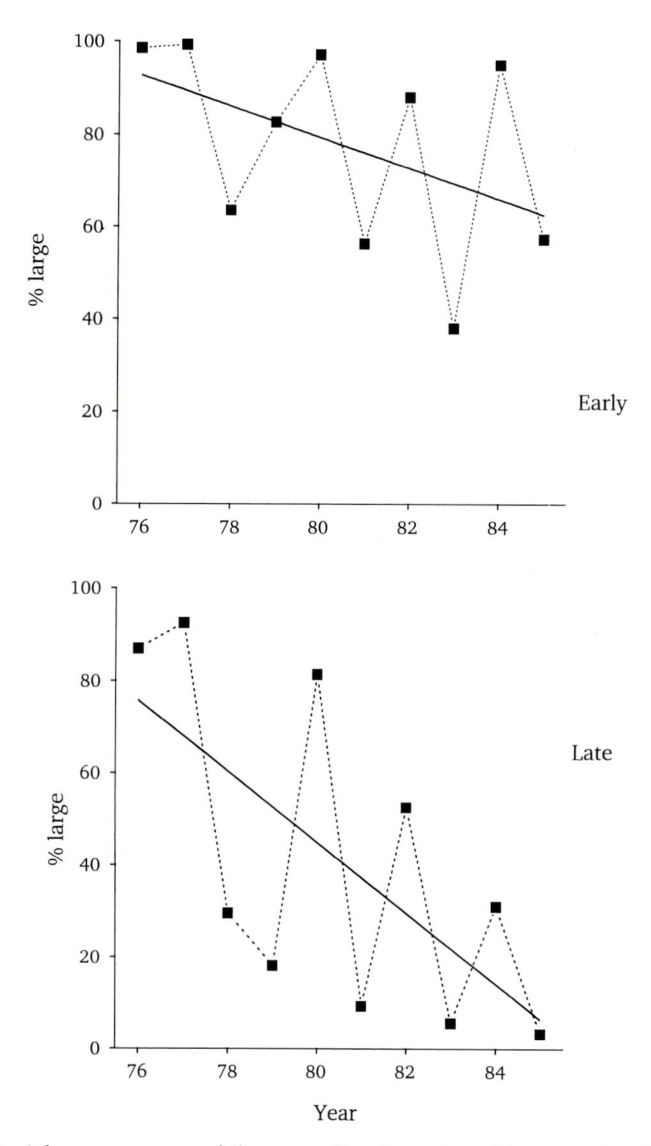

Fig 13.5 The percentage of 'large goslings' produced by nests hatching relatively early or late each year. See text for definitions.

number of small late-hatching goslings will lead to an overall decrease in first-year survival of goslings pooled over all hatch periods, as was done in the annual recovery and recapture analyses of first year survivorship. This analysis strongly supports slower growth rates, especially of late-hatching goslings, as a likely cause of the long term decrease in first year survivorship.

Age-specific breeding propensity

Davies *et al.* (1988) found that the mean body size of two- and three-year-old birds was smaller than that of older birds. They interpreted this as showing that smaller birds breed at younger ages. This analysis was confounded by the long term decline in body size, which was not considered in the analysis. Re-analysis showed that the pattern could be explained because more recent cohorts, which were smaller, were under-represented in the older age classes (Cooch *et al.* 1992).

We can test for body size effects on breeding propensity using local survivorship models. The recapture parameter estimates in these models tell us how likely we are to encounter surviving birds of a given age and size. We tested recapture models which allowed breeding propensity parameters to vary by age (2–5+) and by size class as a gosling against those with uniform parameters across size classes. We followed the same principles of parsimony and model selection as described in the preceding section on fledgling to adult survival. There was no significant improvement in the model fit when size-specific parameters were included for either early ($\chi^2_4 = 2.16$, $P > 0.5$) or late-hatching birds ($\chi^2_4 = 3.31$, $P > 0.4$). This analysis provides no evidence of size differential age-specific breeding propensity.

Adult survival

Do adults of different sizes survive equally well? Ideally, the survival estimates available from band recovery data should be used to answer this question, but the recovery samples of measured geese are too small to test across body sizes. Instead, we have asked whether the patterns of local survival vary among size classes. Because of the long term decline in body size (Fig. 13.3), we can only use those birds which were initially banded as goslings, and therefore are of known cohort. To avoid biases, we restricted the analysis to birds measured at the same age; our largest sample was of birds measured at age 3. A reasonably large sample was available for cohorts banded 1978–1988. The sample of 643 females was divided into three categories, large, medium, and small on the basis of their adult weight. As a direct test of differences among groups, we began by using Test 1 of program RELEASE (following Lebreton *et al.* 1992). There was no indication of any differences overall among adult size classes using the fully time-specific (Cormack–Jolly–Seber) model ($\chi^2_{36} = 24.3$, $P > 0.9$). Further testing with more parsimonious models did not change this result. Thus, we have no good evidence that local survival rates differed among the three size categories of breeding adults. This result differs from that of Davies *et al.* (1988) who found smaller females were disproportionately frequent in older age classes. Our direct analysis of survivorship rates provides no evidence for size-related differences.

13.7.2 Fecundity components

Body size and clutch size have declined during the study. It is tempting to conclude that changes in body size result in smaller clutch sizes. Larger geese can and do carry more grams of nutrient reserves when they depart from migratory staging grounds for the arctic breeding colonies (Alisauskas 1988). Since there is no relationship between body size and egg size (Chapter 12), larger birds potentially have greater discretionary reserves to allocate to egg production than do smaller birds. Thus larger females carrying larger nutrient reserves should be able to lay more eggs (Alisauskas 1988; Alisauskas and Ankney 1990).

However, while the above arguments may appear to be sound, alternative predictions are also possible. The clutch size decline appears to be due to reduced nutrient availability prior to laying, whereas the reduced body size reflects reduced growth rate of goslings prior to fledging. Thus both declines may be attributable to a third common causal factor, reduced food at the colony, rather than themselves being causally related. In addition to an ability to carry larger reserves, larger birds have greater metabolic costs, including transportation costs to the breeding grounds. The magnitude of net discretionary reserves depends on the relative rates of increase of both reserves and costs with body size, which are unknown. Thus, the inference that larger birds have greater reserves to allocate to egg production because they carry larger reserves when leaving migratory staging areas is not necessarily true.

What do our data show? Davies *et al.* (1988) investigated the relationship between structural size and fitness components in the La Pérouse Bay population using females measured in the years 1978–1981. They divided the adult birds into three approximately equal sized groups defined as small, medium, and large on the basis of the first principal component (*PC*1) of a Principal Component Analysis comprised of body weight, tarsus, culmen, and head length. There were no differences in fecundity among the three size categories in terms of total clutch size, clutch size at hatch, goslings leaving nest, or brood size at fledging. Despite methodological problems with this analysis (Alisauskas and Ankney 1990; Cooke *et al.* 1990; Cooch *et al.* 1992), subsequent re-analysis confirmed these negative results.

The components of fitness for which we might expect body size to have its largest effect are clutch size, total clutch weight, and laying date. Cooch *et al.* (1992) tested whether clutch size and laying date correlated with body size at La Pérouse Bay. They performed the analysis within years, to circumvent coincident covariation due to the long term declines in body size and clutch size which confounded earlier analyses. To provide a robust sample, the clutch size measure used was clutch size at hatch, and laying dates were back-dated based on hatching dates and clutch size. A *PC*1 variable composed of culmen, head length, and tarsus was used as a measure of body size. Despite large samples, no significant relationship was found in 15 years of data for adult birds ($n = 962$, 1976–1990).

The only analysis we have done which has found a relationship between adult body size and a fecundity variable is that weight at banding, controlling for the number of days between hatch and capture, correlated negatively with hatch date, although tarsus and culmen did not (Cooch *et al.* 1991*a*). Assuming that body weight at this time correlate positively with weights at laying, this suggests an effect of body condition on the timing of laying consistent with the effect on clutch size. Females in better condition, regardless of body size, bred earlier in the season.

13.8 *Conclusions and speculations*

Body size is a complex character. We have found it helpful to consider separately the selective dynamics of three aspects of body size: initial size, growth rate, and adult size. The major results of our fitness analyses, summarized in Table 13.5, are: (1) within the normal range of variation, there is no evidence for selection on hatchling size. (2) The selection regime on growth rate between hatch and fledging in our population appears to have changed from one slightly favouring faster growing males to one slightly favouring more slowly growing females. If this difference between sexes is due to differential genetic growth trajectories, such selection, and a genetic response, may also have occurred within sexes. Although genes for slower growth may have been favoured in recent years, small individuals late in the season have a lower survival rate. This selection probably has little impact on the genetics of adult body size distribution of the population, since it is primarily an effect of timing of breeding which is itself uncorrelated with adult structural size. (3) Adult females have equivalent fecundity and survival over a wide range of sizes. The reasonably high heritability values for body size characters are con-

Table 13.5 Summary of fitness-body size analyses. Results from Davies *et al.* (1988), Cooke *et al.* (1990), Cooke and Francis (1993), Cooch *et al.* (1991*a*,*b*), Cooch *et al.* (1993), Francis *et al.* (1992b), and this chapter

Survivorship as a function of	
Hatchling size	no difference, perhaps weak fluctuating effects annually
Gosling growth rate	directional selection for faster or slower growth varies with environmental conditions
Fledgling size	covaries with cohort-specific juvenile survival rate
	positive spatial correlation with brood size and recovery rate with years
	positive temporal correlation with seasonal timing
	no effect of body size among early hatchers
	selection against smaller birds later in the year
Adult size	no difference
Fecundity with respect to adult structural size	
Clutch size	no difference
Egg size	no difference
Timing of breeding	no difference (positive mass effect)
Hatching success	no difference
Fledging success	no difference
Breeding propensity	no difference

sistent with this finding, although several other potential explanations are available, and, as always, gene flow may play an important role.

Variation in different aspects of body size has been invoked as the causal mechanism responsible for variation in Snow Goose demography, including demographic changes which have occurred at the La Pérouse Bay colony during the course of our study. Starting with hatchling weight, we have no evidence that variation within the range of sizes observed is related to gosling survival to fledging. Ankney (1980) showed that larger hatchlings could withstand starvation longer, but such severe conditions apparently have not been of selective importance in the 22 years of our study. They may be so more frequently at more northerly colonies, and we must remember that we are not studying a genetically isolated population.

The slower average growth rate of late hatching goslings, and the long term decline in pre-fledging and adult body size, are due to changes in the growth rates of goslings which can be directly attributed to changes in the availability of suitable forage on the traditional brood rearing areas of La Pérouse Bay. There may be population genetic consequences of this change through selection of genes influencing growth rate. We used the differential survival of males and females as a model for what might be occurring more generally in the population. Genetically, the direction of selection for growth rates may have been towards favouring more slowly growing birds in more recent years.

On the other hand, as with clutch size, we face something of a paradox: although smaller fledglings recruit at lower rates, the mean body size in the population is declining. As with clutch size, this implies that much of the selection on body size at this stage may operate independently of genotype, reflecting instead selection against poorly growing goslings regardless of genotype. We appear to have lower recruitment among birds which are small for environmental reasons, as argued for several other species of birds (van Noordwijk *et al.* 1988). Price *et al.* (1988) provided a theoretical framework for this when they explained the dilemma that although laying date was under strong directional selection in many birds, there was not necessarily any genetic change in mean laying date as a result. This model may explain our results too. Selection is operating strongly on environmentally based differences in body size, such that no genetical change in mean body size is predicted, heritability of body size can remain high, and any long term change in body size such as we have detected has an environmental rather than genetical explanation. Selection on this environmental component of growth rate variance may in this case be acting in an opposite direction to that which is favouring more slowly growing goslings at the genetic level.

Variation in adult size appears to have few short term demographic consequences. The long term decline in clutch size appears not to be caused by the long term decline in adult female body size, since there is no relationship between body size and number or size of eggs laid. Similarly, adult body size is not related to the slight decrease in egg size. The recent decline in pre-fledging survivorship of young does not appear to be the result of poorer parental care by smaller parents.

We clearly are lacking an essential part of the story, since we have no explanation for the sexual dimorphism of Snow Geese. We have documented potential costs of size dimorphism in terms of survivorship changes among growing goslings, but we have no idea of the magnitude of potential benefits for males or females. Our limited view of the consequences of pairing behaviour, namely our basically negative findings with respect to assortative mating within cohorts, provides no support for an essential role of body size variation in this process. Our lack of knowledge with regard to the consequences of body size variation for pairing and perhaps anti-predator behaviour is a real gap.

13.9 *Summary*

1. We examined body size variation in hatchlings, fledglings and adults using body weight, tarsus and culmen length. Fledglings and adults are weighed and measured during the annual banding in late July, after nesting is finished and adults are moulting.
2. Males and females differ in fledgling and adult size but not at the egg or hatchling stage.
3. There are strong correlations between fledgling and adult body size, but not between hatchling and fledgling weight. The growth rate during the first five weeks of life accounts for much of the variation in final adult body size.
4. Goslings which hatch later in the season grow more slowly and reach a smaller adult size than those which hatch earlier.
5. There has been a long term decline in gosling growth rate, and therefore adult body size, during the course of the study.
6. Gosling growth rates are slower in years with cold, wet, post-hatching weather.
7. There is no correlation between measures of fecundity and body size of adult females, when effects of cohort are taken into account.
8. Age of first breeding is uncorrelated to body size.
9. Adult annual survival is uncorrelated with body size.
10. The ability of a gosling to survive to fledging does not, in general, differ with hatching weight. In an extremely late year, however, known fledged goslings hatched heavier, while in a benign year, survivors hatched differentially as lighter goslings.
11. A long term decline of male gosling survivorship coincident with decreased foraging opportunities suggests an evolutionary response to selection against genotypes with faster growth rates.
12. Among seasons, there is a close correlation between mean fledgling size and juvenile survival or recruitment.
13. Within seasons, fledglings which hatched early in the year show no differential survivorship with respect to size, however, smaller goslings had lower survival rates among late-hatching nests.
14. There is significant mother/daughter and father/daughter heritability of body size characters.
15. Selection acts against small fledglings, but mainly on an environmental component of the variation related to date of hatch. Despite directional selection, we expect no evolutionary response with respect to body size.

14 *Synthesis and conclusions*

About half a century ago, biologists reached a broad consensus that four fundamental processes could account for evolutionary change within populations. Natural selection, gene flow, genetic drift, and the generation of new variability through mutation were recognized as the major processes moulding the genetic variation of populations. Since that consensus, ecological geneticists and others have tried to determine precisely how these sources of genetic stability and change have interacted to determine the current evolutionary status of populations and might operate in the future to determine the trajectory of particular populations in their environments. The long term goal of these studies has been to generate robust statements about the relative importance or domains of evolutionary processes in nature, and to suggest and test refinements to theoretical population genetics models which predict how evolution should proceed under particular conditions.

This research programme has had mixed results. As field workers showed that the simplifying assumptions of population genetic models rendered them inapplicable in the real world, theoretical population geneticists modelled more realistic ecological situations and considered more complex aspects of genetic systems. As these models become more realistic, testing their applicability to specific organisms becomes more difficult, since discrepancies between a model's predictions and real data may be due to an increased number of violated assumptions or simplifications about genetic or ecological systems. While this makes validation of a model's applicability in nature more difficult, such work is necessary if we are to claim to have a science of microevolutionary biology in the wild.

We studied natural selection and, where possible, microevolutionary change, on certain aspects of the Lesser Snow Goose, an iteroparous, relatively low fecundity vertebrate. Many plants, invertebrates, or even higher fecundity vertebrates such as Guppies (*Poecilia reticulata*) offer tremendous advantages for studying evolutionary processes, since one may experiment and/or sample on time scales which enable one to obtain more complete answers within a reasonable time period. There is a danger, however, in building the empirical study of natural selection and microevolutionary biology entirely on data from such organisms. Evolutionary biology is a robust science when

Table 14.1 A summary of selection patterns on phenotypic traits of Snow Geese

Trait	Mode of selection	Heritability	Genetic status
Plumage color	flat over observed range	high	changing due to gene flow from other populations, impeded by sexual imprinting
Clutch size	directional for larger clutches, but may covary with laying date	low	continually adapting to internal genetic background
Timing of breeding	directional for earlier, but may covary with clutch size	low	continually adapting to internal genetic background
Egg size	flat over observed range	moderate	stable, variation determined by selection on correlated traits?
Sexual dimorphism in gosling growth rate	directional for slower growth rate, differs between sexes?	???	changing with external environment
Adult female body size	flat over observed range	moderate	stable?

we can test the breadth of applicability of generalizations and population genetic models across taxa. As far as possible, we should determine whether or not different processes are more or less important in organisms with a variety of life histories. We follow with a summary of the patterns of natural selection on the characters studied and the implications we draw from them both for theory and for Snow Geese (Table 14.1).

14.1 *Selection on polymorphic traits*

14.1.1 Plumage colour

The most important demonstration from our study of the plumage polymorphism in Lesser Snow Geese is that several population genetic processes jointly influence gene and genotype frequencies in a wild population, and that an animal's behaviour can in some cases be out of equilibrium with the local fitness consequences of its actions.

Present day gene frequencies reflect at least the processes of natural selection, gene flow and non-random mating. Early attempts by F. G. Cooch (1958, 1961) to explain the temporal and spatial distribution of the plumage colours of the Lesser Snow Goose emphasized the importance of directional selection favouring the allele for blue plumage. Cooch argued this from the increase in the frequency of the blue allele in several populations. Our more detailed examination of one colony failed to detect such directional selection despite large samples. There were no differences among colour genotypes for any component of fitness. Despite this, we confirmed a continuing increase in the relative frequency of blue phase birds that Cooch had earlier reported. How can this discrepancy be explained?

Endler (1986) suggests that evolutionary biologists should strive for predictive theories of natural selection. Since within-generation studies suggest no overall selective difference between the two colour phases, we predicted that there would be no changes in the frequency of the two phases over time. This prediction was falsified by our findings and necessitated a revision of our hypothesis. The resolution lies not in selection, but rather in patterns of gene flow and non-random mating (Chapters 3, 5, and 9). Studying and modelling patterns of mate choice and immigration of birds allowed us to predict the overall changes in gene frequency that occur. If we had concentrated our study on selection alone, we would have overlooked the major mechanisms responsible for gene frequency change in the population.

Once we identified the main reasons for gene frequency change within the La Pérouse Bay population, we were able to show that similar processes were at work in other colonies, leading to increasing similarity in phase ratios among colonies with time. Gene flow was reducing the genetic differences in plumage types among colonies. This led to a further prediction that the two colour phases had previously been more isolated from one another in the relatively recent past, and this was confirmed by an examination of the historical record. Blue phase geese formerly bred in the eastern Canadian arctic and wintered in Louisiana, while white phase geese bred in the more western parts of Hudson Bay and wintered in Texas. Populations of birds are mobile and present day patterns may only be understood in terms of the selective and chance events of their past history. Clearly, a detailed analysis of a simple genetic marker provided a much more complex picture of evolutionary processes at work than envisaged by the pioneering studies of Graham Cooch.

Although geese show a strong pattern of assortative mating with respect to morph, there is no evidence that this behaviour enhances the local fitness of discriminatory geese. Morph discrimination may be an incidental by-product of a general adaptive mechanism promoting species recognition, but it does carry some theoretical cost by restricting mate choice, and serves no apparent purpose in this current evolutionary context. This example should be taken as a cautionary tale to those who adopt a 'Panglossian' adaptationist view of animal behaviour.

14.1.2 Sex

It may strike some as odd to consider sex in the context of selection on a genetic polymorphism, but in birds, sex ratios reflect the relative survival of W and Z chromosomes, a chromosomal polymorphism. We often assume that the equilibrium morph frequencies cannot deviate too strongly from 1 : 1, but this is not necessarily the case, both theoretically and empirically, as many animal and plant species demonstrate, albeit not yet in any bird. At the one stage at which we can gather sufficient survival data for both sexes, namely their growth as goslings, we detected an intriguing fitness difference as environmental circumstances changed. Male goslings, which grow more rapidly and achieve larger body sizes than females, had a lower probability of survival towards the end of the study, when foraging conditions were deteriorating (Chapter 13; Cooch *et al.*, submitted). Selection at the gosling growth stage selected against males.

Because of the strongly sex-biased pattern of natal philopatry in Snow Geese, we have limited ability to quantify the subsequent fitnesses of males produced at our colony relative to those from elsewhere. From a broader perspective, the level of sexual dimorphism seen in this species reflects not only the unknown potential payoffs for males of larger size, or to females for smaller size, but also the costs of development for the two sexes. These costs appear to have changed with resource availability when goslings grew. Thus, the current selection against males may in fact be a correlated response to selection operating on the quantitative trait 'growth rate'. To be more specific, selection may be operating on variation in the presumably quantitative trait 'sexual dimorphism in gosling growth alleles', passed on to offspring by parents, which will affect the magnitude of sexual size dimorphism observed in the species.

14.2 *Selection on quantitative traits*

Our studies of selection acting on quantitative variation in Snow Geese fall logically into two categories. Clutch size and the timing of nesting are traits under strong directional selection and display low heritability. On the other hand, there is little evidence of selection affecting egg size and adult body size, which have relatively high heritability. Ever since Fisher (1930) first described his 'Fundamental Theorem', it has been assumed that traits closely related to fitness should, by definition, be under strong directional selection, and that there should be a continual erosion of additive genetic variance. There is a consensus that genetic variation will be continually regenerated by mutation for characters with a multi-locus basis (Lande 1975), and that such mutations will tend to intrinsically degrade the genome, leaving a constantly changing arena within which selection can act (Pomiankowski *et al.* 1991). None the less, there are still expectations that traits associated with fitness will have lower heritabilities, while traits not related to fitness, and therefore not under strong directional selection, should have higher heritabilities. Although these expectations require, among other things, that non-heritable variation be constant, this result has been broadly supported in a large number of traits in a variety of organisms (Mousseau and Roff 1987; Chapter 8). In the bulk of those studies, however, the degree of directional selection was assumed given an inferred relation of the traits to fitness. Our results certainly conform to the expectations, but we have actually demonstrated the presence or absence of selection for the traits displaying, respectively, low and high relative heritability.

14.2.1 Clutch size and laying date

Our analysis of clutch size variation shows that those birds laying the most eggs recruit on average more offspring into the breeding population. We find no evidence for a phenotypic trade-off between fecundity, as measured by clutch size, and annual survival. The closest evidence of an intra-individual microevolutionary trade-off (*sensu* Sterns 1992) is the apparently higher tendency among birds breeding for the first time at a younger age to 'opt out' the following year (Chapter 7; Viallefont *et al.* in press). Our general lack of a phenotypic trade-off is consistent with other studies, among which no general pattern of either negative or positive intra-individual trade-off can be found (see Stearns 1992; Rockwell *et al.* 1987; Barrowclough and Rockwell 1993). The low heritability of clutch size suggests that most of the variation among females in their ability to produce a clutch relates to differences in their environment. In general, those birds that can acquire, transport or mobilize more nutrients produce larger clutches. Much of the variation among females likely relates to variability in food availability and/or opportunities for the females to acquire the food. This does not mean that genetic variation never exists. Rather, strong directional selection would favour alleles that maximize food gathering or assimilation efficiency. Birds carrying alleles that result in them having a competitive disadvantage in communal foraging situations, for example, should be rapidly eliminated from the population while new advantageous alleles that gave their carriers an edge would quickly spread through the population. What we envisage is a continual genetic improvement as directional selection gradually adapts the geese to their food gathering environment.

Since there is a small heritable component to the variation in clutch size, we might predict a phenotypic response to the directional selection according to the classical response to selection formula $R = h^2S$, where R is the response to selection, h^2 is the heritability and S is the selection differential. After all, even with low heritability, clutch size might be expected to increase given strong directional selection. Rockwell *et al.* (1987) calculated that given the heritability and selection differentials measured in this population, we could expect the mean clutch size to increase by 0.2 eggs over a 15 year period. As shown in Chapter 6, however, mean clutch size has declined.

Several resolutions of this paradox are possible (Chapter 11). All of them are related to the fact that the standard response to selection formula is often used outside its correct ecological framework by assuming that the environment remains constant. For example, if the environment degrades such that there is less forage available, then arriving females may simply not have sufficient nutrients to produce the clutch size for which they are genetically predisposed (Rockwell 1988; Cooch *et al.* 1989).

In this context, however, it is also critical to understand that any genetic change brought about by selection is itself part of the changed environment. Fisher (1930) recognized this in his original formulation of the role of selection but his explanation was rather obscure and has been widely ignored or misunderstood by population geneticists (Frank and Slatkin 1992). If we imagine, for example, that part of a female's success at producing a large clutch results from her ability to compete for a limited food resource and that there is some genetic variation for that ability, then selection will increase competitive ability in the population. But since successful competition is by its very nature a relative behaviour and if the entire population is becoming more competitive, then absolute performance, and hence clutch size, may not increase (Cooke *et al.* 1990*b*; also see Frank and Slatkin 1992).

There is an important lesson here for population biologists. Our study of the relative fitness of birds of different clutch sizes led us to the conclusion that birds with a higher clutch size left more offspring. Given at least some genetic variability for this character, we legitimately predicted that clutch size should gradually increase in the population as more favourable phenotypes were selected. However, our data displayed the opposite trend. Conversely, if we had simply examined the long term decline from a population point of view, we might have erroneously inferred that selection was favouring those birds that laid smaller clutches. The general take-home message from our analyses of fitness and clutch size is that when natural selection is being examined in field situations, the direct as well as the selective effects of the environment need to be considered.

We find a very similar pattern in our analysis of the timing of nesting. There is low heritability and a distinct advantage to laying early in the season, although birds laying smaller clutches may delay until after those laying larger clutches have begun. Despite strong directional selection, however, there has been no long term change in mean laying date during the course of the study. The disappearance of snow from the breeding grounds is the proximate factor that usually determines the beginning of nesting, and it would clearly be maladaptive to nest before nest sites became available. However, selection seems to favour those birds that can establish nests as soon as sites become available. The pattern of nest establishment is largely determined by the rate at which sites become available. Considering the low heritability, most of the variation among females is a result of non-genetic differences.

We have found evidence for a phenotypic 'behavioural' trade-off between timing of laying and clutch size. Female Snow Geese could obtain approximately equivalent fitnesses by laying n eggs at the start of the laying period or $n + 1$ eggs later, and the length of the laying period corresponded roughly to the time needed to acquire the nutrients needed to lay an additional egg. This result reinforces the notion that selection is acting jointly on mechanisms which control timing of laying and clutch size. Thus the environmental factors that affect variation in clutch size may also influence variation in laying date. This could explain the strong negative covariance between clutch size and laying date in which birds that lay large clutches lay early in the season.

Selection operates strictly on phenotypic differences among individuals. It can not discriminate whether the basis of the difference is genetic or environmental. Van Noordwijk and his colleagues (for example van Noordwijk *et al.* 1988) have shown how in some cases fitness differences displayed by segments of a population result strictly from environmental differences, even though there is genetic variation within the population for the trait. Price *et al.* (1988) developed a model based on this notion that explains the lack of response of a population to directional selection for laying date. Such a model could explain how directional selection can be observed within generations and yet result in no long term change of the mean laying date. However, our model to resolve the paradox related to clutch size also adequately explains the data on laying date. The essential difference between the two hypotheses is in the predictions about the pattern of selection operating on the genetic component of the variation. In the Price *et al.* (1988) hypothesis, normalizing selection is assumed; in the Cooke *et al.* (1990*b*) hypothesis, directional selection is allowed. We know of no way to discriminate between these possible explanations for laying date variation.

14.2.2 Egg size

Our analysis of egg size variation also presented a paradox. The high heritability was consistent with our failure to detect selective differences between eggs of different sizes, but we were puzzled by the lack of directional selection and the lack of a trade-off between egg size and clutch size. High egg size heritability is commonly found in bird species, but there are few studies that have examined fitness differences in such detail. Those that have been carried out do not lead to any clear generalizations. The main conclusion from our study is that selection is weak or absent. However, this is a surprising conclusion for a bird whose reproductive output is thought to be limited by the amount of nutrient that can be acquired prior to annual reproduction. If a fixed amount of nutrient is available for reproduction, and, as we showed earlier, birds with large clutches are most successful, then one would predict (1) that by laying small eggs, a female could distribute her limited nutrient into more eggs and therefore birds that laid smaller eggs should be favoured or (2) that there would be an advantage to laying larger eggs and some genotypic trade-off between egg size and clutch size.

Lessells *et al.* (1989) examined the covariance of egg size and clutch size and found no evidence for such a trade-off, nor have phenotypic trade-offs been found in other species of birds. In Chapter 12, we asked why egg size was not under intense directional selection, favouring genotypes that lay small eggs? We argued that egg size may be pleiotropically tied to other morphological or physiological characters that themselves display higher levels of genetic variance. Body size, or at least the aspects we have examined, is uncorrelated with egg size and is, thus, not likely the correlated character. Alternatively, physiological characters, such as basal metabolic rate, could strongly influence egg size and could be variable in birds. Since the measurement of physiological variability in a population presents a considerable challenge, it may be some years before we know the answer.

14.2.3 Body size and growth rates

Body size was one of the most complex traits we analysed and is perhaps the one that most interests biologists, judging by the number of studies on the topic. Like egg size, it displayed reasonably high heritability and there was little evidence of a selection differential over the normal range of body sizes and environments of the population. Adult body size is influenced by two factors, the genetic constitution of the bird and the growth rate of the birds in early life. Females are generally smaller than males, but even within sexes there is a considerable range of variation, much of which is heritable. An apparent change in the relative survivorship rates of males and females as feeding conditions deteriorated suggests that selection shifted during our study against the faster-growing males.

Within females, the range of body sizes in part reflects a range of genotypes of roughly equivalent fitness, since there is little evidence that size affects any component of fitness. However, adult body size is also strongly influenced by the environment through its effects on gosling growth, especially during the first few weeks of life (Chapter 13). When the food supply is not limiting, individuals grow rapidly and their fledging (and adult) size reflects their genetic potential. Under less suitable conditions, growth rate and size may fall well below each individual's genetic potential. Within each of those cases, however, some portion of the variation among individuals reflects heritable variance while any mean difference between the cases would result solely from environmental differences. In Chapter 13, we described how differences in the timing of hatch, annual quantity of forage and spatial extent of the brood rearing area can affect growth rates and lead to what are, in effect, 'segments' of the La Pérouse Bay population that differ in size due strictly to environmental causes. At the same time heritable variation would still exist within each segment. This situation seems to be a prime candidate for the operation of the type of model proposed by Price *et al.* (1988) in which a selection gradient is related to environmental differences among phenotypes. Under such a system, both selection and substantial heritability for a trait persist.

14.3 *Selection in a changing environment*

When we began our study, we did not expect the long term demographic and morphological changes we observed. From the population genetic viewpoint, these changes complicated our study, while from the ecological perspective they provided valuable insights. We correlated the changes with either declines in food availability or changes in the behaviour of the main predator, *Homo sapiens*. Because the study has involved individually marked birds, we were often able to rule out alternative explanations for the long term changes, such as selection, gene flow or age related changes. For example, the long term decline in clutch size occurred within marked individuals as well as over the population as a whole. This ruled out natural selection or gene flow as a complete explanations for the decline. The phenomenon could be studied among known-aged birds and we ruled out age-related effects. The long term changes also allowed us to tease apart the sources of environmental and genetic variance, *sensu* van Noordwijk (1987). A study that integrates both population genetics and population ecology concepts can yield an understanding of evolutionary processes that would be impossible with a narrower focus.

Many may find the ecological relationships between demographic and morphological changes and environmental factors the most compelling part of our story. This viewpoint reflects the intuitive ease with which such relationships may be drawn, despite our caution in accepting causal interpretations in the absence of experimental controls. It is more difficult to consider and document the equally profound effects of population genetics, but it is essential to do so to understand the population biology of the species.

14.4 *Practical applications*

Although the main emphasis of this book has been an examination of the potential role of natural selection, the approach we have taken provides practical applications for both management and conservation of geese and other species. The Lesser Snow Goose is a species that provides both recreation and food for large numbers of North Americans. Wildlife agencies in Canada, Mexico and the United States commit large resources to ensure its continuation for future generations. The North American Waterfowl Management Plan sets objectives for the conservation of the species. A fitness components approach to studying a species or population helps identify more exactly the stages of the life cycle at which the birds are most vulnerable and those segments of the population that are most at risk.

Ecological constraints during nutrient acquisition prior to nesting limit the fecundity of individual females and hence, ultimately, the productivity of the breeding population. Any management techniques that allow geese to improve their intake of nutrient in the staging areas and during the northward migration should enhance production. Such techniques could include measures to decrease disturbance of grazing geese during spring staging or to increase the availability of nutrients at this time. In this context it is disturbing that there is evidence of habitat destruction by geese along the Hudson Bay coastline on the northward migration (Kerbes *et al.* 1990). Loss of this salt marsh habitat may well have a negative long term impact on productivity.

Post-fledging was identified as an important stage for mortality. This has previously been suspected, but little documented in birds because it is difficult to study directly. Other studies of birds, however, are now reaching similar conclusions (for example Sullivan 1989).

The approach we have taken has a much wider application than to Snow Geese and, for example, has been successfully applied to a study of the endangered Bahama Parrot, *Amazona bahamensis* (Gnam and Rockwell 1991). By combining information from several such studies, it may be possible to discover general demographic principles that may help us make wiser conservation decisions. There are many endangered and threatened species, where the causes of decline are unknown and there is too little time for detailed studies of many of them. Yet it is critical to discover, for example, which stages of the life cycle present the major threat to survival. It is insufficient to protect the breeding area of a threatened species if high mortality on the wintering grounds is the main reason for population decline. While there is no substitute for detailed studies of endangered species themselves, reasonable predictions made by comparison with well-studied species may be the only option open to us. By providing detailed accounts of both our methods and findings, we hope to encourage others to approach their study organisms using the tools of both ecology and genetics.

References

Abraham, K. F. (1980*a*): Breeding site selection of lesser snow geese. PhD dissertation, Queen's University, Kingston, Ontario.

Abraham, K. F. (1980*b*). Moult migration of lesser snow geese. *Wildfowl*, **31**, 89–93.

Alatalo, R. V. and Lundberg, A. (1986). Heritability and selection on tarsus length in the pied flycatcher *Ficedula hypoleuca*. *Evolution*, **40**, 574–83.

Alatalo, R. V., Gustafsson, L., and Lundberg, A. (1990). Phenotypic selection on heritable size traits: environmental variance and genetic response. *The American Naturalist*, **135**, 464–71.

Aldrich, T. W. and Raveling, D. G. (1983). Effects of experience and body weight on incubation behavior of Canada geese. *Auk*, **100**, 670–79.

Alexander, S. A. (1983). Variation in egg weights of lesser snow geese. BSc thesis, Queen's University, Kingston, Ontario.

Alisauskas, R. T. (1988). Nutrient reserves of lesser snow geese during winter and spring migration. PhD dissertation, University of Western Ontario.

Alisauskas, R. T. and Ankney, C. D. (1990). Body size and fecundity in lesser snow geese. *Auk*, **107**, 440–3.

Anderson, M. G., Rhymer, J. M., and Rohwer, F. C. (1992). Philopatry, dispersal and the genetic structure of waterfowl populations. In *Ecology and management of breeding waterfowl*, (ed. B. D. G. Batt *et al.*), pp. 365–95. University of Minnesota Press, Minneapolis.

Andersson, M. (1984). Brood parasitism within species. In *Producers and scroungers*, (ed. C. J. Barnard), pp. 195–228. Chapman and Hall, New York

Ankney, C. D. (1974). The importance of nutrient reserves to breeding blue geese, *Anser caerulescens*. PhD thesis, University of Western Ontario, London.

Ankney, C. D. (1977). The use of nutrient reserves by breeding male lesser snow geese *Chen caerulescens caerulescens*. *Canadian Journal of Zoology*, **55**, 1984–87.

Ankney, C. D. (1980). Egg weight, survival, and growth of lesser snow goose goslings. *Journal of Wildlife Management*, **44**, 174–82.

Ankney, C. D. (1984). Nutrient reserve dynamics of breeding and molting brant. *Auk*, **101**, 361–70.

Ankney, C. D. and Bisset, A. R. (1976). An explanation of egg-weight variation in the lesser snow goose. *Journal of Wildlife Management*, **40**, 729–34.

Ankney, C. D. and MacInnes, C. D. (1978). Nutrient reserves and reproductive performance of female lesser snow geese. *Auk*, **95**, 459–71.

Ankney, C. D., Afton, A. D., and Alisauskas, R. T. (1991). The role of nutrient reserves in limiting waterfowl reproduction. *Condor*, **93**, 1029–32.

AOU (American Ornithologists' Union) (1983). *Checklist of North American birds*, (6th edn). Allen Press, Lawrence, KS.

Arnold, S. J. (1986). Limits on stabilizing, disruptive and correlational selection set by the opportunity for selection. *The American Naturalist*, **128**, 143–6.

Arnold, S. J. and Wade, M. J. (1984*a*). On the measurement of natural and sexual selection: applications. *Evolution*, **38**, 720–34.

Arnold, S. J. and Wade, M. J. (1984*b*). On the measurement of natural and sexual selection: theory. *Evolution*, **38**, 709–19.

Arnold, T. W. and Rohwer, F. C. (1991). Do egg formation costs limit clutch size in waterfowl?—a skeptical view. *Condor*, **93**, 1032–8.

Aubin, A. E., Dunn, E. H., and MacInnes, C. D. (1986). Growth of lesser snow geese on arctic breeding grounds. *Condor*, **88**, 365–70.

Avise, J. C., Alisauskas, R. T., Nelson, W. S., and Ankney, C. D. (1992). Population genetic structure in an avian species with female natal philopatry. *Evolution*, **46**, 1084–96.

Barnston, G. (1860). Recollections of the swans and geese of Hudson Bay. *Ibis*, **2**, 153–259.

Barrowclough, G. and Rockwell, R. F. (1993). Variance of lifetime reproductive success: estimation based on demographic data. *The American Naturalist*, **141**, 281–95.

Barry, T. W. (1962). Effect of late seasons on Atlantic brant reproduction. *Journal of Wildlife Management*, **26**, 19–26.

Barry, T. W. (1967). Geese of the Anderson River Delta, Northwest Territories PhD thesis, University of Alberta, Edmonton.

Barton, N. H. and Turelli, M. (1989). Evolutionary quantitative genetics: how little do we know? *Annual Review of Genetics*, **23**, 337–70.

Bazely, D. R. (1988). Assessing the impact of goose grazing on vegetation in the arctic. *Ibis*, **130**, 301–02.

Bazely, D. R. and Jefferies, R. L. (1985). Goose faeces: a source of nitrogen for plant growth. *Journal of Applied Ecology*, **22**, 693–703.

Bazely, D. R. and Jefferies, R. L. (1986). Changes in the composition and production of salt-marsh plant communities in response to the removal of a grazer. *Journal of Ecology*, **74**, 693–706.

Beasley, B. A. and Ankney, C. D. (1988). The effect of plumage color on the thermoregulatory abilities of lesser snow goose (*Chen caerulescens caerulescens*) goslings. *Canadian Journal of Zoology*, **66**, 1352–58.

Bent, A. C. (1925). *Life histories of North American wildfowl. Part II*. US Government Printing Office, Smithsonian Institute, US National Museum Bulletin 130, Washington, DC.

Birkhead, T. R. and Biggins, J. D. (1987). Reproductive synchrony and extra-pair copulation in birds. *Ethology*, **74**, 320–34.

Birkhead, T. R. and Møller, A. P. (1992). *Sperm competition in birds: evolutionary causes and consequences*. Academic Press, London.

Black, J. M. and Owen, M. (1988). Variations in pair bond and agonistic behaviors in barnacle geese on the wintering grounds. In *Waterfowl in winter*, (ed. M. W. Weller), pp. 39–57. University of Minnesota Press, Minneapolis.

Black, J. M., Deerenberg, C., and Owen, M. (1991). Foraging behaviour and site selection of Barnacle geese *Branta leucopsis* in a traditional and newly colonised spring staging habitat. *Ardea*, **79**, 349–58.

Blankert, J. J. (1980). Lesser snow goose from Canada in Netherlands. *Dutch Birding*, **2**, 52.

Blokpoel, H. (1974). Migration of lesser snow and blue geese, Part 1: Distribution, chronology, directions, numbers, heights and speeds. Canadian Wildlife Service Reports Series 28.

Blokpoel, H. and Gauthier, M. C. (1975). Migration of lesser snow geese in spring across southern Manitoba, Part 2: influence of the weather and prediction of major flights. Canadian Wildlife Service Reports Series 32.

Boag, P. T. (1987). Effects of nestling diet on growth and adult size of zebra finches (*Poephila guttata*). *Auk*, **104**, 155–66.

Boag, P. T. and van Noordwijk, A. J. (1987). Quantitative genetics. In *Avian genetics: a population and ecological approach*, (ed. F. Cooke and P. A. Buckley), pp. 45–78. Academic Press, London.

Boyd, H., Smith, G. E. J., and Cooch, F. G. (1982). The lesser snow geese of the eastern Canadian Arctic: their status during 1964–79 and their management from 1981 to 1990, Occasional Paper. Canadian Wildlife Service, Ottawa.

Bray, R. (1943). Notes on the birds of Southampton Island, Baffin Island and Melville Pennisula. *Auk*, **60**, 509–12.

Bromley, R. G. and Jarvis, R. L. (1993). The energetics of migration and reproduction of dusky Canada geese. *Condor*, **95**, 193–210.

Brownie, C. D., Anderson, D. R., Burham, K. P., and Robson, D. S. (1985). Statistical inference from band recovery data—a handbook. US Department of Interior, Fish and Wildlife Service Resource Publication No. 156.

Bungaard, J. and Christiansen, F. B. (1972). Dynamics of polymorphisms: I. Selection components in an experimental population of *Drosophila melanogaster*. *Genetics*, **71**, 439–60.

Cabot, D. and West, B. (1983). Studies on the population of Barnacle geese *Branta leucopsis* wintering on the Inishkea Islands, County Mayo: 1. Population dynamics 1961–1983. *Irish Birds*, **2**, 318–36.

Campbell, R. R. (1979). Ecophysiological studies in lesser snow geese (*Anser caerulescens caerulescens*) of the La Perouse Bay colony. PhD thesis, University of Guelph, Ontario.

Cargill, S. M. (1979). Parental investment and hatching success in clutches of the lesser snow goose (*Anser caerulescens caerulescens*). BSc thesis, Queen's University.

Cargill, S. M. and Cooke, F. (1981). Correlation of laying and hatching sequences in clutches in the Lesser Snow Goose (*Anser caerulescens caerulescens*). *Canadian Journal of Zoology*, **59**, 1201–4.

Cargill, S. M. and Jefferies, R. L. (1984). The effects of grazing by lesser snow geese on the vegetation of a sub-arctic salt marsh. *Journal of Applied Ecology*, **21**, 669–86.

Caswell, H. (1989). *Matrix population models*. Sinauer, Sunderland.

Cavelli-Svorza, L. and Feldman, M. (1981). *Cultural transmission and evolution: a quantitative approach*. Princeton University Press.

Charlesworth, B. (1980). *Evolution of age-structured populations*. Cambridge University Press.

Charlesworth, (1987). The heritability of fitness. In *Sexual selection: testing the alternatives*, (ed. J. W. Bradbury and M. B. Anderson), pp. 21–40. John Wiley, New York.

Charnov, E. L. and Krebs, J. R. (1974). On clutch size and fitness. *Ibis*, **116**, 217–19.

Choudhury, S., Black, J. M. and Owen, M. (1992). Do barnacle geese pair assortatively? Lessons from a long-term study. *Animal Behavior*, **44**, 171–73.

Choudhury, S., Jones, C. S., Black, J. M., and Prop, J. (1993). Adoption of young and intraspecific nest parasitism in barnacle geese. *Condor*, **95**, 860–78.

Christiansen, F. B. and Frydenberg, O. (1973). Selection component analysis of natural polymorphisms using population samples including mother–child combinations. *Theoretical Population Biology*, **4**, 425–45.

Clinchy, M. and Barker, I. K. (1994). Effects of parasitic infections on clutch size of lesser snow geese from a northern breeding colony. *Canadian Journal of Zoology*, **72**, 541–4.

Clutton-Brock, T. H., Guinness, F. E., and Albon, S. D. (1982). *Red Deer: behaviour and ecology of two sexes*, Wildlife Behaviour and Ecology Series. University of Chicago Press.

Cole, L. C. (1954). The population consequences of life history phenomena. *Quarterly Review of Biology*, **29**, 103–37.

Cole, R. W. (1979). The relationship between weight at hatch and survival and growth of wild lesser snow geese. MS thesis, University of Western Ontario, London.

Collias, N. E. and Jahn, L. R. (1959). Social behavior and breeding success in Canada geese (*Branta canadensis*) confined under semi-natural conditions. *Auk*, **76**, 478–509.

Collins, M. (1993). Nest depredation in lesser snow geese. MSc thesis, Queen's University, Kingston, Ontario.

Cooch, E. G. (1990). Demographic change in a snow goose population. PhD thesis, Queen's University, Kingston, Ontario.

Cooch, E. G. and Cooke, F. (1991). Demographic changes in a snow goose population—biological and management implications. In *Bird population studies: their relevance to conservation and management*, (ed. C. M. Perrins, J. D. Lebreton, and G. Hirons), pp. 168–89. Oxford University Press.

Cooch, E. G., Lank, D. B., Rockwell, R. F., and Cooke, F. (1989). Long-term decline in fecundity in a snow goose population: evidence for density dependence? *Journal of Animal Ecology*, **58**, 711–26.

Cooch, E. G., Lank, D. B., Dzubin, A., Rockwell, R. F., and Cooke, F. (1991*a*). Body size variation in lesser snow geese: environmental plasticity in gosling growth rates. *Ecology*, **72**, 503–12.

Cooch, E. G., Lank, D. B., Rockwell, R. F., and Cooke, F. (1991*b*). Long-term decline in body size in a snow goose population: evidence of environmental degradation? *Journal of Animal Ecology*, **60**, 483–96.

Cooch, E. G., Lank, D. B., Rockwell, R. F., and Cooke, F. (1992). Is there a positive relationship between body size and fecundity in lesser snow geese? *Auk*, **109**, 667–73.

Cooch, E. G., Jefferies, R. L., Rockwell, R. F., and Cooke, F. (1993). Environmental change and the cost of philopatry: an example in the Lesser Snow Goose. *Oecologia*, **93**, 128–38.

Cooch, E. G., Lank, D. B., Robertson, R. J., and Cooke, F. (1994). The cost of being larger: natural selection on sexually dimorphic growth in lesser snow geese. *Evolution*, submitted.

Cooch, F. G. (1958). The breeding biology and management of the blue goose *Chen caerulescens*. PhD thesis, Cornell University.

Cooch, F. G. (1961). Ecological aspects of the blue-snow goose complex. *Auk*, **78**, 72–89.

Cooch, F. G. (1963). Recent changes in the distribution of color phases of *Chen c. caerulescens. Proceedings of the International Ornithological Congress*, **13**, 1182–94.

Cooch, F. G. and Beardmore, J. A. (1959). Assortative mating and reciprocal difference in the blue-snow goose complex. *Nature*, **183**, 1833–34.

Cooke, F. (1978). Early learning and its effect on population structure. Studies of a wild population of snow geese. *Zeitschrift für Tierpsychologie*, **46**, 344–58.

Cooke, F. (1988). Genetic studies of birds—the goose with blue genes. *Proceedings of the International Ornithological Congress*, **19**, 189–214.

Cooke, F. and Abraham, K. F. (1980). Habitat and locality selection in lesser snow geese: the role of previous experience. *Proceedings of the International Ornithological Congress*, **17**, 998–1004.

Cooke, F. and Cooch, F. G. (1968). The genetics of the polymorphism in the goose *Anser caerulescens. Evolution*, **22**, 289–300.

Cooke, F. and Davies, J. C. (1983). Assortative mating, mate choice and reproductive fitness in snow geese. In *Mate choice*, (ed. P. Bateson), pp. 279–95. Cambridge University Press, New York

Cooke, F. and Francis, C. M. (1993). Challenges in the analysis of recruitment and spatial organization of populations. In *Marked individuals in the study of bird populations*, (ed. J.-D. Lebreton and P. M. North), pp. 295–308. Birkhauser, Basel, Switzerland.

Cooke, F. and Harmsen, R. (1983). Does sex ratio vary with egg sequence in Lesser Snow Geese? *Auk*, **100**, 215–17.

Cooke, F. and McNally, C. M. (1975). Mate selection and colour preferences in Lesser Snow Geese. *Behaviour*, **53**, 151–70.

Cooke, F. and Mirsky, P. J. (1972). A genetic analysis of Lesser Snow Goose families. *Auk*, **89**, 863–71.

Cooke, F. and Rockwell, R. F. (1988). Reproductive success in a lesser snow goose population. *Reproductive success. Studies of individual variation in contrasting breeding systems*, (ed. T. H. Clutton-Brock), pp. 237–47. University of Chicago Press, IL.

Cooke, F. and Sulzbach, D. S. (1978). Mortality, emigration and separation of mated Snow Geese. *Journal of Wildlife Management*, **42**, 271–80.

Cooke, F., Mirsky, P. J., and Seiger, M. B. (1972). Colour preferences in the Lesser Snow Goose and their possible role in mate selection. *Canadian Journal of Zoology*, **50**, 529–36.

Cooke, F., MacInnes, C. D., and Prevett, J. P. (1975). Gene flow between breeding populations of Lesser Snow Geese. *Auk*, **93**, 493–510.

Cooke, F., Finney, G. H. and Rockwell, R. F. (1976). Assortative mating in Lesser Snow Geese. *Behavioral Genetics*, **6**, 127–40.

Cooke, F., Bousfield, M. A., and Sadura, A. (1981). Mate change and reproductive success in the lesser snow goose. *Condor*, **83**, 322–7.

Cooke, F., Findlay, C. S., Rockwell, R. F., and Abraham, K. F. (1983). Life history studies of the lesser snow goose (*Anser caerulescens caerulescens*). II. Colony structure. *Behavioral Ecology and Sociobiology*, **12**, 153–9.

Cooke, F., Findlay, C. S., and Rockwell, R. F. (1984). Recruitment and the timing of reproduction in Lesser Snow Geese. *Auk*, **101**, 451–58.

Cooke, F., Findlay, C. S., Rockwell, R. F., and Smith, J. A. (1985). Life history studies of the lesser snow goose (*Anser caerulescens caerulescens*). III. The selective value of plumage polymorphism: net fecundity. *Evolution*, **39**, 165–77.

Cooke, F., Parkin, D. T., and Rockwell, R. F. (1988). Evidence of former allopatry of the two colour phases of Lesser Snow Goose (*Chen caerulescens caerulescens*). *Auk*, **105**, 467–79.

Cooke, F., Davies, J. C., and Rockwell, R. F. (1990). Body size and fecundity in lesser snow geese: response to Alisauskas and Ankney. *Auk*, **107**, 444–46.

Cooke, F., Taylor, P. D., Francis, C. M., and Rockwell, R. F. (1990*b*). Directional selection and clutch size in birds. *The American Naturalist*, **136**, 261–67.

Cooke, F., Rockwell, R. F., and Lank, D. B. (1991). Recruitment in long-lived birds: genetic considerations. *Proceedings of the International Ornithological Congress*, **20**, 1666–77.

Cooper, J. A. (1978). The history and breeding biology of the Canada geese of Marshy point, Manitoba. *Wildlife Monographs*, **61**.

Crespi, B. J. (1989). Causes of assortative mating in arthropods. *Animal Behavior*, **38**, 980–1000.

Curio, E. (1983). Why do young birds reproduce less well? *Ibis*, **125**, 400–4.

Darwin, C. (1859). *On the origin of species by means of natural selection, or the preservation of favoured races in the struggle for life*. John Murray, London.

Davies, J. C. and Cooke, F. (1983*a*). Annual nesting productivity in snow geese: prairie drought and arctic springs. *Journal of Wildlife Management*, **47**, 291–96.

Davies, J. C. and Cooke, F. (1983*b*). Intraclutch hatch synchronization in the lesser snow goose. *Canadian Journal of Zoology*, **61**, 1398–401.

Davies, J. C., Rockwell, R. F., and Cooke, F. (1988). Body size variation and fitness components in lesser snow geese *Chen caerulescens caerulescens*. *Auk*, **105**, 639–48.

Delacour, J. and Mayr, E. (1945). The family Anatidae. *Wilson Bulletin*, **57**, 3–55.

Diamond, J. M. (1987). A Darwinian theory of divorce. *Nature*, **329**, 765–66.

Drent, R. H. and Daan, S. (1980). The prudent parent: energetic adjustments in avian breeding. *Ardea*, **68**, 225–52.

Dzubin, A. (1974): Snow and blue goose distribution in the Mississippi and Central Flyways: compendium of results of bird recovery analyses. Mimeo Canadian Wildlife Service, prepared for Central and Mississippi Flyway Councils, June 1974, University of Saskatchewan, Saskatoon.

Dzubin, A. and Cooch, E. G. (1992). *Measurements of geese: general field methods*. California Waterfowl Association, Sacramento, CA.

Ebbinge, B. S. (1985). Factors determining the population size of arctic-breeding geese, wintering in western Europe. *Ardea*, **73**, 121–28.

Ebbinge, B. S. (1991). The impact of hunting on mortality rates and spatial distribution of geese wintering in the western Palearctic. *Ardea*, **79**, 197–210.

Emlen, S. T. and Oring, L. W. (1977). Ecology, sexual selection, and the evolution of mating systems. *Science*, **197**, 215–23.

Endler, J. A. (1986). *Natural selection in the wild*. Princeton University Press.

Falconer, D. S. (1989). *Introduction to quantitative genetics*, (3rd edn). Longman, London.

Findlay, C. S. and Cooke, F. (1982*a*). Breeding synchrony in the Lesser Snow Goose (*Anser caerulescens caerulescens*). I. Genetic and environmental components of hatch date variability and their effects on hatch synchrony. *Evolution*, **36**, 342–51.

Findlay, C. S. and Cooke, F. (1982*b*). Synchrony in the Lesser Snow Goose. (*Anser caerulescens caerulescens*) II. The adaptive value of reproductive synchrony. *Evolution*, **36**, 786–99.

Findlay, C. S. and Cooke, F. (1983). Genetic and environmental components of clutch size variance in a wild population of Lesser Snow Geese (*Anser caerulescens caerulescens*). *Evolution*, **37**, 724–34.

Findlay, C. S. and Cooke, F. (1987). Repeatability and heritability of clutch size in lesser snow geese. *Evolution*, **41**, 453.

Findlay, C. S., Rockwell, R. F., Smith, J. A., and Cooke, F. (1985). Life history studies of the lesser snow goose (*Anser caerulescens caerulescens*). VI. Plumage polymorphism, assortative mating and fitness. *Evolution*, **39**, 904–14.

Finney, G. and Cooke, F. (1978). Reproductive habits in the snow goose: the influence of female age. *Condor*, **80**, 147–58.

Fisher, R. A. (1930). *The genetical theory of natural selection*. Oxford University Press.

Flint, P. L. and Sedinger, J. S. (1992). Reproductive implications of egg-size variation in the black brant. *Auk*, **109**, 896–903.

Ford, E. B. (1975). *Ecological genetics*, (4th edn). Chapman and Hall, London.

Forslund, P. and Larsson, K. (1991). The effect of mate change and new partner's age on reproductive success in the barnacle goose, *Branta leucopsis*. *Behavioral Ecology*, **2**, 116–22.

Forslund, P. and Larsson, K. (1992). Age-related reproductive success in the barnacle goose. *Journal of Animal Ecology*, **61**, 195–204.

Foster, B. J. (1957). Snow and blue geese nesting in the southern arctic. *Ontario Field Biologist*, **11**, 22.

Francis, C. M. and Cooke, F. (1992*a*). Migration routes and recovery rates of lesser snow geese from southwestern Hudson Bay. *Journal of Wildlife Management*, **56**, 279–86.

Francis, C. M. and Cooke, F. (1992*b*). Sexual differences in survival and recovery rates of lesser snow geese. *Journal of Wildlife Management*, **56**, 287–96.

Francis, C. M. and Cooke, F. (1993). A comparison of survival rate estimates from live recaptures and dead recoveries of lesser snow geese. In *Marked individuals in the study of bird population*, (ed. J.-D. Lebreton and P. M. North). Birkhauser, Basel, Switzerland.

Francis, C. M., Richards, M. H., Cooke, F., and Rockwell, R. F. (1992*a*). Changes in survival rates of lesser snow geese with age and breeding status. *Auk*, **109**, 731–47.

Francis, C. M., Richards, H. M., Cooke, F., and Rockwell, R. F. (1992b). Long term changes in survival rates of lesser snow geese. *Ecology*, **73**, 1346–62.

Frank, S. A. and Slatkin, M. (1992). Fisher's fundamental theorem of natural selection. *Trends in Ecology and Evolution*, **7**, 92–95.

Frey, I. D., Clauss, M. J., Hik, D. S., and Jefferies, R. L. (1993). Growth responses of arctic graminoids following grazing by lesser snow geese. *Oecologia*, **93**, 487–92.

Ganter, B. and Cooke, F. (1993). Reaction of lesser snow geese to early nest failure. *Wildfowl*, **44**, 170–73.

Gauthier, G. and Tardif, J. (1991). Female feeding and male vigilance during nesting in greater snow geese. *Condor*, **93**, 701–11.

Gauthier, G., Giroux, J.-F., and Bedard, J. (1992). Dynamics of fat and protein reserves during winter and spring migration in greater snow geese. *Canadian Journal of Zoology*, **70**, 2077–87.

Gebhardt-Henrich, S. G. and van Noordwijk, A. J. (1991). Nestling growth in the Great tit. I. Heritability estimates under different environmental conditions. *Journal of Evolutionary Biology*, **4**, 341–62.

Geramita, J. M. and Cooke, F. (1982). Evidence that fidelity to natal breeding colony is not absolute in female Snow Geese. *Canadian Journal of Zoology*, **60**, 2051–56.

Geramita, J. M., Cooke, F., and Rockwell, R. F. (1982). Assortative mating and gene flow in the Lesser Snow Goose: a modelling approach. *Theoretical Population Biology*, **22**, 177–203.

Giroux, J.-F. and Bedard, J. (1986). Sex-specific hunting mortality of greater snow geese along firing lines in Quebec. *Journal of Wildlife Management*, **50**, 416–19.

Gnam, R. and Rockwell, R. F. (1991). Reproductive potential and output of the Bahama Parrot *Amazona leucocephala bahamensis*. *Ibis*, **133**, 400–5.

Gochfeld, M. (1980). Mechanisms and adaptive value of reproductive synchrony in colonial seabirds. In *Behaviour of marine animals*, (ed. J. Burger, L. Olla, and H. E. Winn), pp. 207–65. Plenum Press, New York.

Grafen, A. (1988). On the uses of data on lifetime reproductive success. In *Reproductive success*, (ed. T. H. Clutton-Brock), pp. 454–71. University of Chicago Press, Chicago.

Graham, A. (1769). *Diary written between 1768 and 1769. Observations on Hudsons Bay*, Book 2. Hudsons Bay Co. Archives, Winnipeg, Manitoba.

Grant, B. R. and Grant, P. R. (1989). *Evolutionary dynamics of a natural population: the large cactus finch of the Galapagos*. The University of Chicago Press.

Grau, C. R. (1976). Ring structure of avian egg yolk. *Poultry Science*, **55**, 1418–22.

Gregoire, P. E. and Ankney, C. D. (1990). Agonistic behavior and dominance relationships among lesser snow geese during winter and spring migration. *Auk*, **107**, 550–60.

Gurtovaya, E. N. (1990). Ethological demonstrations and other forms of behaviour of snow geese (*Anser caerulescens caerulescens*) during nesting. (Translated from Russian.) *Zoologichesky Zhurnal*, **69**, 86–98.

Hamann, J. (1983). Intra-seasonal clutch size reduction in lesser snow geese. MSc thesis, Queen's University, Kingston, Ontario.

Hamann, J. and Cooke, F. (1987). Age effects on clutch size and laying dates of individual females in Lesser Snow Geese. *Ibis*, **129**, 527–32.

Hamann, J. and Cooke, F. (1989). Intra-seasonal decline in clutch size in lesser snow geese. *Oecologia*, **79**, 83–90.

Hamann, J., Andrews, B., and Cooke, F. (1986). The role of follicular atresia in inter- and intra-seasonal clutch size variation in lesser snow geese (*Anser caerulescens caerulescens*). *Journal of Animal Ecology*, **55**, 481–9.

Hanson, H. C., Lumsden, H. G., Lynch, J. J., and Norton, H. W. (1972). Population characteristics of three mainland colonies of the blue and lesser snow geese nesting in the southern Hudson Bay region. Research Department (Wildlife) 93. Ontario Ministry of Natural Resources.

Hardy, J. D. and Tacha, T. C. (1989). Age-related recruitment of Canada geese from the Mississippi valley. *Journal of Wildlife Management*, **53**, 97–8.

Harvey, J. M. (1971). Factors affecting blue goose nesting success. *Canadian Journal of Zoology*, **49**, 223–34.

Hawkins, A. S. (1949). Waterfowl populations and breeding conditions. Special Scientific Report No. 2. US Department of the Interior, Fish and Wildlife Service.

Hawkins, A. S., Gollop, J. B., and Wellein, E. G. (1951). Waterfowl populations and breeding conditions, Summer 1951. Special Scientific Report No. 13. US Department of the Interior, Fish and Wildlife.

Hedrick, P. and Murray, R. (1983). Selection and measures of fitness. In *The genetics and biology of Drosophila*. Vol. 3 (eds. M. Ashburner, H. Carson and J. Thompson). Academic Press, New York.

Heinroth, O. (1922). Die Beziehungen zwischen Vogelgewicht, Eigewicht, Gelegegewicht und Brutdauer. *Journal für Ornithologie*, **70**, 172–285.

Henny, C. J. (1967). Estimating band-reporting rates from banding and crippling loss data. *Journal of Wildlife Management*, **31**, 533–38.

Hik, D. S. (1986). Local philopatry and dispersal in the lesser snow goose: the role of reproductive success. BSc thesis, Queen's University, Kingston, Ontario.

Hik, D. S. and Jefferies, R. L. (1990). Increases in the net above-ground primary production of a salt-marsh forage grass: a test of the predictions of the herbivore-optimization model. *Journal of Ecology*, **78**, 180–95.

Hik, D. S., Sadul, H. A., and Jefferies, R. L. (1991). Effects of the timing of multiple grazings by geese on net above-ground primary production of swards of *Puccinellia phryganodes*. *Journal of Ecology*, **79**, 715–30.

Hik, D. S., Jefferies, R. L., and Sinclair, A. R. E. (1992). Foraging by geese, isostatic uplift and asymmetry in the development of plant communities. *Journal of Ecology*, **80**, 395–406.

Howarth, B. (1974). Sperm storage as a function of the female reproductive tract. In *The oviduct and its function*, (ed. A. D. Johnson and C. W. Foley), pp. 237–70. Academic Press, New York.

Huxley, J. S. (1927). On the relation between egg-weight and body-weight in birds. *Zoological Journal of the Linnean Society*, **36**, 457–66.

Jackson, S., Hik, D. S., and Rockwell, R. F. (1988). The influence of nesting habitat on reproductive success of the lesser snow goose. *Canadian Journal of Zoology*, **66**, 1699–703.

James, F. C. (1983). Environmental component of morphological differentiation in birds. *Science*, **221**, 184–6.

Jefferies, R. L. (1988*a*). Pattern and process in arctic coastal vegetation in response to foraging by Lesser Snow Geese. In *Plant form and vegetational structure, adaptation, plasticity and relationship to herbivory*, (ed. M. J. A.

Weger, P. J. M. van der Aart, H. J. During and J. T. A. Verhoeven), pp. 341–69. Academic Publishers, The Hague.

Jefferies, R. L. (1988*b*). Vegetational mosaics: plant-animal interactions, and resources for plant growth. In *Plant evolutionary biology*, (ed. L. D. Gottlieb and S. K. Jain), pp. 340–61. Chapman and Hall, London.

Jefferies, R. L., Jensen, A., and Abraham, K. F. (1979). Vegetational development and the effect of geese on vegetation at La Perouse Bay, Manitoba. *Canadian Journal of Botany*, **57**, 1439–50.

Jones, I. L. and Hunter, F. M. (1993). Mutual sexual selection in a monogamous seabird. *Nature*, **362**, 238–9.

Jones, P. J. and Ward, P. (1976). The level of reserve protein as the proximate factor controlling the timing of breeding and clutch-size in the red-billed Quelea, *Quelea quelea*. *Ibis*, **110**, 547–74.

Kear, J. (1973). The magpie goose *Anseranas semipalmata* in captivity. *International Zoo Yearbook*, **13**, 28–32.

Kerbes, R. H. (1969). Biology and distribution of nesting blue geese on Koukdjuak Plain, Baffin Island, N.W.T. MS thesis, University of Western Ontario, London.

Kerbes, R. H., Kotanen, P. M. and Jefferies, R. L. (1990). Destruction of wetland habitats by lesser snow geese: a keystone species on the west coast of Hudson Bay. *Journal of Applied Ecology*, **27**, 242–58.

Kettlewell, H. B. D. (1973). *The evolution of melanism: the study of a recurring necessity*. Oxford University Press.

Kirkpatrick, M., Price, T. and Arnold, S. J. (1990). The Darwin–Fisher theory of sexual selection in monogamous birds. *Evolution*, **44**, 180–93.

Klomp, H. (1970). The determination of clutch-size in birds. *Ardea*, **58**, 1–124.

Lack, D. (1947). The significance of clutch size. *Ibis*, **89**, 302–52.

Lack, D. (1967). The significance of clutch-size in waterfowl. *Wildfowl*, **18**, 125–28.

Lack, D. (1968). *Ecological adaptations for breeding in birds*. Methuen, London.

Lamprecht, J. (1986). Social dominance and reproductive success in a goose flock (*Anser indicus*). *Behaviour*, **97**, 50–65.

Lande, R. (1975). The maintenance of genetic variability by mutation in a polygenic character with linked loci. *Genetics Research* (*Cambridge*), **26**, 221–35.

Lande, R. and Arnold, S. J. (1983). The measurement of selection on correlated characters. *Evolution*, **37**, 1210–26.

Lank, D. B., Cooch, E. G., Rockwell, R. F., and Cooke, F. (1989*a*). Environmental and demographic correlates of intraspecific nest parasitism in lesser snow geese. *Journal of Animal Ecology*, **58**, 29–45.

Lank, D. B., Mineau, P., Rockwell, R. F., and Cooke, F. (1989*b*). Intraspecific nest parasitism and extra-pair copulation in lesser snow geese. *Animal Behavior*, **37**, 74–89.

Lank, D. B., Rockwell, R. F., and Cooke, F. (1990). Frequency-dependent fitness consequences of intraspecific nest parasitism in snow geese. *Evolution*, **44**, 1436–53.

Lank, D. B., Bousfield, M. A., Cooke, F., and Rockwell, R. F. (1991). Why do snow geese adopt eggs? *Behavioral Ecology*, **2**, 181–7.

Larsson, K. (1993). Inheritance of body size in the Barnacle goose under different environmental conditions. *Journal of Evolutionary Biology*, **6**, 195–208.

Larsson, K. and Forslund, P. (1992). Genetic and social inheritance of body and egg size in the barnacle goose (*Branta leucopsis*). *Evolution*, **46**, 235–44.

Leblanc, Y. (1987). Intraclutch variation in egg size of Canada geese. *Canadian Journal of Zoology*, **65**, 3044–7.

Lebreton, J.-D. and Clobert, J. (1986). *User's guide to program SURGE*, Version 2.0. CEFE/CNRS, Montpellier, France.

Lebreton, J.-D., Burnham, K. P., Clobert, J., and Anderson, D. R. (1992). Modeling survival and testing biological hypotheses using marked animals: a unified approach with case studies. *Ecological Monographs*, **62**, 67–118.

Le Maho, Y. (1977). The emperor penguin: a strategy to live and breed in the cold. *American Scientist*, **65**, 680–93.

Lemieux, L. (1959). The breeding biology of the greater snow goose on Bylot Island, Northwest Territories. *The Canadian Field-Naturalist*, **73**, 117–28.

Lessells, C. M. (1982). Some causes and consequences of family size in the Canada goose *Branta canadensis*. PhD thesis, Oxford University.

Lessells, C. M. and Krebs, J. R. (1989). Age and breeding performance of European bee-eaters. *Auk*, **106**, 375–82.

Lessells, C. M., Cooke, F., and Rockwell, R. F. (1989). Is there a trade-off between egg weight and clutch size in wild lesser snow geese (*Anser caerulescens caerulescens*)? *Journal of Evolutionary Biology*, **2**, 457–72.

Lewis, H. F. and Peters, H. S. (1941). Notes on birds of the James Bay region in the autumn of 1940. *The Canadian Field-Naturalist*, **55**, 111–17.

Lewontin, R. C. (1974). *The genetic basis of evolutionary change*, Columbia Biological Series No. 25. Columbia University Press, New York.

Lindauer, W. (1967). *The hatchability of chicken eggs as influenced by environment and heredity*. Monograph 1, Storrs Agricultural Experiment Station, University of Connecticut, Storrs, CT.

Lynch, M. and Arnold, S. J. (1988). The measurement of selection on size and growth. In *Size-structured populations—ecology and evolution*, (ed. B. Ebenman and L. Persson), pp. 47–59. Springer, Berlin.

Manning, T. H. (1942). Blue and lesser snow geese on Southampton and Baffin Islands. *Auk*, **59**, 158–75.

Martin, K., Cooch, F. G., Rockwell, R. F., and Cooke, F. (1985). Reproductive performance in lesser snow geese: are two parents essential? *Behavioral Ecology and Sociobiology*, **17**, 257–63.

Martinson, R. K. and McCann, J. A. (1966). Proportion of recovered goose and brant bands that are reported. *Journal of Wildlife Management*, **30**, 856–58.

McAtee, W. L. (1910). Notes on *Chen caerulescens, Chen rossii* and other waterfowl in Louisiana. *Auk*, **27**, 337–9.

McAtee, W. L. (1911). Winter ranges of geese on the Gulf Coast. Notable bird records from the same region. *Auk*, **28**, 272–4.

McCleery, R. H. and Perrins, C. M. (1989). Great tit. In *Lifetime reproduction in birds*, (ed. I. Newton), pp. 35–53. Academic Press, London.

McDonald, D. B. (1989). Cooperation under sexual selection: age-graded changes in a lekking bird. *The American Naturalist*, **134**, 709–30.

McIlhenny, E. A. (1932). The blue goose in its winter home. *Auk*, **49**, 279–306.

Mineau, P. (1978). The breeding strategy of a male snow goose *Anser caerulescens caerulescens*. MS thesis, Queen's University, Kingston, Ontario.

Mineau, P. and Cooke, F. (1979*a*). Rape in the Lesser Snow Goose. *Behavior*, **70**, 280–91.

Mineau, P. and Cooke, F. (1979*b*). Territoriality in Snow Geese or the protection of parenthood: Ryder's and Inglis' hypotheses reassessed. *Wildfowl*, **30**, 16–19.

Mitchell-Olds, T. and Shaw, R. G. (1987). Regression analysis of natural selection: statistical inference and biological interpretation. *Evolution*, **41**, 1149–61.

Mousseau, T. A. and Roff, D. A. (1987). Natural selection and the heritability of fitness components. *Heredity*, **59**, 181–97.

Mulder, R. S., Williams, T. D., and Cooke, F. (1994). Dominance brood size and foraging behaviour during brood-rearing in the lesser snow goose: an experimental study. *Condor*, in press.

Nei, M., Chakravarti, A., and Tateno, Y. (1977). Mean and variance of FST finite number of incompletely isolated populations. *Theoretical Population Biology*, **11**, 291–306.

Newell, L. C. (1988). Causes and consequences of egg weight variation in the lesser snow goose (*Chen caerulescens caerulescens*) MS thesis, Queen's University, Kingston, Ontario.

Newton, I. (1977). Timing and success of breeding in tundra-nesting geese. In *Evolutionary ecology*, (ed. B. Stonehouse and C. M. Perrins), pp. 113–26. University Park Press, Baltimore.

Newton, I. (1986). *The Sparrowhawk*. T. & A.D. Poyser, Calton.

Newton, I. (1989*a*). Introduction. In *Lifetime reproduction in birds*, (ed. I. Newton), pp. 1–11. Academic Press, London.

Newton, I. (ed.) (1989*b*). *Lifetime reproduction in birds*. Academic Press, London.

Nichols, J. D. (1992). Capture–recapture models. *BioScience*, **42**, 94–102.

Nichols, J. D., Blohm, R. J., Reynolds, R. E., Trost, R. E., Hines, J. E., and Bladen, J. P. (1991). Band reporting rates for mallards with reward bands of different dollar values. *Journal of Wildlife Management*, **55**, 119–26.

Nur, N. (1988). The cost of reproduction in birds: an examination of the evidence. *Ardea*, **76**, 155–68.

Nur, N. (1990). The cost of reproduction in birds: evaluating the evidence from manipulative and non-manipulative studies. In *Population biology of passerine birds: an integrated approach* NATO Advanced Science Institutes Series, (ed. J. Blondel, A. Gosler, J.-D. Lebreton and R. McCleery), pp. 281–96. Springer, Berlin.

Oberholser, H. C. (1919). *Exanthemops* Elliot an excellent genus. *Auk*, **36**, 562.

O'Connor, R. J. (1980). Energetics of reproduction in birds. *Proceedings of the International Ornithological Congress*, **17**, 306–11.

O'Donald, P. (1980). Sexual selection by female choice in a monogamous bird: Darwin's theory corroborated. *Heredity*, **45**, 201–17.

Ojanen, M., Orell, M., and Vaisanen, R. A. (1979). Role of heredity in egg size variation in the great tit *Parus major* and the pied flycatcher *Ficedula hypoleuca*. *Journal of Avian Biology*, **10**, 22–8.

Otto, C. (1979). Environmental factors affecting egg weight within and between colonies of fieldfare *Turdus pilaris*. *Ornis Scandinavia*, **10**, 111–16.

Owen, M. (1980). *Wild geese of the world. Their life history and ecology.* B. T. Batsford, London.

Owen, M. (1984). Dynamics and age structure of an increasing goose population—the Svalbard Barnacle goose *Branta leucopsis*. *Norsk Polarinstitutt Skriftner*, **181**, 37–47.

Owen, M. and Black, J. M. (1989). Factors affecting the survival of barnacle geese on migration from the breeding grounds. *Journal of Animal Ecology*, **58**, 603–17.

Perrins, C. M. and Birkhead, T. R. (1983). *Avian ecology*, Tertiary Level Biology. Blackie, Glasgow.

Ploeger, P. L. (1968). Geographical differentiation in arctic Anatidae as a result of isolation during the last glacial. *Ardea*, **56**, 1–159.

Pomiankowski, A., Iwasa, Y., and Nee, S. (1991). The evolution of costly mate preferences. I. Fisher and biased mutation. *Evolution*, **45**, 1422–30.

Prevett, J. P. (1972). Family behavior and age-dependent breeding biology of the blue goose, *Anser caerulescens*. PhD thesis, University Western Ontario, London.

Price, T. and Liou, L. (1989). Selection on clutch size in birds. *The American Naturalist*, **134**, 950–9.

Price, T. and Schluter, D. (1991). On the low heritability of life-history traits. *Evolution*, **45**, 853–61.

Price, T., Kirkpatrick, M., and Arnold, S. J. (1988). Directional selection and the evolution of breeding date in birds. *Science*, **240**, 798–9.

Prout, T. (1965). The estimation of fitness from genotypic frequencies. *Evolution*, **19**, 546–51.

Prout, T. (1969). The estimation of fitness from population data. *Genetics*, **63**, 949–67.

Prout, T. (1971). The relation between fitness components and population prediction in *Drosophila*: I. The estimation of fitness components. *Genetics*, **68**, 127–49.

Pugesek, B. H. (1981). Increased reproductive effort with age in the California gull (*Larus californicus*). *Science*, **212**, 822–3.

Quinn, T. W. (1988). DNA sequence variation in the lesser snow goose, *Anser caerulescens caerulescens*. PhD thesis, Queen's University, Kingston, Ontario.

Quinn, T. W. (1992). The genetic legacy of Mother Goose—phylogeographic patterns of lesser snow goose *Chen caerulescens caerulescens* maternal lineages. *Molecular Ecology*, **1**, 105–17.

Quinn, T. W. and White, B. N. (1987). Analysis of DNA sequence variation. In *Avian genetics: a population and ecological approach*, (ed. F. Cooke and P. A. Buckley), pp. 163–98. Academic Press, London.

Quinn, T. W., Quinn, J. S., Cooke, F., and White, B. N. (1987). DNA marker analysis detects multiple maternity and paternity in single broods of the Lesser Snow Goose (*Anser caerulescens caerulescens*). *Nature*, **326**, 392–4.

Quinn, T. W., Davies, J. C., Cooke, F., and White, B. N. (1989). Genetic analysis of offspring of a female–female pair in the lesser snow goose (*Chen c. caerulescens*). *Auk*, **106**, 177–84.

Rainnie, D. J. (1983). Diseases of lesser snow geese (Anser caerulescens caerulescens) on a breeding colony. M. Vet. Sci. thesis, Western College of Veterinary Medicine, University of Saskatchewan.

Ratcliffe, L., Rockwell, R. F., and Cooke, F. (1988). Recruitment and maternal age in lesser snow geese (*Chen caerulescens caerulescens*). *Journal of Animal Ecology*, **57**, 553–63.

Rattray, B. and Cooke, F. (1984). Genetic modelling: an analysis of a colour polymorphism in the snow goose (*Anser caerulescens*). *Zoological Journal of the Linnean Society*, **80**, 437–45.

Raveling, D. G. (1978). The timing of egg laying by northern geese. *Auk*, **95**, 294–303.

Reynolds, C. M. (1972). Mute swan weights in relation to breeding. *Wildfowl*, **23**, 111–18.

Reznick, D. (1985). Costs of reproduction: an evaluation of the empirical evidence. *Oikos*, **44**, 257–67.

Ridley, M. (1983). *The explanation of organic diversity: the comparative method and adaptations for mating*. Clarendon Press, Oxford.

Robertson, G. J., Cooch, E. G., Lank, D. B., Rockwell, R. F., and Cooke, F. (1994). Female age and egg size in the lesser snow goose. *Journal of Avian Biology*, **25**, 149–55.

Rockwell, R. F. (1988). Reproductive fitness in snow geese (*Chen caerulescens*): proximate and ultimate measures. *Proceedings of the International Ornithological Congress*, **19**, 2113–20.

Rockwell, R. F. and Barrowclough, G. F. (1987). Gene flow and the genetic structure of populations. In *Avian genetics: a population and ecological approach*, (ed. F. Cooke and P. A. Buckley), pp. 223–55. Academic Press, London.

Rockwell, R. F. and Cooke, F. (1977). Gene flow and local adaptation in a colonially nesting dimorphic bird: the Lesser Snow Goose. *The American Naturalist*, **111**, 91–7.

Rockwell, R. F., Findlay, C. S., and Cooke, F. (1983). Life history studies of the lesser snow goose (*Anser caerulescens caerulescens*). I. The influence of age and time on fecundity. *Oecologia*, **56**, 318–22.

Rockwell, R. F., Findlay, C. S., and Cooke, F. (1985*a*). Life history studies of the Lesser Snow Goose. V. Temporal effects on age-specific fecundity. *Condor*, **87**, 142–3.

Rockwell, R. F., Findlay, C. S., Cooke, F., and Smith, J. A. (1985*b*). Life history studies of the lesser snow goose (*Anser caerulescens caerulescens*). IV. The selective value of plumage polymorphism: net viability, the timing of maturation, and breeding propensity. *Evolution*, **39**, 178–89.

Rockwell, R. F., Findlay, C. S., and Cooke, F. (1987). Is there an optimal clutch size in Lesser Snow Geese? *The American Naturalist*, **130**, 839–63.

Rockwell, R. F., Cooch, E. G., Thompson, C. B., and Cooke, F. (1993). Age and reproductive success in female lesser snow geese: experience, senescence, and the cost of philopatry. *Journal of Animal Ecology*, **62**, 323–33.

Rohwer, F. C. and Anderson, M. G. (1988). Female-based philopatry, monogamy, and the timing of pair formation in migratory waterfowl. In *Current Ornithology*, Vol. 5 (ed. R. F. Johnston), pp. 187–221. Plenum, New York.

Rohwer, F. C. and Freeman, S. (1989). The distribution of conspecific nest parasitism in birds. *Canadian Journal of Zoology*, **67**, 239–53.

Rose, M. R. (1982). Antagonistic pleiotropy, dominance and genetic variation. *Heredity*, **48**, 63–78.

Rowley, I. (1983). Re-mating in birds. In *Mate choice*, (ed. P. Bateson), pp. 331–60. Cambridge University Press.
Ryder, J. P. (1967). *The breeding biology of Ross' geese in the Perry River region, Northwest Territories*. Canadian Wildlife Service Report Series 3.
Ryder, J. P. (1970). A possible factor in the evolution of clutch size in Ross' goose. *Wilson Bulletin*, **82**, 5–13.
Saether, B.-E. (1990). Age-specific variation in reproductive performance of birds. *Current Ornithology*, 7, 251–82.
Saunders, W. E. (1917). Wild geese at Moose Factory. *Auk*, **34**, 334.
Schluter, D. (1988). Estimating the form of natural selection on a quantitative trait. *Evolution*, **42**, 849–61.
Schluter, D. and Gustafsson, L. (1993). Maternal inheritance of condition and clutch size in the collared flycatcher. *Evolution*, **47**, 658–67.
Schubert, C. A. and Cooke, F. (1993). Egg-laying intervals in the lesser snow goose. *Wilson Bulletin*, **105**, 389–544.
Seber, G. A. F. (1982). *The estimation of animal abundance and related parameters*, (2nd edn). Macmillan, New York.
Seguin, R. J. and Cooke, F. (1983). Band loss from Lesser Snow Geese. *Journal of Wildlife Management*, **47**, 1109–14.
Sequin, R. J. and Cooke, F. (1985). Web tag loss from lesser snow geese. *Journal of Wildlife Management*, **49**, 420–2.
Shields, G. F. and Wilson, A. C. (1987). Calibration of mitochondrial DNA evolution in geese. *Journal of Molecular Evolution*, **24**, 212–17.
Smithey, D. A. (1973). Social organization, behavior, and movement of blue and snow geese wintering in Louisiana. MS thesis, Louisiana State University, Baton Rouge.
Soper, J. D. (1930). Discovery of the breeding grounds of the blue goose. *The Canadian Field-Naturalist*, **44**, 1–11.
Stearns, S. C. (1976). Life history tactics: a review of the ideas. *Quarterly Review of Biology*, **51**, 3–47.
Stearns, S. C. (1992). *The evolution of life histories*. Oxford University Press.
Stutzenbaker, C. D. and Buller, R. J. (1974). *Goose depredation on ryegrass pastures along the Texas Gulf coast*. Special Report, Federal Aid Project W.106R. Parks and Wildlife Department, Austin, TX.
Sullivan, K. A. (1989). Starvation and predation: age-specific mortality in juvenile juncos (*Junco phaeonotus*). *Journal of Animal Ecology*, **58**, 275–86.
Sulzbach, D. S. (1975). A study of the population dynamics of a nesting colony of the lesser snow goose (*Anser caerulescens caerulescens*). MS thesis, Queen's University, Kingston, Ontario.
Sutherland, W. (1987). Random and deterministic components of variance in mating success. In *Sexual selection: testing the alternatives*, (ed. J. W. Bradbury and M. B. Andersson), pp. 209–19. Wiley, Chichester.
Sutton, G. M. (1931). The blue goose and lesser snow goose on Southampton Island, Hudson Bay. *Auk*, **48**, 335–64.
Syroechkovsky, E. V. (1975). Egg weight and its effect upon mortality of nestlings in *Chen caerulescens caerulescens* on Wrangel Island. *Zoologichesky Zhurnal*, **54**, 408–12.
Syroechkovsky, E. V. and Kretchmar, A. V. (1981). Osnovnyye faktory, opredelyayushchiye chislennost' byelogo gusya (The fundamental factors deter-

mining the abundance of the snow goose). In *Ekologiya mlekopitayushchikh i ptits ostrova Vrangelya (The ecology of mammals and birds of Wrangel Island)*, (ed. V. G. Krivosheyev), pp. 3–37. DVNTs Akad. Nauk SSSR, Vladivostok.

Thomas, V. G. and Peach Brown, H. C. (1988). Relationships among egg size, energy reserves, growth rate, and fasting resistance of Canada goose goslings from southern Ontario. *Canadian Journal of Zoology*, **66**, 957–64.

Turelli, M. (1988). Phenotypic evolution, constant variances, and the maintenance of additive variance. *Evolution*, **42**, 1342–7.

van Noordwijk, A. J. (1987). On the implications of genetic variation for ecological research. *Ardea*, **75**, 13–19.

van Noordwijk, A. J. and de Jong, G. (1986). Acquisition and allocation of resources: their influence on variation in life history tactics. *The American Naturalist*, **128**, 137–42.

van Noordwijk, A. J., van Balen, J. H., and Scharloo, W. (1988). Heritability of body size in a natural population of the Great Tit (*Parus major*) and its relation to age and environmental conditions during growth. *Genetics Research (Cambridge)*, **51**, 149–62.

Viallefont, A., Cooke, F., and Lebreton, J.-D. (1994). Age-specific costs of first time breeding. *Auk*, in press.

Wellein, E. G. and Newcomb, W. (1953). Aerial waterfowl breeding grounds in sections of the far north. In *Waterfowl populations and breeding conditions, Summer 1953*, Special Scientific Report, Wildlife No. 25, pp. 16–19. Canadian Wildlife Service.

Welsh, A. H., Peterson, A. T., and Altmann, S. A. (1988). The fallacy of averages. *The American Naturalist*, **132**, 277–88.

Whitehead, P. J., Freeland, W. J., and Tschirner, K. (1990). Early growth of magpie geese, *Anseranas semipalmata*: sex differences and influence of egg size. *Australian Journal of Zoology*, **38**, 249–62.

Williams, G. C. (1957). Pleiotropy, natural selection, and the evolution of senescence. *Evolution*, **11**, 398–411.

Williams, G. (1966). Natural selection, the costs of reproduction and a refinement of Lack's principle. *The American Naturalist*, **100**, 687–90.

Williams, T. D. (1994*a*). Adoption in a precocial species, the lesser snow goose: intergenerational conflict, altruism or a mutually beneficial strategy? *Animal Behavior*, **47**, 101–107.

Williams, T. D. (1994*b*). Intraspecific variation in egg size and egg composition in birds: effects on offspring fitness. *Biological Reviews*, **69**, 35–59.

Williams, T. D., Lank, D. B., and Cooke, F. (1993*a*). Is intraclutch egg-size variation adaptive in the lesser snow goose? *Oikos*, **67**, 250–56.

Williams, T. D., Cooch, E. G., Jefferies, R. L., and Cooke, F. (1993*b*). Environmental degradation, food limitation and reproductive output: juvenile survival in lesser snow geese. *Journal of Animal Ecology*, **62**, 766–77.

Williams, T. D., Lank, D. B., Cooke, F., and Rockwell, R. F. (1993*c*). Fitness consequences of egg-size variation in the lesser snow goose. *Oecologia*, **96**, 331–38.

Williams, T. D., Cooke, F., Cooch, E. G., and Rockwell, R. F. (1993*d*). Effects of declining body condition on gosling survival in mass-banded lesser snow geese. *Journal of Wildlife Management*, **57**, 555–62.

Wobeser, G. (1981). *Diseases of wild waterfowl.* Plenum Press, New York.
Wright, S. (1943). Isolation by distance. *Genetics*, **28**, 114–38.
Wypkema, R. C. P. and Ankney, C. D. (1979). Nutrient reserve dynamics of lesser snow geese staging at James Bay, Ontario. *Canadian Journal of Zoology*, **57**, 213–19.
Yom-Tov, Y. (1980). Intraspecific nest parasitism in birds. *Biological Reviews*, **55**, 93–108.

Author Index

Subject Index

Bold numbers denote references to figures, *italic* numbers denote tables.